Supply Chain and Finance

Series on Computers and Operations Research

Series Editor: P. M. Pardalos *(University of Florida)*

Published

Vol. 1 Optimization and Optimal Control
eds. P. M. Pardalos, I. Tseveendorj and R. Enkhbat

Vol. 2 Supply Chain and Finance
eds. P. M. Pardalos, A. Migdalas and G. Baourakis

Series on Computers and Operations Research Vol. 2

Supply Chain and Finance

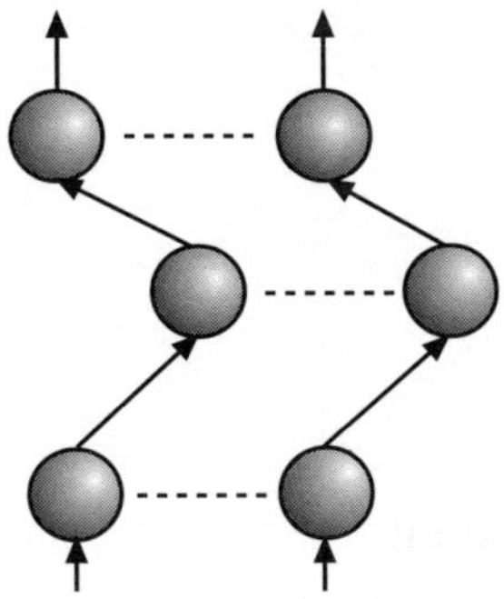

Editors

Panos M. Pardalos
University of Florida, USA

Athanasios Migdalas
Technical University of Crete, Greece

George Baourakis
Mediterranean Agronomic Institute of Chania, Greece

World Scientific

NEW JERSEY • LONDON • SINGAPORE • SHANGHAI • HONG KONG • TAIPEI • BANGALORE

Published by

World Scientific Publishing Co. Pte. Ltd.

5 Toh Tuck Link, Singapore 596224

USA office: Suite 202, 1060 Main Street, River Edge, NJ 07661

UK office: 57 Shelton Street, Covent Garden, London WC2H 9HE

British Library Cataloguing-in-Publication Data
A catalogue record for this book is available from the British Library.

SUPPLY CHAIN AND FINANCE

ISBN 981-238-717-X

Printed in Singapore by World Scientific Printers (S) Pte Ltd

PREFACE

With the globalization of the modern economy, it becomes more and more important to take into account various factors that can affect the economic situation and market conditions in different industries, and a crucial issue here is developing efficient methods of analyzing this information, in order to understand the internal structure of the market and make effective strategic decisions for successful operation of a business.

In recent years, a significant progress in the field of mathematical modelling in finance and supply chain management has been made.

Among these advances, one can mention the development of novel approaches in risk management and portfolio optimization - one of the most popular financial engineering problems first formulated and solved in the famous work by Markowitz in the 50-s. Recent research works in this field have resulted in developing new risk measures that utilize historical information on stock prices and make the portfolio optimization models easily solvable in practice. Moreover, new techniques of studying the behavior of the stock market based on the analysis of the cross-correlations between stocks have been introduced in the last several years, and these techniques often provide a new insight into the market structure.

Another important problem arising in economics and finance is assessing the performance of financial institutions according to certain criteria. Numerous approaches have been developed in this field, and many of them proved to be practically effective.

One more practical research direction that has been rapidly emerging in the last several years is supply chain management, where mathematical programming and network optimization techniques are widely used.

The material presented in the book describes models, methodologies, and case studies in diverse areas, including stock market analysis, portfolio optimization, classification techniques in economics, supply chain optimization, development of e-commerce applications, *etc.* We believe that this book will be of interest to both theoreticians and practitioners working in the field of economics and finance.

We would like to take the opportunity to thank the authors of the chapters, and World Scientific Publishing Co. for their assistance in producing this book.

Panos M. Pardalos
Athanasios Migdalas
George Baourakis

August 2003

CONTENTS

CHAPTER 1

NETWORK-BASED TECHNIQUES IN THE ANALYSIS OF THE STOCK MARKET

V. Boginski
Department of Industrial and Systems Engineering,
University of Florida,
303 Weil Hall, Gainesville, FL 32611, USA
E-mail: vb@ufl.edu

S. Butenko
Department of Industrial Engineering
Texas A&M University
236E Zachry Engineering Center, College Station, TX 77843-3131, USA
E-mail: butenko@tamu.edu

P.M. Pardalos
Department of Industrial and Systems Engineering,
University of Florida,
303 Weil Hall, Gainesville, FL 32611, USA
E-mail: pardalos@ufl.edu

We give an overview of a novel network-based methodology used to analyze the internal structure of financial markets. In the core of this methodology, is a graph representation of the data corresponding to the correlations between time series representing price fluctuations of the financial instruments. The resulting graph is referred to as market graph. Studying properties of the market graph based on data from the U.S. stock market leads to several important conclusions regarding the global structure of the modern stock market. This methodology also provides a new tool for classification of stocks and portfolio selection.

1. Introduction

One of the most important and challenging problems arising in the modern finance is finding efficient ways of summarizing and visualizing the stock market data that would allow one to obtain useful information about the behavior of the market. A large number of financial instruments are traded in the U.S. stock markets, and this number changes on a regular basis. The amount of data generated daily by the stock market is huge. This data is usually visualized by thousands of plots reflecting the price of each stock over a certain period of time. The analysis of these plots becomes more and more complicated as the number of stocks grows. These facts indicate the need for developing new efficient techniques of analyzing this data, that would allow one to reveal the internal structure and patterns underlying the process of stock price fluctuations.

A natural characteristic of the "similarity" or "difference" in the behavior of various stocks is the *correlation matrix* C constructed for a given set of stocks traded in the stock market. If the number of considered stocks is equal to N, then this matrix has the dimension $N \times N$, and each element C_{ij} is equal to the cross-correlation coefficient calculated for the pair of stocks i and j based on the time series representing the corresponding stock prices over a certain period. One can easily see that this matrix is symmetric, *i.e.*, $C_{ij} = C_{ji}, \forall i, j = 1, ..., N$.

The analysis of the correlation matrix gave a rise to several methodologies of studying the structure of the stock market. One of the directions of this type of research deals with analyzing statistical properties of the correlation matrix. These approaches utilize statistical physics concepts and Random Matrix Theory applied to finance. Several works in this area analyze the distribution of eigenvalues of the correlation matrix, which leads to some interesting conclusions.[14,16]

Another approach that extends the techniques of the analysis of the cross-correlation data utilizes *network* representation of the stock market based on the correlation matrix. Essentially, according to this methodology, the stock market is represented as a *graph* (or, a *network*). One can easily imagine a graph as a set of dots (vertices) and links (edges) connecting them. The vertices of this graph represent the stocks, and edges (links) between each pair of vertices are placed according to a certain criterion based on the corresponding correlation coefficient C_{ij}.

It should be noted that representing a certain real-life massive dataset as a large graph with certain attributes associated with its vertices is edges is becoming more and more widely used nowadays, and in many cases it provides useful information about the structure of a dataset it represents.[1,2,7,12]

A recently developed method of representing the stock market as a graph uses the concept of so-called *correlation threshold*. In this case, an edge between two stocks i and j is added to the graph if the corresponding correlation coefficient is greater than the considered correlation threshold. A graph constructed using this procedure is referred to as the *market graph.* Clearly, each value of the correlation threshold defines a different market graph, and studying the properties of these graphs for different correlation thresholds allows to obtain some non-trivial results regarding the internal structure of the stock market.[8,9]

Among the directions of investigating the characteristics of this graph, one can mention the analysis of its *degree distribution*, which represents the global pattern of connections, as well as finding *cliques* and *independent sets* in it. Studying these special formations provides a new tool of classification of the stocks and portfolio selection.

In this chapter, we will discuss these approaches in detail and analyze the corresponding results.

The rest of the chapter is organized as follows. In Section 2, statistical properties of correlation matrices representing real-life stock prices data are discussed. Section 3 presents basic definitions and concepts from the graph theory. Section 4 describes several aspects of the network-based approach of the analysis of the stock market. Finally, Section 5 concludes the chapter.

2. Statistical Properties of Correlation Matrices

As it was pointed out above, the correlation matrix is an important characteristic of the collective behavior of a given group of stocks. As we will see in this section, studying the properties of correlation matrices can provide useful information about the stock market behavior.

The formal procedure of constructing the correlation matrix is as follows. Let $P_i(t)$ denote the price of the instrument i at time t. Then

$$R_i(t, \Delta t) = \ln \frac{P_i(t+\Delta t)}{P_i(t)}$$

defines the logarithm of return of the stock i over the period from (t) to $t + \Delta t$.

The elements of the correlation matrix C representing correlation coefficients between all pairs of stocks i and j are calculated as

$$C_{ij} = \frac{E(R_iR_j) - E(R_i)E(R_j)}{\sqrt{Var(R_i)Var(R_j)}}, \tag{1}$$

where $E(R_i)$ is defined simply as the average return of the instrument i over T considered time periods (*i.e.*, $E(R_i) = \frac{1}{N}\sum_{t=1}^{T} R_i(t)$).[14,15,16]

The first question regarding the properties of this matrix is, what is the distribution of the correlation coefficients C_{ij} calculated for all possible pairs i and j, and how does this distribution change over time? Boginski *et al.*[9] analyzed this distribution for several overlapping 500-day periods during 2000-2002 (with $\Delta t = 1$ day) and found that it has a shape resembling the normal distribution with the mean approximately equal to 0.05 (note, however, that unlike a normal distribution, the distribution of cross-correlations is defined only over the interval [-1,1]). Moreover, the structure of this distribution remained relatively stable over the considered time intervals. This distribution for different time periods is presented in Figure 1.

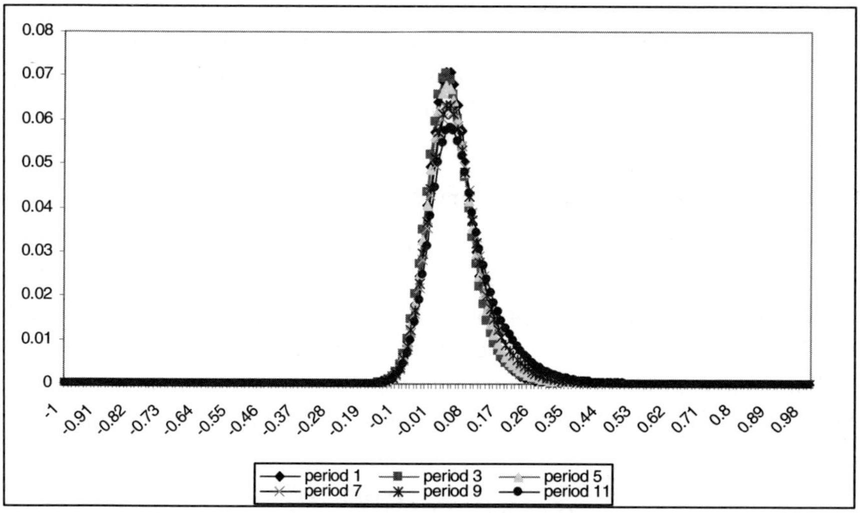

Fig. 1. Distribution of correlation coefficients in the US stock market for several overlapping 500-day periods during 2000-2002 (period 1 is the earliest, period 11 is the latest).

From Figure 1, one can also observe that even though the distributions corresponding to different periods have a similar shape and identical mean, the "tail" of the distribution corresponding to the latest period is significantly "heavier" than for the earlier periods. It means that although the values of correlation coefficients for most pairs of stocks are close to zero (which implies that there is no apparent similarity in the behavior of these stocks), a significant number of stocks have high correlation coefficients and exhibit a similar behavior, and the number of these stocks increases over time.

Similar results were obtained by Laloux *et al.*[14] and Plerou *et al.*[16] using the concepts of random matrix theory (RMT), which was originally developed for modelling the statistics of energy levels in quantum systems [19]. Using RMT, one can either confirm the hypothesis that a given correlation matrix is a "purely random matrix" (*i.e.*, it represents the time series corresponding to completely *uncorrelated* stocks, or find an evidence that there is a deviation from this hypothesis (*i.e.*, there is a significant correlation between some stocks). The methodology of testing this hypothesis is based on the analysis of eigenvalues of the correlation matrix C. According to RMT, all the eigenvalues λ_k of a purely random matrix are expected to belong to a finite interval:

$$\lambda_k \in [\lambda_{min}, \lambda_{max}].$$

The bounds of this interval are determined by the ratio R of the length of the time series (*i.e.*, the number of time periods for which the values of stock prices are considered) to the number of stocks N.[14] Plerou *et al.* [16] present the analysis of the distribution of eigenvalues of the correlation matrix corresponding to prices of stocks of 1000 largest U.S. companies during the years 1994-1995 with $\Delta t = 30$ minutes. The time series for each stock contained 6448 data points, and $R = 6.448$. For this value of R, the bounds of the interval $[\lambda_{min}, \lambda_{max}]$ are estimated to be equal to 0.37 and 1.94 repectively, which means that if all the eigenvalues of the correlation matrix satisfy the condition $0.37 \leq \lambda_k \leq 1.94$, then one would accept the hypothesis that this matrix corresponds to independent time series. However, it turns out that some eigenvalues of this correlation matrix are significantly larger than the upper bound of the interval, and, in fact, the largest eigenvalue of this matrix is more than 20 times larger than λ_{max}.[16] From this discussion, one can conclude that the fluctuations of the stock prices for the considered period are not purely random.

The results described in this section suggest that more and more stocks exhibit similar collective behavior nowadays. As we will see next, this fact is confirmed by the analysis of the stock market from another perspective using graph-theoretical approaches. We will also show how to apply this methodology for classification of stocks and choosing diversified portfolios.

However, before discussing these results, we need to introduce several standard definitions from the graph theory.

3. Graph Theory Basics

To give a brief introduction to the graph theory, we introduce several basic definitions and notations. Let $G = (V, E)$ be an *undirected* graph with the set of n vertices V and the set of edges E.

3.1. *Connectivity and Degree Distribution*

The graph $G = (V, E)$ is *connected* if there is a path through edges from any vertex to any vertex in the set V. If the graph is disconnected, it can be decomposed into several connected subgraphs, which are referred to as the *connected components* of G.

The *degree* of the vertex is the number of edges emanating from it. For every integer number k one can calculate the number of vertices $n(k)$ with the degree equal to k, and then get the probability that a vertex has the degree k as $P(k) = n(k)/n$, where n is the total number of vertices. The function $P(k)$ is referred to as the *degree distribution* of the graph.

3.2. *Cliques and Independent Sets*

Given a subset $S \subseteq V$, by $G(S)$ we denote the subgraph induced by S. A subset $C \subseteq V$ is a *clique* if $G(C)$ is a complete graph, i.e. it has all possible edges. The maximum clique problem is to find the largest clique in a graph.

The following definitions generalize the concept of clique. Namely, instead of cliques one can consider dense subgraphs, or *quasi-cliques.* A *γ-clique* C_γ, also called a *quasi-clique*, is a subset of V such that $G(C_\gamma)$ has at least $\lfloor \gamma q(q-1)/2 \rfloor$ edges, where q is the cardinality (i.e., number of vertices) of C_γ.

An *independent set* is a subset $I \subseteq V$ such that the subgraph $G(I)$ has no edges. The maximum independent set problem can be easily reformulated as the maximum clique problem in the *complementary* graph $\bar{G}(V, \bar{E})$, which is defined as follows. If an edge $(i, j) \in E$, then $(i, j) \notin \bar{E}$, and if $(i, j) \notin E$, then $(i, j) \in \bar{E}$. Clearly, a maximum clique in $\bar{G}$ is a maximum independent set in G, so the maximum clique and maximum independent set problems can be easily reduced to each other.

4. Market Graph: Global Organization and Evolution

In this section, we describe the recently developed methodology utilizing a representation of the stock market as a large graph based on the correlation matrix corresponding to the set of stocks traded in the U.S. stock market. This graph is referred to as the market graph.

The procedure of constructing this graph is relatively simple. Clearly, the set of vertices of this graph corresponds to the set of stocks. For each pair of stocks i and j, the correlation coefficient C_{ij} is calculated according to formula (1). If one specifies a certain *threshold* $\theta, -1 \leq \theta \leq 1$, then an undirected edge connecting the vertices i and j is added to the graph if the corresponding correlation coefficient C_{ij} is greater than or equal to θ. The value of θ is usually chosen to be significantly larger than zero, and in this case an edge between two vertices reflects the fact that the corresponding stocks are significantly correlated.

Boginski *et al.*[8,9] studied the properties of the market graph constructed using this procedure based on the time series of the prices of approximately 6000 stocks traded in the U.S. stock market observed over several partially overlapping 500-day periods during 2000-2002. The interval between consecutive observations were equal to one day (i.e., the coefficients C_{ij} were calculated using formula (1) with $T = 500$ and $\Delta t = 1$ day). These studies produced several interesting results that are discussed in the next subsections.

4.1. *Edge Density of the Market Graph as a Characteristic of Collective Behavior of Stocks*

Changing the parameter θ allows one to construct market graphs where the connections between the vertices reflect different degrees of correlation between the corresponding stocks. It is easy to see that the number of connections (edges) in the market graph decreases as the threshold value θ increases.

The ratio of the number of edges in the graph to the maximum possible number of edges is called the *edge density*. The edge density of the market graph is essentially a measure of the fraction of pairs of stocks exhibiting a similar behavior over time. As it was pointed out above, specifying different values of θ allows one to define different "levels" of this similarity. Figure 2 shows the plot of the edge density of the market graph as a function of θ. As one can see the decrease of the edge density is exponential, which can be easily understood taking into account that the distribution of correlation coefficients is similar ro normal.

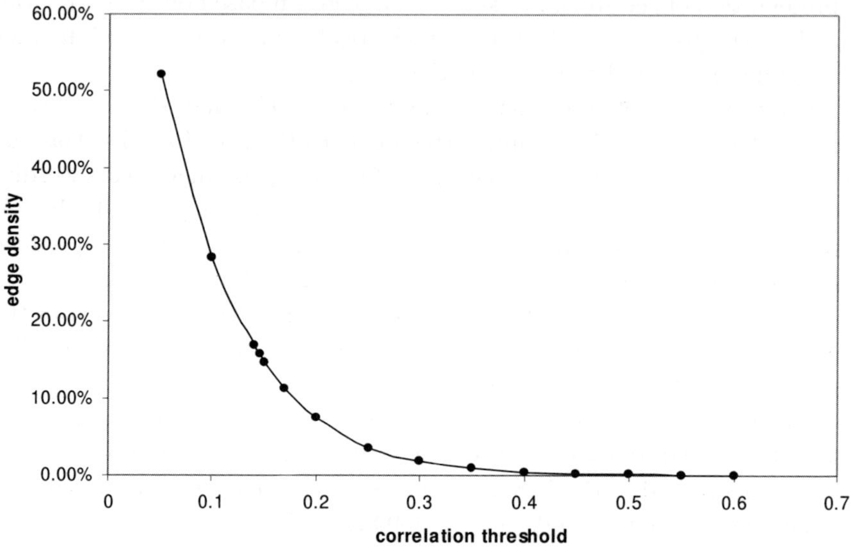

Fig. 2. Edge density of the market graph for different values of the correlation threshold.

On the other hand, one can look at the changes of the edge density of the market graph over time. In Ref. 9 these dynamics were analyzed for 11 overlapping 500-day periods in 2000-2002, where the 1st period was the earliest, and the 11th period was the latest. In order to take into account only highly correlated pairs of stocks, a considerably large value of θ ($\theta = 0.5$) was specified. It turned out that the edge density of the market graph corresponding to the latest period was more than 8 times higher than for the first period. The corresponding plot is shown in Figure 3. These facts are in agreement with the results discussed in the previous section, where we pointed out that the number of stocks demonstrating similar behavior steadily increases. The dramatic jump of the edge density suggests that there is a trend to the "globalization" of the modern stock market, which means that nowadays more and more stocks significantly affect the behavior of the others, and the structure of the market becomes not purely random. However, one may argue that this "globalization" can also be explained by the specifics of the time period considered in the analysis, the later half of which is characterized by a general downhill movement of the stock prices.

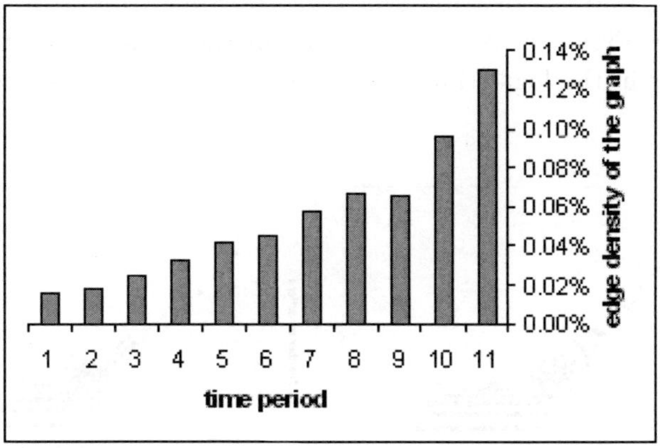

Fig. 3. Evolution of the edge density of the market graph during 2000-2002.

4.2. *Global Pattern of Connections in the Market Graph*

The edge density of the market graph discussed in the previous subsection is a global characteristic of connections between stocks, however, it does not reflect the *pattern* of these connections. For this purpose, the concept of *degree distribution* defined in Section 3 is utilized.

It turns out that the degree distribution of the market graph has a highly specific *power-law* structure, *i.e.*, this graph follows the *power-law model*[3] which states that the fraction $P(k)$ of vertices of degree k in the graph is proportional to some power of k, *i.e.*,

$$P(k) \propto k^{-\gamma}. \tag{2}$$

Equivalently, one can represent it as

$$\log P \propto -\gamma \log k, \tag{3}$$

which demonstrates that this distribution would form a straight line in the logarithmic scale, and the slope of this line would be equal to the value of the parameter γ.

According to Refs. 8, 9, the power-law structure of the market graph is stable for different values of θ, as well as for different considered time periods. Figure 4 demonstrates the degree distribution of the market graph (in the logarithmic scale) for several values of θ. In Ref. 9, the authors considered the degree distribution of the market graph for 11 overlapping time periods, and the distributions corresponding to four of these periods are shown in Figure 5. As one can see, all these plots are approximately straight lines in the logarithmic scale, which coincides with (3).

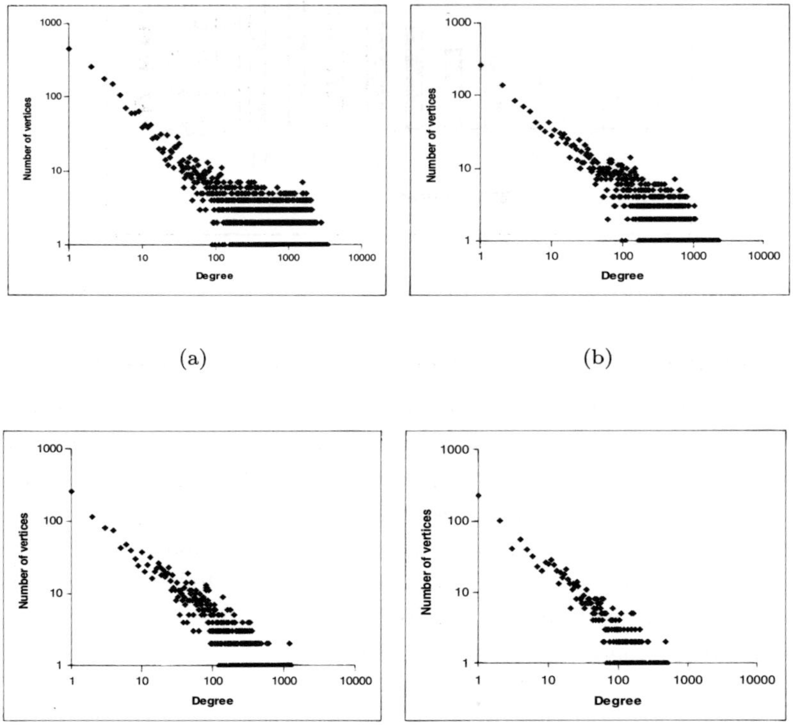

Fig. 4. Degree distribution of the market graph for a 500-day period in 2001-2002 corresponding to (a) $\theta = 0.3$, (b) $\theta = 0.4$, (c) $\theta = 0.5$, (d) $\theta = 0.6$.

The stability of the degree distribution of the market graph implies that there are highly specific patterns underlying the stock price fluctuations. However, an even more interesting fact is that besides the market graph, many other graphs representing real-life datasets arising in diverse application areas also have a well-defined power-law structure.[5,7,10,11,12,13,18,17] This fact served as a motivation to introduce a concept of "self-organized" networks,[4,5,6] and it turns out that this phenomenon also takes place in

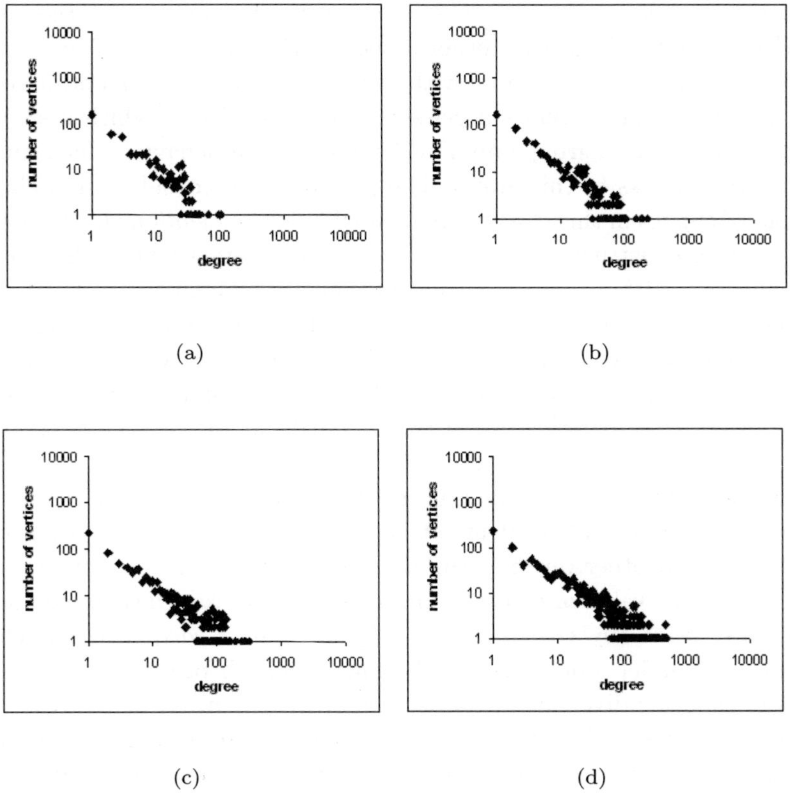

Fig. 5. Degree distribution of the market graph for different 500-day periods in 2000-2002 with $\theta = 0.5$: (a) period 1, (b) period 4, (c) period 7, (d) period 11.

finance.

5. Interpretation of Cliques and Independent Sets in the Market Graph

Another significant result of Ref. 8 is a suggestion to relate some correlation-based properties of the stock market to certain *combinatorial* properties of the corresponding market graph. For example, the authors attacked the problem of finding large groups of highly-correlated stocks by applying simple algorithms for finding large cliques in the market graph constructed using a relatively large value of correlation threshold. As it was mentioned above, a clique is a set of *completely interconnected* vertices, therefore, partitioning the market graph into large cliques defines a natural classification of stocks into dense clusters, where any stock that belongs to the clique is highly correlated with *all* other stocks in this clique. The fact that all stocks in a clique are correlated with each other is very important: it shows that this technique provides a classification of stocks, in which a stock is assigned to a certain group only if it demonstrates a behavior which similar to *all* other stocks in this group. The possibility to consider quasi-cliques instead of cliques in this classification should also be mentioned. This would allow one to construct larger groups of "similar" stocks while the density of connection within these groups would remain high enough.

Interestingly, the size of the maximum clique in the market graph was rather large even for a high correlation threshold. The details of these numerical experiments can be found in Ref. 8. For example, for $\theta = 0.6$ the edge density of the market graph is only 0.04%, however, a large clique of size 45 was detected in this graph.

Independent sets in the market graph are also important for practical purposes. Since an independent set is a set of vertices which are not connected with any other vertex in this set, independent sets in a market graph with a *negative* value of θ correspond to sets of stocks whose price fluctuations are negatively correlated, or *fully diversified portfolios*. Therefore, finding large independent sets in the market graph provides a new technique of choosing diversified portfolios. However, it turns out that the sizes of independent sets detected in the market graph are significantly smaller than clique sizes,[8] which indicates that one would not expect to find a large diversified portfolio in the modern stock market.

The results described in this subsection provide another argument in support of the idea of the globalization of the stock market, which was proposed above based on the analysis of the properties of correlation matrices and the edge density of the market graph.

6. Conclusion

In this chapter, we have discussed a new network-based methodology of the analysis of the behavior of the stock market. Studying the properties of the market graph gives a new insight into the internal structure of the stock market and leads to several important conclusions.

It turns out that the power-law structure of the market graph is quite stable over time; therefore one can say that the concept of self-organized networks, which was mentioned above, is applicable in finance, and in this sense the stock market can be considered as a "self-organized" system.

The results of studying the structural properties of the market graph provide a strong evidence supporting the well-known idea about the globalization of economy which has been widely discussed recently.

All these facts show that the market graph model is practical, and this research direction needs to be further developed.

References

1. J. Abello, P.M. Pardalos, and M.G.C. Resende. On maximum clique problems in very large graphs, *DIMACS Series*, 50, American Mathematical Society, 119-130 (1999).
2. J. Abello, P.M. Pardalos, and M.G.C. Resende, editors. *Handbook of Massive Data Sets.* Kluwer Academic Publishers (2002).
3. W. Aiello, F. Chung, and L. Lu. A random graph model for power law graphs, *Experimental Math.* **10**, 53-66 (2001).
4. R. Albert, and A.-L. Barabasi. Statistical mechanics of complex networks. *Reviews of Modern Physics* **74**: 47-97 (2002).
5. A.-L. Barabasi and R. Albert. Emergence of scaling in random networks. *Science* **286**: 509–511 (1999).
6. A.-L. Barabasi. *Linked.* Perseus Publishing (2002).
7. V. Boginski, S. Butenko, and P.M. Pardalos. Modeling and Optimization in Massive Graphs. In: *Novel Approaches to Hard Discrete Optimization*, P. M. Pardalos and H. Wolkowicz, eds. American Mathematical Society, 17-39 (2003).
8. V. Boginski, S. Butenko, and P.M. Pardalos. On Structural Properties of the Market Graph. In: *Innovation in Financial and Economic Networks*, A. Nagurney, ed. Edward Elgar Publishers (2003).
9. V. Boginski, S. Butenko, and P.M. Pardalos. Power-Law Networks in the Stock Market: Stability and Dynamics. To appear in *Proceedings of 4th WSEAS International Conference on Mathematics and Computers in Business and Economics* (2003).
10. A. Broder, R. Kumar, F. Maghoul, P. Raghavan, S. Rajagopalan, R. Stata, A. Tomkins, J. Wiener. Graph structure in the Web, *Computer Networks* **33**: 309–320 (2000).
11. M. Faloutsos, P. Faloutsos, and C. Faloutsos. On power-law relationships of the Internet topology, *ACM SICOMM* (1999).
12. B. Hayes. Graph Theory in Practice. *American Scientist*, **88**: 9-13 (Part I), 104-109 (Part II) (2000).
13. H. Jeong, B. Tomber, R. Albert, Z.N. Oltvai, and A.-L. Barabasi. The large-scale organization of metabolic networks. *Nature* **407**: 651-654 (2000).
14. L. Laloux, P. Cizeau, J.-P. Bouchad and M. Potters. Noise Dressing of Financial Correlation Matrices. *Phys. Rev. Lett.* **83**(7), 1467–1470 (1999).
15. R. N. Mantegna, and H. E. Stanley. *An Introduction to Econophysics: Correlations and Complexity in Finance.* Cambridge University Press (2000).

16. V. Plerou, P. Gopikrishnan, B. Rosenow, L.A.N. Amaral, and H.E. Stanley. Universal and Nonuniversal Properties of Cross Correlations in Financial Time Series. *Phys. Rev. Lett.* **83**(7), 1471–1474 (1999).
17. D. Watts. *Small Worlds: The Dynamics of Networks Between Order and Randomness*, Princeton University Press (1999).
18. D. Watts and S. Strogatz. Collective dynamics of 'small-world' networks. *Nature* **393**: 440-442 (1998).
19. E.P. Wigner. *Ann. Math.* **53**, 36 (1951).

CHAPTER 2

ON THE EFFICIENCY OF THE CAPITAL MARKET IN GREECE: PRICE DISCOVERY AND CAUSALITY IN THE ATHENS STOCK EXCHANGE AND THE ATHENS DERIVATIVES EXCHANGE

H. V. Mertzanis

Hellenic Capital Market Commission
Department of Research, Market Surveillance and Int'l Relations
1, Kolokotroni and Stadiou str.
105-62 Athens, Greece
E-mail: h.mertzanis@cmc.ase.gr

The chapter examines the interactions, using high frequency data, between the price series of the share price index of futures contracts traded on the Athens Derivatives Exchange and the underlying spot asset - the FTSE/ASE-20 Index - in the Athens Stock Exchange. This allows conclusions to be drawn on the impact of market structure on informed trading and on the nature of the cost-of-carry model. The usual result of futures leading spot is rejected, with clear bi-directionalcausality, and with many significant lags. This suggests that an electronic market may enhance price discovery. However, price discovery is quite slow. Also, this suggests that there is no preferred market for informed trading in the environment, and that tests for the presence of arbitrage opportunities and the correctness of the cost-of-carry model may be ineffective unless the lag structure is taken into account.

Keywords: Market efficiency, price discovery, stock and derivatives markets.

1. Introduction

The chapter examines the interactions, using high frequency data, between the price series of the share price index of futures contracts traded on the Athens Derivatives Exchange (ADEX) and the underlying spot asset - the FTSE/ASE-20 Index - in the Athens Stock Exchange.

The most important reasons for engaging in this study are the following: Firstly, and most importantly, the question of market efficiency, which underlies a great deal of financial research, is addressed in this area by the uncovering of the price discovery process. Price discovery is two things: the differential reaction of the different markets to new information, and the rate at which the new information is incorporated into price. Semi-strong form market efficiency precludes the possibility of earning excess returns on current public information, so if markets demonstrate this efficiency, the time-lag in both of these two must be sufficiently small to prevent economically significant excess returns. Also, an aim of security market design and regulation is optimal price discovery, so the choice of market structure will depend heavily on the best market for this. The theory behind this is best discussed in O'Hara,[17] and empirical evidence of the speed of price discovery abounds in this literature.[4,25] In Greece there is a fully automated and integrated trading system in both the equity and derivative markets, thus allowing us to comment on the price discovery process in comparison with other studies.

Secondly, the potential causal relationship may indicate to regulators, which of the two markets is most likely to be used by informed traders. Regulators, attempting to detect the presence of traders illegally using price-sensitive information, would wish to know the most likely market for these informed traders, and whether the market structure allows or impedes this detection.

Finally, the implementation of arbitrage trading strategies, which ensure a fair price for futures contracts (that is, with respect to the cost-of-carry model or some variation of it), must take into account the lead-lag relationship of the asset and its derivative security. If this is not done, problems may arise which take the form of apparent mispricing of futures contracts, and violations of the simple cost-of-carry model. Hence, some (but not all) of the mispricing discussed in Brailsford and Hodgson[3] might arise from delayed implementation of arbitrage, purely due to the time lag in reaction of the different markets. Also, the violations of the cost-of-carry model, like those demonstrated in Hearley[11] and others, may be due to the same effect.

Studies, which examine the joint time-series relationship between derivatives and their underlying spot assets, are not uncommon, and in general have similar motivations to those, listed above. An early study is Garbade and Silber.[8] More recent studies concentrate on allowing the most general specification possible for the dynamics of the two series, and testing for the causality or lead-lag relationship. Examples of this include Stoll and Whaley,[22] Tang *et al.*,[23] Wahab and Lashgari,[26] Ghosh,[9] and Tse.[25] Most studies conclude that future lead spot prices.

Note that many studies presume that a test of Granger or Granger-

Sims causality implies that action in one market causes a reaction in the other. This is not true; it may simply react first. For example, Hamilton (pp. 306-307)[10] gives an example in which "Granger causality" may have no economic interpretation; the series which acts first does not necessarily cause a reaction in another series. Note also that recent studies by Engle and Susmel[5] and Arshanapalli and Doukas[2] suggest that a common factor could be driving the relationship (particularly, in their cases, in volatility) and that "causality" that we see is no more than one market reacting more quickly than the other to an outside influence or shock. This is the sense in which we must interpret our results here, because the reaction in both is perceived to be a response to an external information shock. This is argued too by Turkington and Walsh[24] who, by making use of impulse response functions, show evidence of bi-directional causality between the two markets.

This study aims to address the extent and timing of the lead-lag relationship between the FTSE/ASE-20 futures and the underlying spot index. Two issues need to mention in relation to this. Firstly, as noted above, trading in both the equity and the derivatives markets in Greece is executed on a fully automatic and integrated trading system. This makes the institutional setting for this study most unique. Previous studies have either had open outcry in both markets (as in the US studies) or open outcry in the equity market and electronic trading in the futures market (as most European studies). The only real exception is Shyy and Lee,[20] who use the French equity market (electronic) and futures market (open outcry). Nowadays, most equity and derivatives markets are fully electronic, but no recent studies exist, to my knowledge, examining their joint interaction.

Secondly, we use the econometric methodology of Stoll and Whaley[22] and Fleming *et al.*,[6] which is further developed by Turkington and Walsh.[24] The methodology has four parts. Firstly, index values implied from the cost-of-carry model are calculated, so that we have two series: an actual spot and an implied spot. This ensures that any effects found cannot be attributed to non-linearities between future prices and spot prices. Secondly, we test for the presence of cointegration between the two levels series. (This is performed to confirm previous conclusions of the nature of these series). If cointegration were present, any causality test would need to be on first differences using a bivariate vector error-correction (VEC) model. Thirdly, however, following precedent literature, we filter out any microstructural effects from the actual spot and implied spot series by fitting an ARMA (p,q) model. Finally, we test for causality using the innovations from the

ARMA (p,q) process. The innovations series will not demonsrate cointegration (even though the levels were integrated of order 1 and cointegrated) because the innovations should be stationary. As a result, a conventional bivariate vector autoregression (VAR) was run on the innovations to test for causality. Impulse response functions are also plotted.

Unlike previous studies, we find bi-directional causality (feedback) from the FTSE/ASE-20 futures and the index itself using the innovations. The number of significant 5-minute lags was quite large, up to seven for both markets. They demonstrate that a shock in one market causes the other market to continue reacting for many lags, in fact for up to an hour in both series.

Section 2 discusses the institutional structure of the futures market in Greece, the data we use and the method. Section 3 gives tabulated results and discusses them. Unlike many previous studies, we are able to draw conclusions on all three of the initial aims listed above; price discovery, causality and the presence of arbitrage. Section 4 summarizes the results, gives some concluding comments based on these aims, and suggests some directions for future research.

2. Market Structure, Data and Method

2.1. *Structure of the Greek market and the General Index*

Unlike many international derivatives exchanges, the Greek futures market is a separate entity to the stock exchange. The ADEX was established in 1997 and has since grown to the largest electronic exchange in the south-eastern European region.

The ADEX trades in nearly 7 different futures and options contracts, the most heavily traded of which is the FTSE/ASE-20 futures contract. Computerized trading on the ADEX extending from 10:45 am – 16:45 pm facilitates FTSE/ASE-20 futures trading. In total, futures are traded on the ADEX for 6 hours per day, without break for lunch. Thus the market structure of the ADEX in comparison to the underlying stock market also provides a testable environment for the automated trading hypothesis. That is, if we expect different price discovery reactions in the two electronic markets, this may be the result of causes other than electronic efficiency.

FTSE/ASE-20 contracts first traded on the ADEX on August 2000 and have been whole-heartedly embraced by the market (for full information see *http://www.adex.ase.gr*). The FTSE/ASE-20 futures contracts are denominated in terms of the FTSE/ASE-20 Share Price Index, with the value of one futures contract designated as 5 EUR multiplied by the index value. The futures contract is traded in index points, while the monetary value of the contract is calculated by multiplying the futures price by the multiplier 5 EUR per point. For example, a contract trading at 2,185 points has value of 10,925 EUR. The FTSE/ASE-20 is market-capitalization weighted price index of approximately the 20 largest companies traded on the ASE. Hence, it represents over 80% of the total market value of domestically listed stocks, providing a highly satisfactory representation of market movements. The futures contract on the index FTSE/ASE-20 is cash settled in the sense that the difference between the traded price of the contract and the closing price of the index on the expiration day of the contract are settled between the counter-parties in cash. As a matter of fact, as the price of the contract changes daily, it is cash settled on a daily basis, up until the expiration of the contract.

Contracts are written on a quarterly expiry basis with contracts available up to the 3rd Friday of the expiration month. The minimum daily fluctuation of the FTSE/ASE-20 contracts value is 0.25 index point, which is equivalent to 1.25 EUR.

FTSE/ASE-20 futures contracts do not attract statutory stamp duty charges but do require the deposit of collateral (margin) with the Athens Derivatives Exchange Clearing House (ADECH). Today, the initial margin is 12% of the value of one contract, thus affords significant leverage. The margin account is marked to market at the end of each trading day with the settlement of clearing account required by 12 noon the following day. Note that the ADEX offers reduced margin requirements for spread positions in which the trader is long in one contract month and short in another. The spread concession is levied against offsetting positions held in different contract months. The termination of trading on an FTSE/ASE-20 futures contract is the last business day of the contract month whilst cash settlement occurs on the second day following the last trading day.

The trading costs of each market are different. On the equity market, the value-weighted spread for the stocks that compose the FTSE/ASE-20 Index is approximately 0.7%, and additional costs involving stamp duty, taxes and brokerage (which ranges from 1 to 3%) need to be considered. Also, index tracking in the sense of Roll (1992) will incur substantial rebalancing costs. However, the costs of trading on the futures exchange appear to be less. Applying Roll's (1984) estimator of the effective bid-ask spread yields an approximate value of 0.65% for the futures series that we have, and since there is no stamp duty and in general lower brokerage fees, we might suspect that the costs of trading on the futures exchange are relatively less than the equity market.

We can hence conclude that reduced trading costs and greater leverage may induce informed traders to the futures market. However, the speed of price discovery is expected to be greater in an electronic market, in this case, the equity market, perhaps counteracting this benefit.

2.2. *Data and Method*

The chosen sample period is from January 2nd, 2002 to May 31st, 2001, where a sample is drawn every 1 minute. Also, because the trading patterns are quite likely to be materially different from normal daytime trading, trades in the first 15-minute period of each day were also omitted, so the observations on each day start at 11:00am and finish at 16:45pm. This left us with 345 paired observations per day, for the 92 days of the sample, a total of 31,740 pairs of observations. To this data we applied the method described below.

Using this data, we firstly generated an implied spot index price series from the futures prices. This involved using the simple cost-of-carry model with observed futures prices and contract maturity dates, daily dividend yields collected from the issues of the ASE Monthly Bulletin for 2002, and proxy risk-free rates as Treasury notes, collected from the Bank of Greece Monthly Bulletin, with durations that best matched the maturity dates. The implied spot index series was generated using the cost-of-carry model:

$$S(t) = F(t, T)e^{-(r-d)(T-t)} \tag{1}$$

where $S(t)$ is the implied spot price, $F\ (t,\ T)$ is the observed futures price at time t for a contract expiring at time T, r is the risk-free rate of interest and d is the dividend yield. This means that we are assuming that the cost-of-carry model holds instantaneously, or, that the implied spot price reflects the futures price as if no time lag between them existed.

The second step is to test for cointegration of the two series. The approach we use is to first perform the usual augmented Dickey-Fuller test of each levels series to examine whether the series are integrated of the same order. Then, if the series are integrated, we test for cointegration using the Johansen procedure.[12] There are two likelihood ratio tests that we can use, and since our data involves two distinct series, the variables are cointegrated if and only if a single cointegrating equation exists. The first statistic (λ_{trace}) tests whether the number of cointegrating vectors is zero or one, and the other (λ_{max}) tests whether a single cointegrating equation is sufficient or if two are required.

In general, to see if R cointegrating vectors are correct, construct the following test statistics:

$$_{Trace}(R) = -T\Sigma_{i=R+1}^{\mathrm{P}} ln(1 - \lambda_i) \tag{2}$$

$$_{Trace}(R, R+1) = -Tln(1 - \lambda_{R+1}) \tag{3}$$

where P is the number of separate series to be examined, T is the number of useable observations and λ_i are the estimated eigenvalues obtained from the (i+1) x (i+1) "cointegrating matrix".[10]

The first test statistic (λ_{trace}) tests whether the number of distinct cointegrating vectors is less than or equal to R. The second test statistic (λ_{max}) tests the null that the number of cointegrating vectors is R against an R+l alternative. Johansen and Juselius[13] provide the critical values of these statistics.

Thirdly, we fitted an ARMA (p, q) model to each levels series, and collected the residuals. In the same way as in Ref. 27 the "innovations" represent the unexpected component of the prices of implied and actual spot series, purged of short-run market microstructure effects like bid-ask bounce and serial correlation potentially induced by non-synchronous calculation of the index (that is, the one which is induced by stale prices from thinly traded index components).[14] Since we are examining informational effects, these "innovations" are precisely the components of the price series that we wish to examine. This should have the effect of sharpening our inference.

If the levels series were indeed cointegrated, estimation and testing for causality would have to be via the Johansen bivariate vector[12] error correction (VEC) approach. If not, we can use the conventional bivariate vector autoregression (VAR). (We find that the levels were indeed cointegrated, so causality tests for the levels involve the VEC parameterisation. The innovations should not be cointegrated if the ARMA (p, q) models are correctly specified, so the causality tests are through a VAR).

Our results of the causality tests are presented only for the innovations; the levels causality tests produced similar results. The equations estimated are:

$$\Delta(FTSE/ASE{-}20)_t = c_1+\sum_{j=1}^{n}\lambda_\iota\Delta(FTSE/ASE{-}20)_{t-j}+\sum_{j=1}^{n}\lambda_j\Delta F_{t-j}+\lambda_{\iota t} \tag{4}$$

$$\Delta F_t = c_2+\sum_{j=1}^{n}\lambda_\iota\Delta(FTSE/ASE-20)_{t-j}+\sum_{j=1}^{n}\lambda_j\Delta F_{t-j}+\lambda_{2t} \tag{5}$$

where FTSE/ASE-20_t is the actual spot index price change innovation and Δ F_t is the implied spot index price change innovation. Both are generated by the ARMA (p, q) filters for the respective series. Impulse response functions were generated based on a shock of one-tenth of an index point, although this was not crucial to the results.

The causality test applied is simple Granger-causality. We first run equation 4. The regression is repeated with the restriction that all of the exogenous series coefficients (the values of λ_j) are zero. The statistic used is $S = T(RSS_0 - RSS1)/RSS_1$, where p = number of restricted coefficients, T = sample size, RSS = residual sum of squares and the subscripts on the RSS terms are restricted (l) and unrestricted (0).

Equation 5 is then estimated, unrestricted at first (giving RSS_0) and then with the λ_j values constrained to be zero (giving RSS_1). The conclusions that we draw are: (i) If S is not significant for either equation, there is no Granger-causality present, (ii) if S is significant for equation 4 but not equation 5, then innovations in the index are said to Granger-cause the innovations in the futures price, (iii) if S is significant for equation 5 but not for equation 4, the innovations are futures are said to Granger-cause the innovations in the index and (iv) if S is significant for both equations, then there is bi-directional Granger- causality, or feedback.

3. Results and Discussion

The results that we present here are in three brief parts. Firstly, we see results of tests for unit roots and cointegration for the levels series. Then, the estimated values of the innovation VARs are tabulated, together with the causality result determined. Finally, impulse response function graphs for these VARs are given. The results of the augmented Dickey-Fuller (ADF) with no trend or intercept terms, and the ARMA (p, q) results are reported in Table 1. The Johansen tests appear in Table 2.

Table 1. ADF and ARMA Results

Variable	*ADF*	*T-Stat*	*I (1)*	*ARMA*		*D-W (Innovations)*
				p	q	
FTSE/ASE-20	ADF (1)	-0.498	Yes	1	1	1.906
F	ADF (6)	-0.698	Yes	2	1	1.965

Table 2. Johansen Results

Variable	*Statistic*	*Number of Coint. Eq.*	*Likelihood Ratio Stat.*	*1% Critical Value*	*Cointegration Rank*
FTSE/ASE-20 and F	Trace	None	165.3	20.04	1
	Trace	At most 1	0, 67	6.65	

In this table, FTSE/ASE-20 is the actual spot index and F is the implied spot index. The ADF test clearly shows that the actual spot index and the implied spot index are non-stationary in the levels and stationary in the first difference.

Note that the ADF test is quite sensitive to structural change or outliers in the series. Additionally the inclusion or exclusion of an intercept term or deterministic tend in the regression also biases results toward accepting the null. The series are examined for potential outliers and the test is reapplied under the different specifications. The recalculated test statistics change only marginally. Note from Table 2 that for two series to be cointegrated; only one cointegrating equation must exist or equivalently the rank of the cointegrating matrix must be one. This is indeed what we find for the levels.

Fitting the ARMA (p, q) series with the lags illustrated yielded white noise in the innovations, and these innovations are both integrated of order zero. Surprisingly the FTSE/ASE-20 series shows a lower degree of serial correlation (only 1 lag) than the F series (seven lags). This is contrary to expectations, as we would expect a high order of positive serial correlation in the index due to the progressive adjustment process.

We found that the results for the raw series (which are cointegrated) and the innovations series (which are not) were very similar with regard to the causality tests we performed, so only the results for the innovation series are presented. Table 3 gives these results.

We can see from Table 3 that the price discovery process is quite slow, with endogenous lags still significant to lag 4 for the actual index innovations and lag 6 for the implied spot index innovations. The exogenous variables to both series were significant out to lag 7. Note that the values of the λ_i coefficients (those for the actual spot index series in the implied spot index equation) are much larger than the correspondent exogenous series

Table 3. Regression Results

Lag	Coefficient FTSE/ASE-20	Coefficient F
1	- 0.041 (-0.06)**	1.072 (27.63)**
2	- 0.039 (-3.21)**	0.832 (-23.05)**
3	- 0.032 (-2.54)**	0.721 (16.34)**
4	- 0.031 (-3.18)*	0.523 (13.11)**
5	- 0.027 (-1.71)	0.532 (11.89)**
6	- 0.008 (-1.23)	0.493 (9.32)**
7	- 0.014 (-0.67)	0.232 (5.23)**

Note: T-stats in parentheses

coefficients, the λ_i, (those of the implied spot index series in the actual spot index equation). However, both are significant to a similar length, and the causality test results indicate that bi-directional causality (i.e., feedback) is indeed present, with however very strong evidence of future leading spot prices. (The S statistics were both significant, but are not given in Table 3).

It is difficult to interpret the results for this estimation on their own, because there is so much complex feedback between the lags of each equation and the system as a whole. However, a very intuitive way of understanding how the system behaves includes the use of impulse response functions (IRPs). These involve assuming that the system is at a steady state, and then disturbing it using a shock or innovation into the error term of one of the equations. The shock filters back through the lag structure *of both equations simultaneously*, and the value of the dependent variable at each time period that follows the impulse can be calculated from the estimated equations (3) and (4) and then graphed. However, no such methodology is used in this chapter.

4. Summary and Conclusions

This section firstly summarizes the results and then draws conclusions on the topics suggested in the introduction. We have found that FTSE/ASE-20 futures and the spot FTSE/ASE-20 index are integrated of order 1 and cointegrated. This means that causality tests of the changes in each need to be correctly specified by vector error correction (VEC) models. However, we use the approach of firstly filtering out microstructure effects like bid-ask bounce and serial correlation induced by non-synchronous observations in the index components, using an ARMA (p, q) specification. The resulting two series of innovations are integrated of order zero, so testing for causality is made by vector autoregression (VAR). Unlike previous studies, we find strong evidence of bi-directional causality (or feedback) between the two series, with however strong evidence of future leading spot prices.

We motivated this study from three different angles. Firstly, we can draw conclusions on the price discovery process between the futures and the spot index. We have found that the discovery time of the true price, following an information shock, depends on whether the shock is an "own" market shock or an "other" market shock. If there is an information shock in the index, it will presumably be some piece of market wide information that hits the equity market first, and will be rapidly assimilated into the index. However, a shock in the index can take as long as one hour to adjust in the futures market. Almost exactly the reverse applies for a futures market shock. Neither market appears to adjust more quickly than the other; the only factor of importance is which market picks up the news first.

This leads to our second point. If one market or the other dominates the capture of new pieces of information, we could comfortably say that that market is the trading "habitat" of informed traders. The direction of causality would be strongly from one market to the other, as informed traders in one market would commence trading and drive the reaction in the other market. However, we find that there is feedback between the markets; if informed traders do indeed choose a "habitat", it is not along the simple division of the type of instrument they choose.

Taken with previous evidence, we can also say that the open outcry style of market is no more likely to attract informed traders than an electronic system, and may be less likely. This is because previous evidence suggests that futures in an electronic trading system seem to lead the spot asset traded on an open outcry system. However, reversing these (in Greece) does not cause spot to lead futures. It seems that the electronic equity trading may have counteracted the benefits informed traders enjoy in the futures market.

Thirdly, if arbitrage opportunities and deviations cost-of-carry seem to arise at high frequency, as has been seen in recent evidence in Greece and elsewhere, it may be due to a misspecification in the cost-of-carry model. An extended high-frequency cost-of-carry model, that takes into account the lead-lag relationship between futures and spot, may eliminate some or all of these deviations.

One further point that has arisen during this study is that the futures market appears to react much more to index shocks than the other way around. A futures shock causes a small change in the index, but an index shock causes an enormous change in the futures contract, about 25 times the size of the index change. One is tempted to try to explain this by saying that the futures market over-reacts to the spot market, but we have seen that the change is permanent, not temporary, so this is not a valid explanation.

Extending on this chapter could be in three obvious directions. Firstly, the above study is conducted only on FTSE/ASE-20 futures and the underlying index. Repeating the study using a wider range of derivative contracts and their respective underlying assets would broaden our conclusions. However, as noted in the introduction, the FTSE/ASE-20 futures and the FTSE/ASE-20 are one of the few futures-spot pairings that captures economy-wide factors. Other futures contracts are written on spot assets that are more specific commodities, so news in these assets will be more specific to that asset. The same type of study could be extended (with a little more effort) to options and their underlying asset. The use of options would be particularly useful in studying the effect of company specific information, because individual share futures (a recent innovation in Greece) have relatively thin trading compared to the options series on the same stock.

Secondly, testing for the presence of arbitrage opportunities and mispricing in futures contracts could be extended to allow for the price discovery lag that we have found. We may find, at high frequency sampling, that this time lag causes apparent mispricing, which could be captured by allowing the futures price in the cost-of-carry model to reflect lagged as well as contemporaneous spot prices. The same could apply to the spot price; hence the cost-of-carry becomes a bivariate specification not unlike the VAR we have studied here.

Lastly, the "over-reaction" we have noted here, of futures to underlying index shocks, needs to be examined further. The resulting increased volatility in futures prices will have consequences for, among other things, hedging, arbitrage strategies and margin requirements.

References

1. M. Aitken and P. Swan. The cost and responsiveness of equity trading in Australia, SIRCA research paper (1995).
2. B.Arshanapalli and J. Doukas. Common Volatility in S&P 500 Stock Index and S&P 500 Stock Index Futures Prices during October 1997, *Journal of Futures Markets*, Vol. 14, no.8, pp. 915-925 (1994).
3. T. Brailsford and A. Hodgson. Mispricing in stock index futures: A re-examination using the SP. *Australian Journal of Management*, **22**, pp. 2143 (1997).
4. F. de Jong and T. Numan. High frequency analysis of lead-lag relationships between financial markets, *Journal of Empirical Finance*, **4**, pp. 259-277 (1997).
5. R. Engle and R. Susmel. Common Volatility in International Equity Markets. *Journal of Business & Economic Statistics*, **11**(2), pp. 167-176 (1993).
6. J. Flemming, B. Ostdiek, and R.E. Whaley. Trading costs and the relative rates of price discovery in stock, futures and options markets. *Journal of Futures Market*, **16**, pp. 353-387 (1996).
7. M. Forster and T.J. George. Anonymity in securities markets. *Journal of Financial Intermediation*, **2**, pp. 168-206 (1992).
8. K.D.Garbade and W.L. Silber. Price movement and price discovery in the futures and cash markets. *Review of Economics and Statistics*, **64**, pp. 289-297 (1982).
9. A. Ghosh. Cointegration and error correction models: Intertemopral causality between index and futures prices. *Journal of Futures markets*, **13**, pp. 193-198 (1993).
10. J.D. Hamilton. *Time Series Analysis.* Princeton University Press, Princeton, NJ (1994).
11. R. Heaney. A test of the cost-of-carry relationship using 96 day bank accepted bills and the All-Ordinaries share price index. *Australian Journal of Management*, **20**, pp. 75-104 (1995).
12. S. Johansen. Estimation and hypothesis testing for cointegrating vectors in Gaussian vector autoregressive models. *Econometrica*, **59**, pp. 1551-1580 (1991).
13. S. Johansen and K. Juselius. Maximum likelihood estimation and inference on cointegration with application to the demand for money. *Oxford Bulletin of Economics and Statistics*, **47**, pp. 169-209 (1990).
14. A. Lo and A.C. Mackinlay. An econometric analysis of non-synchronous trading. *Journal of Econometrics*, **45**, pp. 181-211 (1990).
15. A. Madhavan. Trading Mechanisms in Securities Markets, *Journal of Finance*, **47**, pp. 607-641 (1992).
16. L. Meulbroek. An Empirical Analysis of Illegal Insider Trading. *Journal of Finance*, **147**, pp. 1661-1699 (1992).
17. M. O'Hara. *Market Microstructure Theory.* Blackwell Publishers, Cambridge MA (1995).
18. R. Roll. A simple implicit measure of the effective bid/ask spread in an

efficient market. *Journal of Finance*, **39**, pp. 347-350 (1984).
19. R. Roll. A mean/variance analysis of tracking error. *Journal of Portfolio Management*, **118**, pp. 13-22 (1992).
20. G. Shyy and J.H Lee. Price transmission and information asymmetry in Bund Futures markets: LIFFE vs. DTB. *Journal of Futures Markets*, **15**, pp. 87-99 (1995).
21. J.A. Stephan and R.E.Whaley. Intraday price change and trading volume relations in the stock and option markets. *Journal of Finance*, **45**, pp. 191-220 (1990).
22. H. R. Stoll and R.E.Whaley. The dynamics of stock index and stock index futures returns. *Journal of Financial and Quantitative Analysis*, **25**, pp. 441-468 (1990).
23. G.N. Tang, S.C. Mak, and D.F.S Choi. The causal relationship between stock index futures and cash index prices in Hong Kong. *Applied Financial Economics*, **2**, pp. 187-190 (1992).
24. J. Turkington and D. Walsh. Price discovery and causality in the Australian share price index future market, working paper, University of Western Australia (1997).
25. Y.K. Tse. Lead-lag relationship between spot index and futures price of the Nikkei Stock average. *Journal of Forecasting*, **14**, pp. 553-563 (1995).
26. M. Wahab and M. Lashgari. Price dynamic and error correction in stock index and stock index futures: A cointegration approach. *Journal of Futures Markets*, **13**, pp. 711-742 (1993).
27. D. M. Walsh. Price reaction to order flow 'news' in Australian equities. *Pacific Basin Journal of Finance*, **7**, pp. 1-23 (1997).

CHAPTER 3

ASSESSING THE FINANCIAL PERFORMANCE OF MARKETING CO-OPERATIVES AND INVESTOR OWNED FIRMS: A MULTICRITERIA METHODOLOGY

G. Baourakis

Mediterranean Agronomic Institute of Chania
Dept. of Economic Sciences, Management / Marketing / Finance
P.O.BOX 85, 73 100 Chania, Crete, Greece.
E-mail: baouraki@maich.gr

N. Kalogeras

Marketing and Consumer Behaviour Group
Dept of Social Sciences, Wageningen University
Hollandseweg 1, 6706
KN Wageningen, The Netherlands
E-mail: Nikos.Kalogeras@Alg.MenM.WAU.NL

C. Zopounidis

Technical University of Crete
Dept. of Production Engineering and Management, Financial Engineering Laboratory
University Campus, 73100, Chania, Crete, Greece.
E-mail: kostas@ergasya.tuc.gr

G. Van Dijk

Marketing and Consumer Behaviour Group
Dept of Social Sciences, Wageningen University
De Leeuwenborch, Hollandseweg 1, 6706 KN Wageningen, The Netherlands
E-mail: coop99@xs4all.nl

This chapter examines the evaluation of economic and financial viability of marketing cooperatives (MCs) and Investor Owned Firms (IOFs). The analysis covers the periods from 1993-98 for MCs and from 1994-98 for IOFs. The data is based on the financial characteristics of 10 MCs and 2 IOFs established and operating in Crete (the largest Greek island) and

8 food processing and marketing companies operating in Greece, but chosen exclusively for their vast familiarity to Greek consumers. The assessment procedure includes data analysis techniques in combination with a multicriteria analysis method (PROMETHEE II). The analysis results in an overall ranking of the examined firms' performance. It further indicates the strengths and weaknesses of the involved firms with regard to their financial behavior, thus contributing to the identification of market imperfections of the examined firms. Therefore, relevant conclusions are drawn concerning the revision of corporate strategies.

Keywords: Marketing co-operatives, investor-owned-firms, financial ratio analysis, data analysis, multicriteria decision aid, strategies.

1. Introduction

The empirical domain of this study is the agri-food industry in Greece. The food sector is undergoing a structural change in terms of internationalization, network relationships, and concentration. Nowadays, a range of organizational choices - joint ventures, long term contacts, and strategic alliances - increases a firm's interdependence and ensures its ability to produce to specifications.[3] Shifts in customer requirements, merging competitors, technical and organizational innovations make markets fluid and complex, compelling agribusiness firms to become market oriented in order to anticipate change, sense and respond to trends, and to act faster than the competitors.[9,23]

However, it is reasonable to expect the entire food industry to respond to policy reforms, environmental changes, technological progress and rapid changes in consumer demand. Of particular importance in this respect is the financial and market behavior of MCs and IOFs established and operating in rural areas. Van Dijk [26] argues that these firms are operating under different conditions in the industrialized market economies as compared with conditions when co-operatives were still social and business innovations. The agribusiness sector of the Netherlands and Denmark co-operatives are for instance characterized by export-orientation, an increasingly internationalized industry and a pursuit of direct investments.

Moreover it is worth mentioning, that MCs and IOFs could be distinguished as two different hierarchies, where MCs have less freedom in their choice of financial structure than IOFs. According to Hendrikse and Veerman,[16] this occurs due to the fact that MCs require member control, which precludes the design of an efficient number of contingencies regarding the allocation of decision power.

More specifically, concerning the financial structure of MCs and IOFs, the driving force of their financial instruments and viability is that the impact of wealth constraint of entrepreneurs differs for each financial instrument.[1] This seems rational if we consider the totally different organizational structure that these two types of firms hold. Financial managers and analysts should try to loosen the above mentioned constraint by designing financial instruments which would maintain special organizational forms; make them comparable with others; reduce their risk of failure; and at the same time eliminate their inefficiencies.[16]

This chapter is organized as follows. After the introductory part, both the IOFs and Co-op Firms will be presented in the subsequent two sections as well as a brief market outlook of where the examined firms operate. The methodological framework of this study is presented in detail in Section 4. Section 5 presents the results of the study and a relevant discussion is made. Finally, at the end, the study's limitations and conclusions are drawn along with suggestions for further empirical research.

2. Co-ops vs IOFs: A Literature Overview

Contrary to IOFs, co-operative (co-op) firms are generally regarded as a separate form of business organization. Engelhardt[10] distinguished the following general characteristics of co-op firms: there is a real co-operation between economic units which consists of more than just mutual co-ordination (i.e involvement between conventional companies in a cartel); it is a personal co-operation and not a collectivistic co-operation imposed by public power; the co-op members are separate economic units which are legally and economically independent; co-operation involves the integration of one or more functions performed by the co-operative economic unit.

Caves and Peterson[6] support that traditional values and principles of co-op firms give rise to a financial performance that may differ significantly from that of IOFs. According to Getzloglannis,[14] the theoretical and economic analysis demonstrate that the performance of co-op firms, measured in terms of profitability, leverage, solvency, liquidity and efficiency, may be entirely different from the one of IOFs. A number of reasons have been laid down to explain this phenomenon. The difference in objectives seems to be the most important. Co-op firms are generally considered to be service-to-members maximizers subject to a profit constraint, while IOFs are rate of return to equity (at a given risk level) maximizers. Moreover, Kyriakopoulos[18] summarizes the main distinctive characteristics between

IOFs and co-op firms. These characteristics are presented in Table 1.

Table 1. Distinctive Characteristics of Co-op Firms vs. IOFs

	Co-op Firms	**IOFs**
Dividend	None or Limited	Unlimited
Equity	Membership Certificates	Transferable Shares
Location	Patrons' Place	Profit Criterion
Management	Patron Controlled	Autonomous
Objective	Max. Patron Income	Max. Shareholder Value
Owners	Patrons	Investors
Price Policy	In Patrons benefit	To Increase Profits
Profit Allocation	Patronage	Shares
Services	Extension, Education	None at all
Taxation	Tax Free Patronage Income	Corporate Taxation
Voting	Democratic	In Proportion of Shares.

Source: Kyriakopoulos, 1997.

In addition, Van Dijk[26] more distinctly specifies the main differences in the two examined organizational and managerial systems (see Table 2).

Table 2. Organisation and Management Systems in Co-op Firms/IOFs.

Type	**Capital Suppliers**	**Buyer/Seller of Goods & Services**	**Profit**
Co-op Firms	Members	Members	Condition
IOFs	Shareholders	Clients	Goal

Source: Van Dijk , 1997.

Furthermore, Hendrikse and Veerman[15] support that the relationship between the financial structure of MCs and the undertaking of control by the members of the co-op is a main issue to be taken under consideration for the strategic planning and the designing of managerial organizational structure. They also argue that the shortage of agricultural and horticultural markets poses a serious threat to the survival of the MC. Comparative static results demonstrate the relevant financial instruments (i.e personal liability, financial contributions and bank relationships), organizational form (i.e democratic decision making and internal control systems) and economic systems.

In the same work, evidence is provided depicting that in the Netherlands and the USA, MCs have less freedom in their choice of financial structure than IOFs, because their charter requires member control, which precludes the design of an efficient number of contingencies regarding the allocation of decision power.[15] MCs are restricted to the use of non-voting equity and debt as sources of funds, because MC members feel strongly that the integrity of the MC is destroyed when control has to be shared with non-members. However, internal financial constraints may force them to acquire outside funds. This poses a problem in the competition with other organizations, because the domination of control requirements will most likely result in a higher premium for outside funds. Along the same argument, Van Dijk[26] mentions that, in essence, IOFs create new market opportunities for the co-op firms under the conditions of investor driven membership, diversified membership and market fragmentation. This new model of co-op is the so-called *New Generation Co-operative (NGCs).*

Thus, theoretical financial and economic analysis concerning co-op firms' performance indicate that it may be greatly determined by the co-op principles of risk sharing and mutual responsibility and may affect productive and economic efficiencies in a manner such that financial performance would be different from the one realised by IOFs. An empirical evaluation of the performance should be of high value to creditors, lenders, financial analysts, and firms' managers/marketers as well as to governments and to those who are interested in the financial and economic performance of co-op firms in comparison with IOFs.[14]

3. A Brief Market Outlook

The reorganization of food and agri-business is not linear, involving only the scale of business operations, as well as the transformation of the market place. Global proliferation of technology and managerial know-how, reorganization and international economic boundaries, deregulation of the markets, and heterogeneity in consumer behavior mark a major economic shift from production to market-oriented competition.[3] Likewise, Nilsson[19] supports that the horizontal integration strategy alone, though necessary in many instances, is not sufficient enough to provide a competitive edge because the main drive is to exploit economies of scale assuming a commodity type of business.

The Greek food processing and marketing companies face this stiff and rapidly changing market environment. Most of them try to revise their strategies and proceed towards new organizational arrangements. It can be said that an additional reason for choosing the food sector as a subject of study is that it is the most profitable one in Greek manufacturing.

Simultaneously, with the structural changes which were held in the international agri-business sector, the co-op movement flourished around the world despite some setbacks and many continuing challenges occurring during the 20th century. Almost every country in the world possesses co-op organizations.[26] It is interesting to note that 28 out of 50 countries (56%) earn revenues exceeding $1 billion.[12] Europe, at the continental level holds the first position and provides 49% of the countries with more than $1 billion in sales. Asia (22%) and the USA (15%) rank second and third, respectively, while Africa and Australia share the fourth position with 7% (see Figure 1).

A noticeable point is that European countries' turnover is much higher than that of Asian ones. Where countries with more than 500,000 co-op members are concerned, Asia holds a larger share (46%) than Europe (27%).[12] This fact reveals the dynamic role and the high importance of European co-op movements in the global food marketing environment. More specifically,[11] by considering the country of origin of the top 30 co-ops in the EU countries, the Netherlands holds the first position (31%), followed by, in descending order, France (23%), Denmark (13%), Germany (10%), Ireland (10%), Sweden (7%), Finland (3%) and the UK (3%).

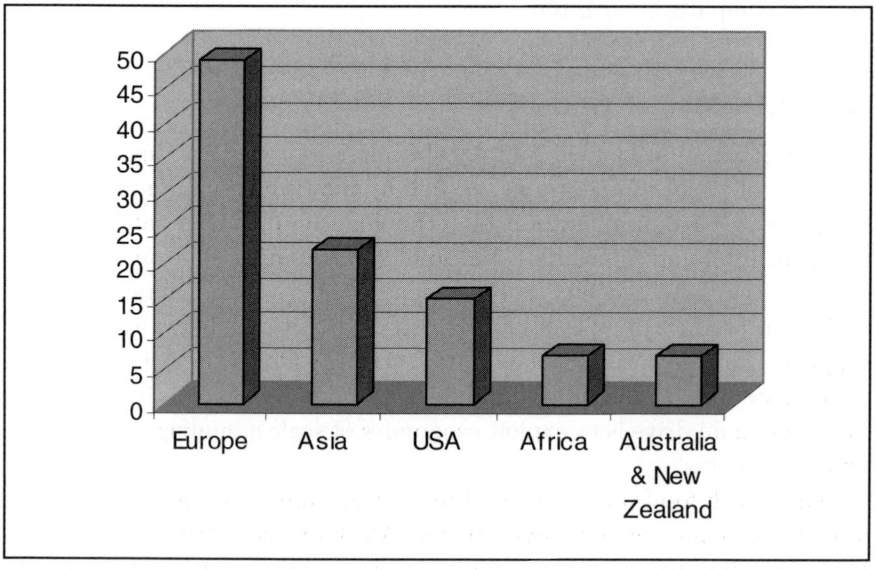

Fig. 1. Percentage Distribution of Countries with More than $1 Billion Sales Per Continent. *Source:* Eurostat, 1996

In Europe, the largest marketing co-op firms, in terms of turnover and number of employees, are the German *BayWa,* the Finnish *Metsalito* and the Dutch *Campina Melkunie.* By ranking them by type of activity, the Dutch *Campina Melkunie* and French *Sodiaal* are the leader co-ops in the dairy sector The German *BayWa* and the Dutch *Cebeco Handelsraad* are the largest multi-purpose co-ops, the French *Socopa* and *UNCAA* dominate in the meat and farm supply sector respectively, and the Dutch *Greenery/VTN* and *Bloemenveiling* are the leaders in the fruit & vegetable and flower auction sectors.[25]

Co-op firms mainly cater to the interests of people who live and work in the rural areas of Greece and are organized along three levels (local co-ops, Union of co-ops and Central Unions). The main categories of local co-ops are: multi-purpose, selling, production, requisite and diverse. The majority of the local co-ops are multi-purpose. The small economic size of the local co-ops (with an average of 55 local co-ops and 6000 farmer-members) have led to the formation of Unions whose activities are related basically to the marketing of food products. Nowadays, these amount to 130. The Central Union of Co-ops was formed by the 185 Unions and 23 local co-ops and they carry out the marketing activities of one product or similar products at the national and international level.[18]

The number of agricultural co-op firms in Greece adds up to 6,920 and the number of memberships is approximately 784,000. The turnover of co-ops adds up to 0.8 billion EUROs and Greece is ranked sixth in terms of co-op turnover and first in terms of number of co-op firms. This is somehow ambiguous but very easily explained if we take into consideration the general agricultural statistics in Greece. For instance, the active population that is mainly occupied in agriculture in Greece is 669,000 people (18,8%) of the total labor force. Twelve percent (12%) of the Gross Domestic Product (GDP) is derived from the agricultural sector and 30% from total exports.[12]

4. Methodological Framework

4.1. *Characteristics of Examined Firms & Sampling Procedure*

The source of the empirical research is derived from 10 MCs and 2 IOFS established in Crete, and from the 8 most known food processing and marketing companies (IOFs) in Greece. The common characteristic of the two groups is, mainly, that they both belong to the same industrial sector and produce similar products. The sample selection was also made based on the idea that the application of the methodology suggested in the current work, can be applied to different business forms that function more or less in the same economic and business environment and face almost the same levels of financial risk and market uncertainty.

For the 8 food processing and marketing companies, appropriate financial information was gathered for the period 1994-98 from the database of ICAP Hellas, a Greek company, which provides integral financial information and business consulting. The examined firms mainly concentrate on processing agricultural raw foods. Some of them process a large amount of Greek seasonal and frozen vegetables and others concentrate on dairy products. Their average size, in terms of numbers of employees, can be characterized as large and almost all of them have an annual revenue of more than 60 million euros while their own capital is maintained at very high levels from year to year.

Additionally, in order to gain a general idea about the examined MC activities, the financial data/structure and the way that they were organized, personal interviews were conducted with their accounting managers and staff. The island of Crete, which is located on the southern border of Europe, was selected. A number of first-degree co-ops, formed from 16 agricultural unions of co-ops and a central union were chosen. The primary products produced by these unions include olive-oil, cheese, wine, fruits and vegetables.

By conducting personal interviews with the managers of Cretan MCs, it was discovered that most of them were established many decades ago, thus, having a history based on old-fashioned co-op values and principles. Therefore, the administration of the central unions, agricultural unions, and local co-ops usually face problems in adapting to the new situation and to the rapidly changing marketing environment. Many of them are well-known to Greek and some European consumers because of their high quality produce.

The basic problem which Cretan co-ops face nowadays is their negative financial performance (as it appears in their balance sheets). Most of them do not properly use or cope with their invested capital in the most efficient way. They always have to face high overhead costs, and there is a general imbalance in the invested capital structure. Size is also a limiting factor acting as an obstacle in their expansion in operational activities (processing and direct marketing applications, strategic management, *etc.*).[2]

More specifically, the sample selection was made by taking the following specific criteria described below into account:

The MCs selected were the Unions of co-ops located in the capital areas of the Cretan prefectures (*Heraklion*, *Chania* and *Rethymnon*). Some other Unions were selected from all over Crete (*Ierapetra*, *Apokorona* & *Sfakia*, *Kolimvari* and *Kissamos*) according to financial and economic size. Moreover, two first degree co-ops (*Koutsoura* and *Archanes*) which provided us with an interesting 5 year entrepreneurship profile were also chosen. Finally, two investor owned firms (IOF's), *VIOCHYM* and *AVEA*, were selected to be examined. Both firms are located in the prefecture of Chania, are totally market-oriented and operate in their own sector, respectively (*VIOCHYM* in the juice-sector and *AVEA* in the olive-oil sector).

On the other hand, concerning the case of food processing and marketing companies (IOFs), it was not possible to include all producers (for example, co-ops which are not obliged to publish information, highly diversified firms whose main activity is not the one under examination in this study and very small-sized firms) due to information obstacles. All the selected firms are of similar size in terms of total assets. Also, the eight selected IOFs are the most renowned to Greek consumers and, during the last decade, have developed a significant exporting activity.

4.2. *Principal Component Analysis*

In the first step of the financial performance evaluation and viability of the considered co-ops and IOFs, a multivariate statistical analysis was conducted, namely: principal components analysis. Principal components analysis is applied to select a limited set of financial ratios that best describe the financial performance of the sample throughout the considered time period. Principal components analysis was applied separately through the years 1993-98 for the co-ops and 1994-98 for the IOFs, in order to determine the most important financial ratios for every examined one-year period of this study.

4.3. *Financial Ratio Analysis*

Ratio analysis is widely used to evaluate financial performance. Within the theory of industrial organization there exist formal measures of performance, which are well-established.[13,21] However, their application is difficult to implement because of the unavailability of required data. Despite its limitations, ratio analysis is a solid tool commonly used in corporate finance to provide valuable comparisons between economic and financial analysis. We rely, therefore, on the financial ratios and on further elaboration by using data selection techniques.

A number of ratios have been found to be useful indicators of financial performance and risk bearing ability of the firms and co-ops under examination. These ratios could be grouped into three categories as depicted in Table 3: profitability, solvency and managerial performance ratios.[7]

Table 3. Financial ratios used in the evaluation of MCs and IOFs

Codification	Financial ratios	
NI/NW	Net income/Net worth	
EBIT/TA	Earning before interest and taxes/Total assets	**Profitability**
GP/SALES	Gross profit/Sales	
NI/SALES	Net income/Sales	
TL/TA	Total liabilities/Total assets	
CA/CL	Current assets/Current liabilities	**Solvency**
QA/CL	Quick assets/Current liabilities	
LTD/(LTD+NW)	Long term debt/(Long term debt+Net worth)	
INV×360/SALES	Inventory×360/Sales	
ARC×360/SALES	Accounts receivables×360/Sales	**Managerial Performance**
CL×360/CS	Current liabilities×360/Cost of sales	

4.4. *Multicriteria Method*

The evaluation of the financial performance and viability of the selected firms and co-ops has been carried out using the PROMETHEE II multicriteria method (Preference Ranking Organization Method of Enrichment Evaluations).[4] This method is the most appropriate for the decision-maker in order to provide with tools enabling him to advance in solving a decision problem where several, often conflicting multiple criteria must be taken into consideration. The PROMETHEE II method is known to be one of the most efficient and simplest multicriteria methods. It is based on the outranking relations' concept, which was found and developed by Bertrand Roy.[22] Roy defined the outranking relation as a binary relation **S** between alternatives a and b in a given set of alternatives A, such that in $a\mathbf{S}b$, a outranks b. However, there is no essential reason to refute the statement that a is at least as good as b.

The construction of the outranking relation through the PROMETHEE II method involves the evaluation of the performance of the alternatives in a set of criteria. Each criterion is given a weight p depending on its importance. The weight increases with the importance of the criterion. The criteria's weights constitute the basis for the assessment of the degree of preference for alternative a over alternative b. This degree is represented in the preference index $\pi(a,b)$:

$$\pi(a,b) = \sum_{j=1}^{n} p_j H_j(d) \Big/ \sum_{j=1}^{n} p_j$$

The preference index for each pair of alternatives (a,b) ranges between 0 and 1. The higher it is (closer to 1) the higher the strength of the preference for a over $b$is. $H_j(d)$ is an increasing function of the difference d between the performances of alternatives a and b on criterion j. $H_j(d)$ is a type of preference intensity.[27] The H_j function can be of various different forms, depending upon the judgment policy of the decision maker. Generally, six forms of the H function are commonly used (see Figure 2).

For the purposes of this study the Gaussian form of the $H_j$was used for all financial ratios. The use of the Gaussian form requires only the specification of the parameter σ. This function is a generalization of all the other five forms, whereas the fact that it does not have discontinuities contributes to the stability and the robustness of the obtained results.[5]

The results of the comparisons for all pairs of alternatives (a,b) are organized in a value outranking graph. The nodes of the graph represent the alternatives under consideration (firms, co-ps, etc.), whereas the arcs between nodes a and b represent the preference of alternative a over alternative b (if the direction of the arc is $a \rightarrow b$) or the opposite (if the direction of the arc is $a \leftarrow b$). Each arc is associated with a flow representing the preference index $\pi(\alpha,\beta)$. The sum of all flows leaving a node a is called the leaving flow of the node, denoted by $\phi^+(\alpha)$. The leaving flow provides a measure of the outranking character of alternative a over all the other alternatives. In a similar way, the sum of all flows entering a node a is called the entering flow of the node, denoted by $\phi^-(\alpha)$. The entering flow measures the outranked character of alternative a compared to all the other alternatives. The difference between the leaving and the entering flow $\phi(\alpha)=\phi^+(\alpha)-\phi^-(\alpha)$ provides the net flow for the node (alternative) a which constitutes the overall evaluation measure of the performance of the alternative a. On the basis of their net flows the alternatives are ranked from the best (alternatives with high positive net flows) to the worst ones (alternatives with low net flows).

By using the methodology that is described above, the PROMETHEE

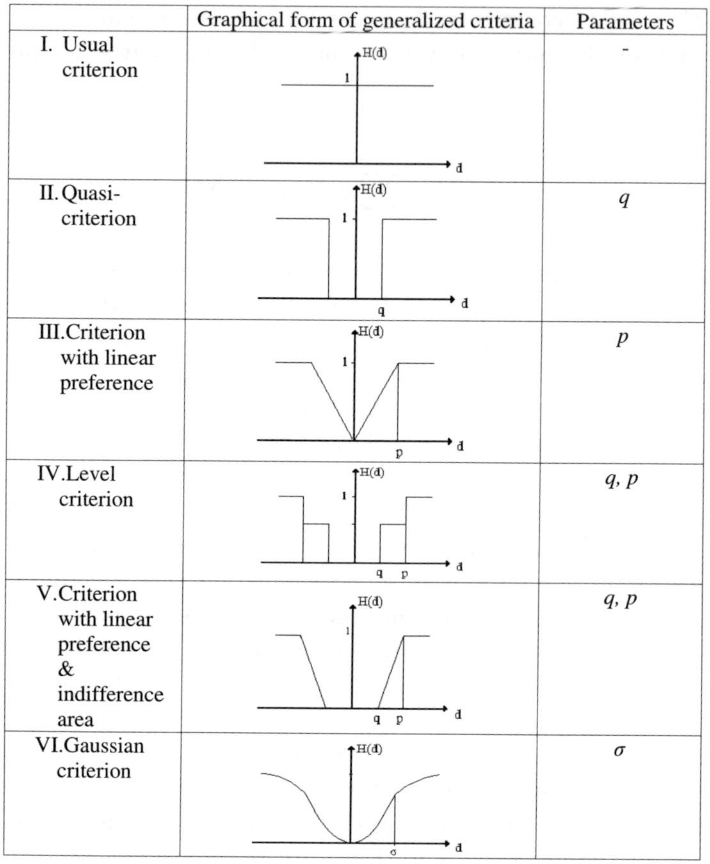

Fig. 2. Forms of the Preference Function (Source: Brans et al., 1986)

II contributes significantly towards making an integrated and rational evaluation and assessment of the performance and viability of the co-op firms and IOFS examined in this study, by specifying the impact of all those factors (financial ratios) on them.

5. Results and Discussion

5.1. *Firms' Attitude through Principal Component Analysis*

Concerning the results of principal component analysis, evidence is provided that in each year, three to four principal components corresponding to eigenvalues higher than one were extracted. In all cases, the cumulative percentage of the total variance explained by the extracted components is at least 70%. In most cases the initial principal components (those that explain most of the variance) involve the profitability and the solvency (including liquidity) of the firms under consideration thereby highlighting the significance of these two factors in characterizing their financial status.

The component loadings of each ratio were used for the selection of the most significant financial ratios. In particular, the most frequently appearing ratios were those finally selected for further investigation during the assessment of the performance and viability of the considered co-op firms and IOFs. The summarized outcomes of this analysis are presented in Tables 4 and 5. These tables present the ratios found to have the highest principal components loadings developed for each year (the corresponding ratios are marked with "+"). Ratios with high loadings are the ones with the higher explanatory power with regard to the financial characteristics of the considered MCs and IOFs. The last column of both figures illustrates the frequency of each ratio selected as a significant explanatory factor according to the results of the principal components analysis. On the basis of this frequency, a limited set of financial ratios is selected to perform the evaluation of the MCs and IOFs (the selected ratios are underlined).

Table 4. Significant Financial Ratios Selected Through Principal Components Analysis for 10 MCs and 2 IOFs

	1993	1994	1995	1996	1997	1998	Frequency
NI/NW			+	+	+		3
EBIT/TA	+			+		+	3
GP/SALES		+	+			+	3
NI/SALES		+			+		2
TL/TA			+	+			2
CA/CL	+				+	+	3
QA/CL				+		+	2
LTD/(LTD+NW)		+		+	+		3
INV×360/SALES		+	+		+		3
ARC×360/SALES	+		+	+			3
CL×360/CS		+					1

5.2. *Overall Ranking of the Examined Firms*

Taking into consideration the limited number of ratios derived from the above procedure, an assessment procedure through the PROMETHEE II

Table 5. Significant Financial Ratios Selected Through Principal Components Analysis for the Processing and Marketing IOFs

	1994	1995	1996	1997	1998	Frequency
NI/NW		+	+	+	+	4
EBIT/TA		+	+	+		4
GP/SALES	+	+	+	+		4
NI/SALES	+	+	+	+	+	5
TL/TA	+			+	+	3
CA/CL	+	+	+	+	+	5
QA/CL	+	+		+	+	4
LTD/(LTD+NW)	+	+	+	+	+	5
INV×360/SALES	+	+	+	+	+	5
ARC×360/SALES	+	+	+	+	+	5
CL×360/CS	+	+	+	+	+	5

method was also carried out. As previously mentioned, this application requires the determination of the appropriate evaluation criteria (the financial ratios which were selected through the principal components analysis) as well as the shape of the H_j function for each selected evaluation criterion j. The shape of the H_j function selected for every financial ratio j, is the Gaussian form (Gaussian criterion) defined as follows: $H_j(d) = 1 - \exp(-d^2/2\sigma^2)$, where d is the difference among the performance level of MCS and IOFs a and b for the financial ratio g_j $[d = g_j(a) - g_j(b)]$, and σ is the standard deviation of the ratio g_i.

Different scenarios were examined to discern the significance of the selected ratios tested. The seven scenarios investigated covered representative examples of the weighting schemes that one could apply in considering the significance of profitability, solvency and managerial performance, during the corporate assessment process. These scenarios presented in Table 6, take into consideration the categorization of the selected financial ratios following the scheme described above.

For each weighting scenario, a different evaluation of the considered MCs and IOFs is obtained using the PROMETHEE II method. More specifically, for each scenario the MCs and IOFs are ranked in descending order starting from the ones with the highest financial performance to the ones with the lowest financial performance. The ranking is determined on the basis of the net flows obtained through the PROMETHEE II method (high net flow corresponds to high financial performance and vice versa). Table 7 illustrates the average ranking of the MCs obtained for each year of the analysis along all seven weighting scenarios (smaller values for the ranking indicate better firms). To measure the similarities of the results obtained

Table 6. Weighting Scenarios for the Application of the PROMETHEE II Method. *Note: Within each category of financial ratios the corresponding ratios are considered of equal importance*

	Profitability	Solvency	Managerial performance
Scenario 1	50.0%	33.3%	16.7%
Scenario 2	16.7%	33.3%	50.0%
Scenario 3	16.7%	50.0%	33.3%
Scenario 4	50.0%	16.7%	33.3%
Scenario 5	33.3%	50.0%	16.7%
Scenario 6	33.3%	16.7%	50.0%
Scenario 7	33.3%	33.3%	33.3%

for each scenario, the Kendall's coefficient of concordance (Kendall's W) is used. The possible values for the Kendall's W are positioned in the interval 0 to 1. If the Kendall's W is 1, this means that the rankings for each weighting scenario are exactly the same. As presented in Table 7, the Kendall's W was depicted to be very high throughout all the years (in all cases above 0.9 with a 1% significance level). This indicates that the results obtained from the PROMETHEE II are quite robust for different weight scenarios, thus increasing the confidence in the results that are obtained from this analysis. The Kendall's W is also employed to measure the similarities between the rankings along all years. The MC result was 0.732, which is significant at the 1% level indicating that the evaluations obtained in each year are quite similar.

Table 7. Average Rankings of the MCs and 2 IOFs throughout the Years and All Scenarios

	1993	1994	1995	1996	1997	1998	Average ranking
HERAKLION	11.42	10.00	5.85	10.00	8.57	8.42	10th
VIOCHYM	7.71	4.28	3.42	4.14	2.71	1.00	4^{rth}
A VEA	8.85	9.14	11.28	10.28	12.00	3.14	11th
KOUTSOURAS Co-op	2.71	1.00	1.42	2.14	2.28	4.00	1st
CHANIA	1.42	2.42	2.85	1.42	3.42	*	2nd
KOLIMVARI	6.00	7.85	7.28	6.42	7.28	7.14	6th
KISSAMOU	3.85	7.42	6.85	10.00	10.28	10.00	8th
APOKORONAS & SFAKIA	5.14	2.57	3.14	3.14	1.57	3.28	3rd
PEZA	3.14	11.00	10.00	9.85	10.14	11.00	12th
IERAPETRA	8.42	6.57	10.00	7.28	5.28	6.00	7th
RETHYMNON	8.42	4.85	8.00	5.00	6.57	3.85	5th
ARXANES Co-op	10.85	10.85	7.57	8.28	7.85	8.28	9th
Kendall's W	**0.918**	**0.920**	**0.943**	**0.944**	**0.903**	**0.915**	

Notes: * Data not available.
The last column (average ranking) refers to the whole time period.

The results from the above table reveal that the best firm, ranked first, throughout the years and for all the scenarios is the MC of Koutsouras. This result is quite interesting because the Koutsouras co-op is much smaller in financial terms, in comparison with the majority of the examined businesses in this category, and is involved in the production and marketing of greenhouse products. Its operations are mainly aimed at the production, distribution and trade of the members' products in both domestic and foreign (West European) markets. In other words, this co-op presents an integrated dynamic and flexible business scheme, while at the same time it succeeds in keeping its overhead costs at low levels.

Except for the Union of Agricultural Co-ops of Chania, which is ranked second, the remaining Unions, which are located in the capital area of each Cretan prefecture, are ranked in much lower positions. These results indicate low degree of entrepreneurship, flexibility and financial and managerial performance during the examined period.

The corresponding results obtained by examining the juice producing & marketing companies are presented in Table 8. The company which ranked first is EVGA S.A. This seems quite reasonable if we consider EVGA's market in the Greek market during the last fifteen years. Ranking second is the General Food Company, "Uncle Stathis." This firm has achieved a very healthy financial performance within the five year period of examination because it expanded its production, differentiated its products (by using modern marketing strategies, i.e. well-designed packaging, high advertising expenditure, etc) and proceeded to export to several Balkan and Mediterranean countries. As in the case of MCs, the Kendall's W for the annual rankings of IOFs for the obtained weighting scenarios is, in all cases, quite high (higher than 0.7) at a 1% significance level.

The above ranking is of high managerial importance if we consider, as previously mentioned, that the ranked IOFs are those of firms which are most familiar to both Greek and non – Greek consumers. Also, they hold very high market shares in the Greek food-manufacturing market and own a well established brand name.

6. Conclusion

This study attempts to provide evidence that the multicriteria decision aid methodology adopted and utilized in the analysis constitutes a major scientific tool that significantly contributes towards this aim. Future research should be oriented to the design of financial instruments, which

Table 8. Average Rankings of the Food Processing and Marketing IOFS throughout the Years and All Scenarios

	1994	1995	1996	1997	1998	Average ranking
CHIPITA INTERNATIONAL S.A	2.71	3.42	4.71	6.28	3.71	3rd
DELTA DAIRY S.A	*	*	*	3.71	4.28	5th
DODONI S.A	6.71	3.14	2.28	3.14	5.14	4th
ELAIS S.A	4.71	2.57	2.14	3.00	3.00	8th
FAGE S.A	3.14	6.57	2.00	4.00	2.42	7th
KNORR CPC HELLAS S.A	1.00	4.28	4.42	3.28	6.14	6th
EVGA S.A	4.00	5.71	6.28	7.14	7.42	1st
UNCLE STATHIS S.A	5.71	2.14	6.14	4.57	4.00	2nd
Kendall's W	**0.714**	**0.788**	**0.880**	**0.741**	**8.240**	

Notes: *Data not available.
The last column (average ranking) refers to the whole time period.

maintains the special agri-business character and eliminates the inefficiencies associated with their organizational nature. A multicriteria DSS for the assessment of the agribusiness firms is within the immediate research plans.

The results of the current study enhance our understanding concerning a firm's market behavior and orientation. The movement from the production philosophy to the market-oriented questions enhances the ability of agri-business to rapidly process market information across their production, processing and marketing chain. Therefore, the agri-business sector can revise its strategies and move forward. Through increased financial performance evaluation both up and downstream, a firm may raise the specter of the traditional Greek producing, processing and marketing companies' role in counterbalancing market power.

References

1. P. Aghion and P. Bolton. An Incomplete Contracts Approach to Financial Contracting. *Review of Economic Studies*, **59**, 473-494 (1992).
2. G. Baourakis and K. Oustapassidis. Application of Strategies for the Increase of the Competitiveness and the Strategic Restructuring of the Cretan Marketing Co-ops. Final Working Paper, Regional Operational Programme, Crete 1994-98, Phase II, Mediterranean Agronomic Institute of Chania, Crete, Greece (2000).
3. M. Boehje, J. Akridge and D. Downey. Restructuring Agribusiness for the 21st century, *Agribusiness: An International Journal,* **11**, 493-500 (1995).
4. J.P. Brans and Ph. Vincke. A Preference Ranking Organization Method: The PROMETHEE Method for Multiple Criteria Decision-Making, *Management Science*, **31**, 647-656 (1985).

5. J.P. Brans and Ph. Vincke, and B. Mareschal. How to Rank and How to Select Projects: The PROMETHEE Method *European Journal of Operational Research*, **24**, 228-238 (1986).
6. R.E. Caves and B.C. Peterson. Co-operatives Shares in Farm Industries: Organization and Policy Factors. *Agribusiness: An International Journal*, **2**, 1-19 (1986).
7. J.K. Courtis. Modelling a Financial Ratios Categoric Framework. *Journal of Business Finance and Accounting*, **5**, 371-386 (1978).
8. G.S. Day. The Capabilities of Market-Driven Organizations, *Journal of Marketing,* **58**, 37-52 (1994).
9. A.P. DeGeus. Planning as Learning. *Harvard Business review,* **66** (March-April), pp. 70-74 (1988).
10. W.W. Engelhardt. *Der Functionswandel der Genossenschaften in Industrialisierten Marktwirtsschaften.* Berlin, Dunker und Humblot (1971).
11. Eurostat, Yearbook of Agricultural Statistics, Luxembourg (1995).
12. Eurostat, Yearbook of Agricultural Statistics, Luxembourg (1996).
13. G.D. Ferrier and P.K. Porter. The Productive Efficiency of US Milk Processing Co-operatives. *Journal of Agricultural Economics,* **42**, 1-19 (1991).
14. A. Getzloganis. Economic and Financial Performance of Cooperatives and Investor Owned Firms: An Empirical Study. In *Strategies and structures in the Agro-Food Industries*, J. Nilsson and G. Van Dijk (eds.), Van Gorcum, Assen, 171-180 (1997).
15. G.W.J. Hendrikse and C.P. Veerman. Marketing Co-operatives and Financial Structure.' Tilburg University, The Netherlands, Center Discussion Paper 9546 (1994).
16. G.W.J. Hendrikse and C.P. Veerman. Marketing Co-operatives as a System of Attributes. In: *Strategies in the Agro-food Industries,* J. Nillson and Gert van Dijk, eds., Van Gorsum, Assen, 111 – 129 (1997).
17. Icap Hellas, Business Directory: Annual Financial Reports for Greek Companies, 1994-1998 (series).
18. K. Kyriakopoulos. Co-operatives and the Greek agricultural economy". In: *Agricultural co-operatives in the European Union*, O-F. Van Bekkum and G. Van Dijk, eds., EU Commission – DG XXIII, Van Gorcum, 62-72 (1997).
19. J. Nilsson. The mergence of New organizational Models for Agricultural Co-operatives. *Swedish Journal of Agricultural Research,* **28**, 39-47 (1998).
20. C. Parliament, Z. Lerman and J. Fulton. Performance of Co-operatives and Investor Owned Firms in The Dairy Industry, *Journal of Agricultural Co-operation,* **4**, 1-16 (1990).
21. P. K. Porter and G.W. Scully. Economic Efficiency in Co-operatives. *Journal of Low and Economics,* **30**, 489-512 (1987).
22. B. Roy. Classement et choix en présence de points de vue multiples: La méthode ELECTRE. *R.I.R.O*, **8**, pp. 57-75 (1968).
23. P.M. Senge. *The Fifth Discipline: The Art and Practice from the Learning Organization,* New York: Doubleday (1990).
24. R.J. Sexton and J. Iskow. What do know about the Economic Efficiency of Co-operatives: An Evaluating Survey. *Journal of Agricultural Co-operation*

1, 15-27 (1993).
25. O.F. Van Bekkum and G. Van Dijk, eds. *Agricultural co-operatives in the European Union,* EU Commission – DG XXIII, Van Gorcum (1997).
26. G. Van Dijk. Implementing the Sixth Reason for Co-operation: New Generation Co-operatives in Agribusiness. In: *Strategies and Structures in the Agro-food Industries,* J. Nillson and Gert van Dijk, eds., Van Gorsum, Assen, 171-182 (1997).
27. P. Vincke. *Multicriteria Decision Aid*, John Wiley & Sons Ltd, New York (1992).

CHAPTER 4

ASSESSING COUNTRY RISK USING MULTICRITERIA CLASSIFICATION APPROACHES

E. Gjonca
Mediterranean Agronomic Institute of Chania
Dept. of Economics, Marketing and Finance
73100 Chania, Greece

M. Doumpos
Mediterranean Agronomic Institute of Chania
Dept. of Economics, Marketing and Finance
73100 Chania, Greece

G. Baourakis
Mediterranean Agronomic Institute of Chania
Dept. of Economics, Marketing and Finance
73100 Chania, Greece

C. Zopounidis
Technical University of Crete
Dept. of Production Engineering and Management
Financial Engineering Laboratory
University Campus
73100 Chania, Greece

Country risk evaluation is an important component of the investment and capital budgeting decisions of banks, international lending institutions and international investors. The increased internationalization of the global economy in recent decades has raised the exposure to risks associated with events in different countries. Consequently, substantial resources are now being devoted to country risk analysis by international organizations and investors who realize the importance of identifying, evaluating and managing the risks they face. This study presents the contribution of multicriteria decision aid in country risk assessment. The proposed approach is based on multicriteria decision aid classification methods, namely the UTADIS method (**UT**ilités **A**dditives **DIS**criminantes) and the MHDIS method (**M**ulti-group **H**ierarchical **DIS**crimination). Both methods lead to the development of country risk classification models in the form of additive utility functions that classify a set of countries into predefined risk classes. The efficiency of the proposed methods is illustrated through a case study using data derived by the World Bank. The two multicriteria methods are employed to develop appropriate models for the classification of countries into five risk groups, according to their creditworthiness and risk level. Several validation tests are performed in order to compare the classification results of the two methods with the corresponding results obtained from statistical and econometric analysis techniques.

Keywords: Country risk, multicriteria decision aid, classification.

1. Introduction

Country risk assessment is one of the most important analytical tools used by leading institutions and investors in determining the creditworthiness of a particular country. The rapid growth of the international debt of developing countries in the 70s, the increasing number of debt reschedulings in the early 80s, the two oil crises in 1973 and 1979 and the post-war recessions in 1974/75 led to an unstable and uncertain international economic, political and social environment.

Country risk evaluations concerned scientists, bankers, investors, and financial managers from the early years. However, the systematic study of this problem started at the beginning of the 1970's. Various commonly accepted definitions of country risk have been found. In general, country risk is defined as the probability that a country will fail to generate enough foreign exchange in order to fulfil its obligation towards the foreign creditors.[11] According to Mondt and Despontin,[21] country risk is divided into two different kinds of risks: (a) an economic (financial) risk which shows the capacity of a country to service its debt, and (b) a political risk which indicates that a country is not willing to pay its foreign currency loans. In a broader sense, Calverley[6] defined country risk as potential, economic and financial losses due to difficulties raised from the macro-economic and/or political environment of a country. From the foreign investor's point of view, Nordal[23] defined country risk for a given country as the unique risk faced by foreign investors when investing in that country as opposed to investing in other countries.

The purpose of this chapter is to present the contribution of multicriteria decision aid (MCDA) in country risk assessment. The proposed classification approaches namely the UTADIS method (**UT**ilités **A**dditives **Dis**criminates) and the M.H.DIS method (**M**ulti–group **H**ierarchical **DIS**crimination), combine utility function–based frameworks with the preference disaggregation paradigm. The methods are applied to the country risk assessment problem, in order to develop models that classify a sample of 125 countries into four groups according to their economic performance and creditworthiness. The data used are derived from the World Bank and refer to a five-year period (1995-1999). A comparison with discriminant analysis is also performed to evaluate the relative discriminating performance of UTADIS and M.H.DIS methods as opposed to a well-known multivariate statistical technique with numerous applications in financial decision-making problems (including country risk assessment).

Compared to previous studies on the use of MCDA methods in country risk assessment,[21,30,11,24,15,16] this study considers a richer set of data. In particular the data used in the analysis are the most recent ones that could be obtained, covering not simply a one year period, but a broader range of five years (1995-1999). Using this multi-period sample, the analysis is focused on the investigation of the predictive performance of developed country risk assessment models involving their ability to provide early warning signals for the problems that the countries may face regarding their performance and creditworthiness. The analysis of this significant issue is performed through the development of country risk models on the basis of the most recent data (year 1999) and then by testing the performance models on the data of the previous years (1995-1998).

The rest of the chapter is organized as follows. Section 2 presents a brief overview of the applications of MCDA approaches in country risk assessment, and provides a description of the proposed preference disaggregation methodologies (UTADIS and M.H.DIS methods). Section 3 is devoted to the application of the UTADIS and M.H.DIS methods in the assessment of country risk, and to their comparison with discriminant analysis. Finally, section 4 concludes the chapter and discusses some future research directions.

2. Multicriteria Classification Analysis

Multicriteria analysis, often called multiple criteria decision making (MCDM) by the American School and multicriteria decision aid (MCDA) by the European School, is a set of methods that allow for the aggregation of several evaluation criteria in order to choose, rank, sort or describe a set of alternatives. The flexibility of MCDA methods, their adaptability to the preferences of decision makers and to the dynamic environment of decisions related to country risk, and the subjective nature of such decisions, have already attracted the interest of many researchers in developing more reliable and sophisticated models for country risk assessment. Generally, four different approaches can be distinguished in MCDA:[32] (1) the outranking relations, (2) the multiattribute utility theory, (3) the multiobjective programming, and (4) the preference disaggregation.

The latter two approaches have already been applied in country risk assessment. Mondt and Despontin[21] and Oral *et al.*[24] proposed methodologies based on the multiobjective programming approach. More specifically, in their study Mondt and Despontin[21] used the perturbation method, a variant of the well-known STEM method,[5] in order to develop a portfolio of countries that could be financed by a bank. On the other hand, Oral *et al.*[24] proposed a goal programming formulation in order to estimate the parameters of a generalized logit model for country risk assessment, taking into account economic and political factors, as well as the geographical region of each country. The application of the preference disaggregation approach in country risk assessment was demonstrated in detail by Cosset *et al.*[11] They used the MINORA multicriteria decision support system, which is based on the UTASTAR preference disaggregation method, in order to develop a model for assessing country risk. Another study that applied the multicriteria decision aid framework in country risk assessment is that of Tang and Espinal[30] who used a simple multi-attribute model to assess country risk. Doumpos *et al.*[15] used the preference disaggregation approach in their country risk analysis. The methods applied were the UTASTAR, UTADIS and a variant of the UTADIS method (UTADIS I). Zopounidis and Doumpos[33] went further from their early study applying the UTASTAR method and the three variants of the UTADIS method (UTADIS I, II, III) in order to develop sorting and ranking country risk models. Finally, Doumpos and Zopounidis[34] proposed an alternative approach known as M.H.DIS to measure financial risks. The proposed approach based on MCDA was applied to the country risk problem to develop a model that classifies the countries into four groups based on their economic performance and creditworthiness.

During the last decade there have been significant changes in the world economic and political environment, which have directly affected the risk of each country. Consequently, new country risk models should be developed in order to consider the new conditions that govern the world economy. Furthermore, the advances in several scientific fields and more specifically, in MCDA provide new powerful tools in the study of complex decision problems including country risk assessment. The exploitation of the capabilities that these advances provide could result in the development of more reliable country risk models that can be used in real word cases by economic analysts of banks as well as from governmental officers, to drive real time estimations.

This is the basic motivation of the research presented in this chapter. The aim is to provide an integrated analysis of the country risk of 125 countries from the most economically developed ones to the less economically developed countries, by classifying them in classes according to their economic performance. A brief description of the proposed methods, UTADIS and M.H.DIS is presented below.

2.1. *The UTADIS Method*

The UTADIS method is a variant of the well-known UTA method (UTilites Additives) proposed by Jacquet-Lagrèze and Siskos.[20] The objective of the UTADIS method is to develop a criteria aggregation model used to determine the classification of alternatives in predefined homogeneous classes $C_1, C_2, \ldots, C_q$.[14] The groups are assumed to be defined in an ordinal way, such that group C_1 includes the countries with the highest performance/creditworthiness and group C_q includes the countries with the lowest performance/creditworthiness.

The method operates on the basis of a non-parametric regression-based framework that is similar to the one commonly used in traditional statistical and econometric classification techniques (e.g., discriminant analysis, logit, probit, etc.). Initially, using a training sample a classification model is developed. If the classification accuracy of the model in the training sample is satisfactory, then it can be used to any other sample for extrapolating purposes.

Formally, the classification model (criteria aggregation model) developed through the UTADIS method has the form of an additive utility function:

$$U'(\mathbf{g}) = \sum_{i=1}^{m} p_i u'_i(g_i) \in [0, 1] \tag{1}$$

where:

- $\mathbf{g}$ is the criteria vector $\mathbf{g}=(g_1, g_2, \ldots, g_m)$. In country risk analysis the criteria vector $\mathbf{g}$ consists of the country risk indicators used to measure the performance and creditworthiness of the countries.
- $p_i \in [0, 1]$ is the weight of criterion g_i (the criteria weights p_i sum up to 1).
- $u'_i(g_i)$ is the corresponding marginal utility function normalized between 0 and 1.

Conceptually, the global utility $U'(\mathbf{g}_j)$ of a country $\mathbf{x}_j$ is an aggregate index of the overall performance of the country on the basis of all criteria. The higher the global utility, the higher is the overall performance and creditworthiness of the country.

The aggregation made through the additive utility function considers both the performance of the countries on each criterion (country risk indicator) and the weight of the criterion (the higher the weight the more significant is the criterion). The performance of the country on each criterion is considered through the marginal utility functions $u'_i(g_i)$. The marginal utility functions provide a mechanism for transforming the criteria's scale into a utility/value scale ranging between 0 and 1. This enables the expression of the performance of the countries on each criterion in utility/value terms according to the intrinsic preferential/value system of the decision maker (country risk analyst). The higher the marginal utility of an alternative on a criterion (closer to 1), the higher is the performance of the country. Generally, the marginal utility functions are non-linear monotone functions defined on each criterion's range. These functions are increasing for criteria whose higher values indicate performance and decreasing in the opposite case (criteria of decreasing preference).

The problem with the use of the additive utility function (1) is that both the criteria weights p_i and the marginal utilities $u'_i(g_i)$ are unknown variables. Therefore the estimation of this utility function requires non-linear techniques, which are usually computationally intensive. This problem is addressed using the transformation $u_i(g_i) = p_i u'_i(g_i)$. Since $u'_i(g_i)$ is normalized between 0 and 1, it is clear that $u_i(g_i)$ ranges in the interval $[0, p_i]$. Thus, estimating the marginal utility function $u_i(g_i)$ is equivalent to estimating both the criteria weights p_i and the marginal utilities $u'_i(g_i)$. In this way, the additive utility function is simplified to the following form:

$$U(\mathbf{g}) = \sum_{i=1}^{m} u_i(g_i) \in [0, 1] \qquad (2)$$

The global utility defined on the basis of the equation (2) serves as an index used to decide upon the classification of the countries into the predefined classes. The classification is performed through the comparison of the global utilities of the countries to some utility thresholds $u_1 > u_2 > \ldots > u_{q-1}$ that define the lower bound of each class:

$$\left.\begin{array}{l} U(\mathbf{g}_j) \geq u_1 \Rightarrow \mathbf{x}_j \in C_1 \\ u_2 \leq U(\mathbf{g}_j) < u_1 \Rightarrow \mathbf{x}_j \in C_2 \\ \ldots\ldots\ldots\ldots\ldots\ldots\ldots\ldots\ldots \\ U(\mathbf{g}_j) < u_{q-1} \Rightarrow \mathbf{x}_j \in C_q \end{array}\right\} \quad (3)$$

The development of the additive utility function and the specification of the utility thresholds is performed using linear programming techniques so as to minimize the violations of the classification rules (2) by the countries considered during model development (training sample). Details of the model development process can be found in the work by Doumpos and Zopounidis.[14]

2.2. *The M.H.DIS Method*

The M.H.DIS method has been proposed as a non-parametric approach to study discrimination problems involving two or more ordered groups of alternatives.[34] The employment of a hierarchical process for the classification of alternatives to groups using available information and holistic judgments, are the main distinctive features of the M.H.DIS method. A second major difference between the two methods involves the mathematical programming framework used to develop the classification models. Model development in UTADIS is based on a linear programming formulation followed. In M.H.DIS, the model development process is performed using two linear programs and a mixed integer that gradually adjust the developed model so that it accommodates two objectives: (1) the minimization of the total number of misclassifications, and (2) the maximization of the clarity of the classification. These two objectives are pursued through a lexicographic approach, i.e., initially the minimization of the total number of misclassifications is required and then the maximization of the clarity of the classification is performed. The common feature shared by both M.H.DIS and UTADIS involves the form of the criteria aggregation model that is used to model the decision maker's preferences in classification problems. Both methods employ a utility–based framework.

The development of discrimination models through the M.H.DIS method is achieved through a regression procedure similar to the one used in UTADIS. Initially, a training sample consisting of n alternatives $\mathbf{x}_1, \mathbf{x}_2 \ldots \mathbf{x}_n$, classified into q ordered classes C_1, $C_2 \ldots C_q$ is used for model development. The alternatives are described (evaluated) along a set of m evaluation criteria $\mathbf{g}=(g_1,\ g_2\ \ldots .g_m)$. The development of the discrimination model is performed so as to respect the pre-specified classification as much as possible. In that respect, the developed model should be able to reproduce the classification of the alternatives considered in the training sample. Once this is achieved the discrimination model can be used for extrapolation purposes involving the classification of any new alternative not included in the training sample.

The method proceeds progressively in the classification of the alternatives into the predefined classes, starting from class C_1 (best alternatives). The alternatives found to belong in class C_1 (correctly or incorrectly) are excluded from further consideration. In a second stage, the objective is to identify the alternatives that belong in class C_2. Once again, all the alternatives that are found to belong in this class (correctly or incorrectly) are excluded from further consideration, and the same procedure continues until all alternatives are classified into the predefined classes. The number of stages in this hierarchical discrimination procedure is $q-1$ (i.e., for two classes there will be only one stage; for three classes there will be two stages, etc). Throughout the hierarchical discrimination procedure, it is assumed that the decision maker's preferences are increasing monotone functions on the criteria's scale. This assumption implies that as the evaluation of an alternative on a criterion increases, the decision regarding the classification of this alternative into a higher (better) class is more favorable to a decision regarding the classification of the alternative into a lower (worse) class. According to this assumption the following general classification rule is imposed:

The classification of an alternative $\boldsymbol{x}$ into one of the predefined classes $C_1, C_2, ..., C_q$ should be determined on the basis of the utilities of the corresponding alternative decisions regarding the classification of $\boldsymbol{x}$, that is, on the comparison of the utility of classifying $\boldsymbol{x}$ into C_2, etc. The classification decision with the maximum utility is chosen.

The utilities used in the M.H.DIS method are estimated through an additive utility function similar to the ones used in UTADIS:

$$U_k(\mathbf{g}) = \sum_{i=1}^{m} u_{ki}(g_i) \in [0,1]$$

$U_k(\mathbf{g})$ denotes the utility of classifying any alternative into class C_k on the basis of the alternative's evaluations on the set of the criteria $\mathbf{g}$, while $u_{ki}(g_i)$ denotes the corresponding marginal utility function regarding the classification of any alternative into class $C_k$according to a specific criterion.

At each stage k of the hierarchical discrimination procedure ($k = 1, 2, \ldots, q-1$), two utility functions are constructed. The first one corresponds to the utility of the decision to classify an alternative into class C_k [denoted as $U_k(\mathbf{g})$], while the second one corresponds to the utility of the decision to classify an alternative into a class lower than C_k [denoted as $U_{\sim k}(\mathbf{g})$]. Both utility functions apply to all alternatives under consideration. Based on these two utility functions the classification of an alternative x with the evaluation g_x on the criteria is performed using the hierarchical procedure presented in 1.

Details of the model development process used in the M.H.DIS method can be found in the studies by Zopounidis and Doumpos,[34] as well as Doumpos and Zopounidis.[14]

3. Application

The performance of the UTADIS and M.H.DIS methods and their applicability in country risk assessment are explored in this section. The recent economic crises have demonstrated in the clearest way that country risk is a crucial risk factor with significant impact on any corporate entity with an international activity. The significance of the country risk assessment problem, along with its complexity that is due to the plethora of factors of different nature that are involved (i.e., macroeconomic, social, political factors, etc.) makes country risk assessment a challenging research problem where several scientific fields such as statistical analysis and operations research can provide significant contribution.

3.1. *Data Set Description*

This application entails the assessment of the country risk for 125 countries from different geographical regions all over the world. The selection was based on the accessibility of the data of the countries, in order to have an entire sample of data. The data used, are derived from the World Bank,

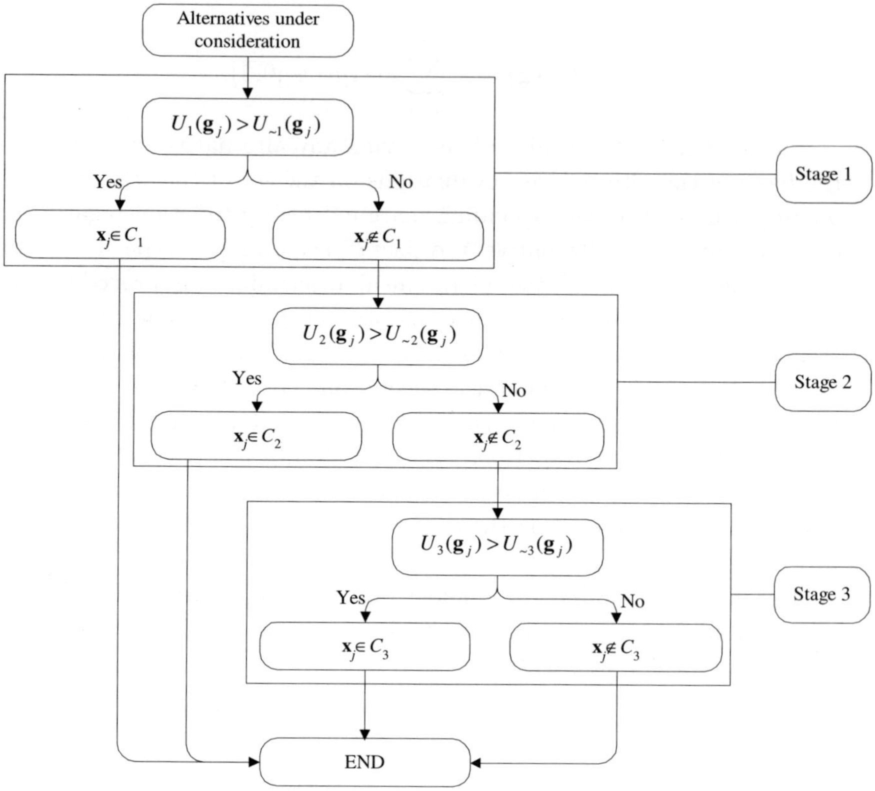

Fig. 1. The hierarchical discrimination procedure in M.H.DIS method (Source: Doumpos and Zopounidis, 2002)

and refer to a five-year period (1995-1999). They involve a significantly 38 indicators relative to country risk assessment including detailed external trade indicators, economic growth indicators, inflation and exchange rates, the balance of payments, tax policies, macroeconomic indicators, indicators upon structural transformation

Obviously, the incorporation of such a number of evaluation criteria would result in the development of an unfeasible country risk assessment model with limited practical value. To overcome this problem, a factor analysis is performed to select the most relevant criteria that best describe the economic performance and the creditworthiness of the countries.

On the basis of the factor analysis results (factor loadings) and the relevance of considered criteria to country risk assessment as reported in the international literature, 12 evaluation criteria are finally selected to be included in the developed country risk assessment model (2).

Table 1. Economic Indicators (Evaluation Criteria)

Gross international reserves in moths of imports
Trade as percentage of GDP
External balance as percentage of GDP
GNP annual growth rate
Total debt service to GDP ratio
Liquid liabilities as percentage of GDP
Inflation, GDP deflator
FDI, net inflows as percentage of GDP
Exports to GNP annual growth rate
Exports to GDP annual growth rate
Exports annual growth rate
Industry, value added as percentage of GDP

According to the World Bank the countries under consideration are grouped into four classes according to their income level:
high-income economies (class C_1): This group includes 28 countries, mostly European countries, United States, Australia, New Zealand, Canada, Japan, Hong Kong, Singapore, etc. These countries are considered as the world's top economies with a stable political and social development;
upper-middle income economies (class C_2): Twenty countries are included in this second group. They represent Europe, South and Eastern Asia, and South America. These countries cannot be considered as developed ones neither from the economic nor from the socio-political point of view. However, they do have some positive perspectives for future development;
lower-middle income economies (class C_3): The third group includes 37 countries from Eastern Europe, Asia, Africa and South Latin America. These countries are facing economic as well as social and political problems, that make their future doubtful and uncertain;
low-income economies (class C_4): This final group consists of 40 countries, mostly from Africa and Asia, who face significant problems from all aspects.

This classification constitutes the basis for the development of the appropriate country risk assessment model using the UTADIS and M.H.DIS methods.

3.2. *Presentation of Results*

Following the methodology that was described above, the UTADIS and M.H.DIS methods were applied in the sample data of 125 countries for five years, to develop classification country risk models according to the grouping and ranking provided by the World Bank. The most recent year is used as the training sample, while the previous years are used to test the generalizing performance of the methods. The obtained results of the two methods are presented in this section.

3.2.1. *Results of UTADIS*

The additive utility model developed through the UTADIS method is consistent with the predefined grouping of the countries according to their economic performance, which is related to the risk and creditworthiness of a country. The classification results of UTADIS are presented in Table 2. The elements $C_1 - C_1$, $C_2 - C_2$, $C_3 - C_3$and $C_4 - C_4$, represent the classification accuracy for each of the four classes, while all the other elements correspond to classification errors.

With regard to the training sample, the overall classification accuracy of UTADIS for the recent year 1999 is 86.83% and it classifies the countries for the previous years (1995-1999) less accurately.

UTADIS classifies almost correctly all the countries belonging to high-income economy group during the five-year period. It performs quite well in identifying the countries belonging to the low-middle income economies. Significant misclassification errors are obtained for upper-middle and lower-middle income economies.

The foreign direct investment as percentage of GDP, was found to be the dominant indicator in the classification of the countries, with a weight of 76.50 %. The rest of the evaluation criteria have rather similar significance in the developed classification model, ranging from 0.41 % for the industry, value added as percentage of GDP, to 6.01% for total debt service to GDP ratio (Table 3).

3.2.2. *Results of M.H.DIS*

Since the sample used involves four classes of countries, the hierarchical discrimination process of the M.H.DIS method consists of three stages. In the first stage, the discrimination among the countries belonging to the high-income economy group and the countries belonging to the rest of the

Table 2. Classification Results of UTADIS

Years	Original Classification	Estimated Classification				Overall Accuracy
		C1	C2	C3	C4	
1999	C1	96.43	3.57	0.00	0.00	86.83
	C2	5.00	80.00	15.00	0.00	
	C3	0.00	10.81	78.38	10.81	
	C4	0.00	0.00	7.50	92.50	
1998	C1	100.00	0.00	0.00	0.00	65.96
	C2	5.00	30.00	60.00	5.00	
	C3	2.70	10.81	51.35	35.14	
	C4	0.00	5.00	12.50	82.50	
1997	C1	100.00	0.00	0.00	0.00	63.61
	C2	5.00	30.00	60.00	5.00	
	C3	2.70	10.81	59.46	27.03	
	C4	0.00	2.50	32.50	65.00	
1996	C1	100.00	0.00	0.00	0.00	64.56
	C2	5.00	40.00	45.00	10.00	
	C3	2.70	13.51	43.24	40.54	
	C4	0.00	5.00	20.00	75.00	
1995	C1	96.43	0.00	3.57	0.00	64.14
	C2	10.00	45.00	30.00	15.00	
	C3	2.70	13.51	35.14	48.65	
	C4	0.00	0.00	20.00	80.00	

classes is performed. In the second stage, the countries belonging to the upper-middle economy group are discriminated from the countries of the lower-middle and the low-income economy groups. Finally, the third stage involves the discrimination among the countries of the lower-middle and the low-income economy group.

Table 3. Significance of Evaluation Criteria for UTADIS (weights in percentage

Evaluation criteria	Weight (%)
Gross international reserves in moths of imports	0.64
Trade as percentage of GDP	3.31
External balance as percentage of GDP	3.43
GNP annual growth rate	1.03
Total debt service to GDP ratio	6.01
Liquid liabilities as percentage of GDP	2.34
Inflation, GDP deflator	1.41
FDI, net inflows as percentage of GDP	76.50
Exports to GNP annual growth rate	0.75
Exports to GDP annual growth rate	1.09
Exports annual growth rate	3.08
Industry, value added as percentage of GDP	0.41

The classification results presented in Table 4 show that M.H.DIS classifies correctly all the countries in the groups they actually belong to for the year 1999, resulting in a classification accuracy of 100%. M.H.DIS performs almost correctly in identifying the countries belonging to the high-income and low-income economy groups for all the period of time under consideration. The classification accuracy for these groups varies from 96.43% to 100% for the first group and 62.50% to 100% for the last group.

Countries belonging to the upper-middle and lower-middle economy groups are assigned to other groups, resulting in significant classification errors. The classification accuracies for these groups range from 45.00% to 50.00% for the middle-income economies and 48.65% to 54.05% for the lower-middle income economies.

Finally, it is clear that the major problem in both methods is to identify the countries belonging to the upper-middle and lower-middle income economy groups. It should be pointed out that most of the countries belonging to upper-middle income economy group are assigned to lower-middle income economy group and vice versa.

Concerning the significance of the evaluation criteria (Table 5) in the classification model developed through the M.H.DIS method, total debt service to GDP ratio is clearly the dominant indicator which best discriminates among the countries belonging to the high-income economies from the rest of the countries. Its weights count for 39.87% and 30.01% for the first pair of utility functions. Inflation, GDP deflator (30.01%) and external balance as percentage of GDP (29.19%) are the most significant indicators able to discriminate the countries belonging to the upper-middle income economies from the rest of countries belonging to the lower-middle and low income economies. And finally, liquid liabilities as percentage of GDP (29.66%) and foreign direct investments (16.22%) are able to provide an accurate classification of the countries in the lower-middle and low-income economies respectively.

3.2.3. *Comparison with DA*

For comparison purposes, discriminant analysis (DA) is also applied in our case study. DA can be considered as the first approach to introduce multiple factors (variables) in the discrimination among different groups of objects. When there are more than two groups, the application of multiple discriminant analysis (MDA) leads to the development of linear discriminant functions that maximize the ratio of among-group to within group

Table 4. Classification results of M.H.DIS

Years	Original Classification	Estimated Classification				Overall Accuracy
		C_1	C_2	C_3	C_4	
1999	C_1	100.00	0.00	0.00	0.00	100.00
	C_2	0.00	100.00	0.00	0.00	
	C_3	0.00	0.00	100.00	0.00	
	C_4	0.00	0.00	0.00	100.00	
1998	C_1	100.00	0.00	0.00	0.00	70.34
	C_2	0.00	50.00	35.00	15.00	
	C_3	0.00	21.62	51.35	27.03	
	C_4	0.00	2.50	17.50	80.00	
1997	C_1	100.00	0.00	0.00	0.00	69.81
	C_2	0.00	60.00	25.00	15.00	
	C_3	2.70	21.62	56.76	18.92	
	C_4	0.00	15.00	22.50	62.50	
1996	C_1	100.00	0.00	0.00	0.00	69.66
	C_2	0.00	60.00	40.00	0.00	
	C_3	5.41	29.73	48.65	16.22	
	C_4	5.00	5.00	20.00	70.00	
1995	C_1	96.43	3.57	0.00	0.00	67.62
	C_2	5.00	45.00	30.00	20.00	
	C_3	8.11	24.32	54.05	13.51	
	C_4	0.00	7.50	17.50	75.00	

Table 5. Significance of Evaluation Criteria for M.H.DIS (weights in %)

Evaluation criteria	Utility functions					
	$U_1(\mathbf{g})$	$U_{\sim 1}(\mathbf{g})$	$U_2(\mathbf{g})$	$U_{\sim 2}(\mathbf{g})$	$U_3(\mathbf{g})$	$U_{\sim 3}(\mathbf{g})$
Gross international reserves	10.01	4.17	0.01	4.92	0.01	5.00
Trade as percentage of GDP	8.32	0.01	4.93	9.73	15.34	3.99
External balance as percentage of GDP	0.01	0.01	18.83	29.19	11.69	10.29
GNP annual growth rate	0.01	8.78	0.01	0.01	1.72	10.01
Total debt service to GDP ratio	39.87	30.01	21.95	22.22	16.26	4.27
Liquid liabilities as percentage of GDP	5.80	0.29	9.51	10.15	29.66	9.71
Inflation, GDP deflator	5.13	12.23	30.01	4.26	5.45	3.34
FDI, net inflows as percentage of GDP	1.69	29.07	4.88	4.92	14.09	16.22
Exports to GNP annual growth rate	7.53	0.79	9.82	4.80	1.72	10.12
Exports to GDP annual growth rate	21.60	12.26	0.04	9.79	0.01	9.69
Exports annual growth rate	0.01	2.35	0.01	0.01	0.01	3.57
Industry, value added as percentage of GDP	0.01	0.01	0.01	0.01	4.05	12.79

variability: this assumes that the variables follow a multivariate normal distribution and that the dispersion matrices of the groups are equal. In this case study MDA is selected for comparison purposes due to its popularity in the field of finance in studying financial decision problems requiring a grouping of set of alternatives.[3] Furthermore, the method is popular among academic researchers in evaluating the performance under new classification approaches. Finally, it should be noted that MDA has already been applied in several studies on country risk assessment.[26] The objective of performing the discriminant analysis is to examine how a different statistical approach could perform in this specific case study compared to the UTADIS and M.H.DIS methods.

Table 6. Classification Results of LDA, UTADIS and M.H.DIS (accuracy in %)

Year	LDA	UTADIS	M.H.DIS
1999	65.92	86.83	100.00
1998	65.34	65.96	70.34
1997	56.32	63.61	69.81
1996	52.37	64.56	69.66
1995	54.55	64.14	67.62

Looking to the results presented in Table 6, the overall classification accuracies of M.H.DIS are significantly higher than the classification accuracies of UTADIS and LDA for the five-year period. These results indicate that M.H.DIS performs better than UTADIS and LDA, although the differences between M.H.DIS and UTADIS are smaller compared to the differences between M.H.DIS and LDA. The higher difference of classification performance occurs for the year 1999.

4. Conclusions and Discussion

This chapter has presented an alternative approach for the analysis and evaluation of country risk. The proposed methodology based on the preference disaggregation approach of multicriteria decision aid, constitutes a flexible tool that can be used by economic analysts, managers of banks and international credit institutions, in order to derive integrated estimations concerning the assessment of country risk.

The country risk problem in this application was studied as a classification problem. The obtained results are very satisfactory since the obtained country risk models are consistent with the classification of the international institution, namely the World Bank. Both methods, UTADIS and M.H.DIS illustrated their ability to identify the countries under consideration in the four predefined classes. M.H.DIS performed more accurately in classifying the countries in their original groups demonstrating a higher efficiency in the analysis of complex real-word decision problems regarding financial risk assessment. The results obtained through the comparison with discriminant analysis and the UTADIS pronounced this remark. Such an approach provides decision makers (financial/credit/stock market analysts, investors, etc.) with a valuable tool to perform real-time evaluations on the financial risks of the considered alternatives.

Based on this approach additional comparative methods such as logistic regression, neural networks and machine learning algorithms could be applied to provide real-time support in the study of decision problems related to country risk assessment. Further research is required using a broader set of data, focusing more on social and political indicators. New combinations of different methods could be made to provide integrated support to analysts in the study of country risk.

References

1. B. Abassi and R.J. Taffler. Country risk: A model of economic performance related to debt servicing capacity. Working paper 36, City University Business School, London (1982).
2. T. Agmon and Deitrich J.K. International lending and income redistribution: An alternative view of country risk. *Journal of Banking and Finance* **7**, 483-495 (1983).
3. El. Altman, R. Avery, R. Eisenbeis, and J. Stinkey. *Application of Classification Techniques in Business, Banking and Finance.* JAI Press, Greenwich (1981).
4. E. M. Balkan. Political instability, country risk and probability of default. *Applied Economics* **24**, 999-1008 (1992).
5. R. Benayoun, J. de Montgolfier, J. Tergny, and O. Larichev. Linear programming with multiple objective function: Stem method (STEM). *Mathematical Programming* **1**, 3, 366-375 (1971).
6. J. Calverly. *Country Risk Analysis.* Butterworth and Co (Publishers) Ltd, Second Edition, 3-4 (1990).
7. J.T. Citron and G. Nickelsburg. Country risk and political instability. *Journal of Development Economics* **25**, 385-392 (1987).
8. W.D. Cook and J.H. Hebner. A multicriteria approach to country risk evaluation: With an example employing Japanese Data. *International Review of Economics and Finance* **2**, 4, 327-348 (1993).
9. J. C. Cosset and J. Roy. The determinants of country risk ratings. Document de Travail 89-43, Universite Laval, Quebec, Canada (1989).
10. J. C. Cosset and J. Roy. Expert judgments of political riskiness: An alternative approach. Document de Travail 88-12, Universite Laval, Quebec, Canada (1998).
11. J.C. Cosset, Y. Siskos, and C. Zopounidis. Evaluating country risk: A decision support approach. Global Finance Journal **3**, 1, 79-95 (1992).

12. P. Dhonte. Describing external debt situations: A roll-over approach. *IMF Staff Papers* **22**, 159-186 (1975).
13. M. Doumpos and C. Zopounidis. Assessing financial risk using a multicriteria sorting procedure: the case of country risk assessment. *Omega: The International Journal of Management Science*, **29**, 97-109 (2001).
14. M. Doumpos and C. Zopounidis. *Multicriteria Decision Aid Classification Methods*. Dordrecht: Kluwer Academic Publishers, (2002).
15. M. Doumpos, C. Zopounidis, and M. Anastassiou. Assessing country risk using multicriteria analysis. In: *Operational Tools in the Management of Financial Risks*, C. Zopounidis (ed.), Kluwer Academic Publishers, Dordrecht 309-326 (1997).
16. M. Doumpos, K. Pendaraki, C. Zopounidis, and C. Agorastos. Assessing country risk using a multi–group discrimination method: A comparative analysis. *Managerial Finance*, **27**, 7-8, 16-34 (2001).
17. G. Feder and R. Just. A study of debt servicing capacity applying logit analysis. *Journal of Development Economics* **4**, 25-38 (1977).
18. G. Feder and L.V. Uy. The determinants of international creditworthiness and their policy implications. *Journal of Policy Modeling* **7**,1, 133-156 (1985).
19. C. R. Frank and R. Cline. Measurement of debt servicing capacity: An application of discriminant analysis. *Journal of International Economics* **1**, 327-344 1971.
20. E. Jacquet-Lagrèze and Y. Siskos. Assessing a set of additive utility functions for multicriteria decision making: The UTA method. *European Journal of Operational Research* **10**, 151-164 (1982).
21. K. Mondt and M. Despontin. Evaluation of country risk using multicriteria analysis. Technical Report, Vrije Universite Brussel (September 1986).
22. J.L. Mumpower, S. Livingston, and T.J. Lee. Expert judgments of political riskiness. *Journal of Forecasting* 6, 51-65 (1987).
23. K.B. Nordal. Country risk, country risk indices and valuation of FDI: a real options approach. *Elsevier: Emerging Markets Review* **2** 197-217(2001).
24. M. Oral, O. Kettani, J.C. Cosset, and D. Mohamed. An estimation model for country risk rating. *International Journal of Forecasting* **8**, 583-593 (1992).
25. F.M. Place. Information quality, country risk assessment, and private bank lending to less-developed countries. UMI Dissertation Services (1989).
26. K. G. Saini and P.S. Bates. Statistical techniques for determining debt-servicing capacity for developing countries: Analytical review of the literature and further empirical results. *Federal Reserve Bank of New York Research Paper No. 7818* (1978).
27. N. Sargen. Use of economic indicators and country risk appraisal. *Economic Review, Federal Reserve Bank of San Francisco*, San Francisco, CA (1977).
28. A.C. Shapiro. Currency risk and country risk in international banking. *Journal of Finance* XL, 3, 881-893 (1985).
29. R.J. Taffler and B. Abassi. Country risk: A model for predicting debt servicing problems in developing countries. *Journal of the Royal Statistical Society* **147**, 4, 541-568 (1984).

30. J.C.S. Tang and C.G. Espinal. A model to assess country risk. *Omega: The International Journal of Management Science* **17**, 4, 363-367 (1989).
31. World Bank. World Development Indicators. World Bank Publications (2001).
32. C. Zopounidis. Multicriteria Decision Aid in financial management. *European Journal of Operational Research*, **119**, 404-415 (1997).
33. C. Zopounidis and M. Doumpos. A multicriteria decision aid methodology for the assessment of country risk. *European Research on Management and Business Economics* **3**, 3, 13-33 (1997).
34. C. Zopounidis and M. Doumpos. Building additive utilities for multi-group hierarchical discrimination: The M.H.DIS method. *Optimization Methods and Software*, **14**, 3, 219-240 (2000).

CHAPTER 5

ASSESSING EQUITY MUTUAL FUNDS' PERFORMANCE USING A MULTICRITERIA METHODOLOGY: A COMPARATIVE ANALYSIS

K. Pendaraki

Technical University of Crete,
Dept. of Production Engineering and Management,
Financial Engineering Laboratory,
University Campus, 73100 Chania, Greece
E-mail: dina@ergasya.tuc.gr

M. Doumpos

Technical University of Crete,
Dept. of Production Engineering and Management,
Financial Engineering Laboratory,
University Campus, 73100 Chania, Greece
E-mail: dmichael@ergasya.tuc.gr

C. Zopounidis

Technical University of Crete,
Dept. of Production Engineering and Management,
Financial Engineering Laboratory,
University Campus, 73100 Chania, Greece
E-mail: kostas@ergasya.tuc.gr

Becoming more and more popular, mutual funds have begun to play an increasingly important role in financial markets. In particular, the evaluation of the performance of mutual funds has been a very interesting research topic not only for researchers, but also for managers of financial, banking and investment institutions. This chapter explores the performance of a non-parametric approach in developing mutual fund's performance models. The proposed approach is based on the UTADIS (Utilités Additives DIScriminates) multicriteria decision aid method. The data set used to examine the mutual funds' performance consists of daily data of the Greek domestic equity mutual funds, and is derived from the Alpha Trust Mutual Fund Management Company S.A. (A.E.D.A.K.). The sample consisting of 33 mutual funds is used to estimate the performance of the method in classifying the funds into two groups. A cross-validation procedure is employed to evaluate the predictive performance of the models and a comparison with linear discriminant analysis is also performed. The results indicate the superiority of the UTADIS method as opposed to the traditional discrimination technique, while the developed models are accurate in classifying the total sample correctly with rate approximately 80% (overall accuracy).

Keywords: Mutual fund's performance, multicriteria decision aid, cross-validation.

1. Introduction

Within the E.U at present 26,512 Mutual Funds operate, with total assets rising to EURO 3,503 bn (data as of 31/12/2001; Association of Greek Institutional Investors). In the same way, the industry of collective investments in Greece is growing rapidly. According to recent data of the Association of Greek Institutional Investors, today, 27 Mutual Fund Management Companies are managing 266 Mutual Funds, with assets rising to 23.86 bn EURO (data as of 29/03/2002). A decade earlier (in 1990s), there were operating only 7 Mutual Fund Management Companies which were managing only 7 mutual funds with assets rising to 431.4 million EURO. The American Investment Company Institute counts more than 8,200 mutual funds when the listed companies in the Stock Exchanges of NYSE and NASDAQ at the end of 1999 were about 7,800.

This situation highlights the great growth of the Mutual Fund Market worldwide. Thus, it is very difficult for investors to choose funds according to their decision policy, the risk levels that are willing to take, and their profitability goals. Today, in USA numerous business magazines, private firms, and financial institutions are specialized in giving regular rankings and ratings of mutual funds. Representative examples are the evaluations of funds given by Morningstar[26] and the two well-known investors services of Moody's[27] and Standard & Poor's,[32] which greatly influence U.S. investor behavior.

In Greece, there are no such institutions regarding the evaluation of mutual fund performance available to the Greek investors. The adoption of the evaluation systems of the foreign markets in the Greek capital market is not feasible, because these systems are based in specific characteristics that is not possible to be complied with the Greek market features. According to Sharpe,[31] such measures, like Morningstar's, are appropriate measures to investors that place all their money in one fund. Morningstar makes the assumption that investors have some other basis for allocating funds and plan to use information provided by Morningstar in the case that they have to come up with a decision regarding which fund or funds to choose from each peer group. Thus, such measures are not appropriate performance measures when evaluating the desirability of a fund in a multifund portfolio, where the relevant measure of risk is the fund's contribution to the total risk of the portfolio.

The analysis of the nature and definition of risk in portfolio selection and management shows that the risk is multidimensional and is affected by a series of financial and stock market data, qualitative criteria and macroeconomic factors which affect the capital market. Many of the models used in the past are based on one-dimensional approaches that do not fit to the multidimensional nature of risk.[5,19]

The empirical literature on the evaluation of the performance of mutual fund portfolios includes the Treynor index,[34] the Sharpe's index,[30] the Jensen's performance index,[22] the Treynor-Mazuy model,[35] the Henriksson-Metron model,[18] the CAPM, and several optimization models. Even though the performance measures proposed in past studies have been widely used in the assessment of portfolio performance, researchers have noted several restrictions in their application, such as the use of a proxy variable of the theoretical market portfolio that can be criticized as inadequate, the evaluation of the performance of an investment manager for long and not short time periods, the acceptance of the assumption of borrowing and lending with the same interest rate, the validity of the Capital Asset Pricing Model, the consistency of the performance of investment managers over time, etc.

The multicriteria decision aid (MCDA) provides the requisite methodology framework in handling the problem of portfolio selection and management through a realistic and an integrated approach.[20] MCDA methods incorporate the preferences of the decision-maker (financial/credit analysts, portfolio managers, managers of banks or firms, investors, etc.) into the analysis of financial decision problems. They are capable of handling qualitative criteria and are easily updated, taking into account the dynamic nature of the decision environment as well as the changing preferences of the decision-maker.

On the basis of the MCDA framework, this chapter proposes the application of a methodological approach, which addresses the mutual funds' performance assessment problem through a classification approach. Precisely, in this chapter: (a) a factor analysis is used for the selection of appropriate variables which best describe the performance of mutual funds, (b) a MCDA classification method (UTADIS) is used to identify high performance mutual funds, (c) a leave-one-out cross-validation approach is employed for model validation, and (d) a comparison with a well-known multivariate statistical technique (discriminant analysis) is performed. On the basis of this approach, the objective is to develop classification models that can be used to support the mutual funds' performance assessment process by classifying 33 Greek domestic equity mutual funds into two groups.

The rest of the chapter is organized as follows. Section 2 reviews the past research on mutual fund appraisal. Section 3 outlines the main features of the proposed multicriteria methodology. Section 4 is devoted to the application of the proposed methodology, underlines the variables and gives a brief description of the data set used, while Section 5 describes the obtained empirical results. Finally, Section 6 concludes the chapter and summarizes the main findings of this research.

2. Review of Past Empirical Studies

According to prior research, consumers pay great attention to the selection of the mutual funds that best accommodate their own financial situation.[25] Thus, it is obvious that mutual funds classes are helping investors to choose funds according to their decision policy, the risk levels that are willing to take, and their profitability needs. Today, there has been a wide variety of studies regarding the development of different models for the evaluation of the performance of mutual funds.

Friend *et al.*[13] presented the first extensive and systematic study of mutual funds. They created an index of five securities with the elements weighted by their representation in the mutual funds sample under consideration. According to their results, there is no strong relationship between turnover rates and performance. In 1966, two papers were written that dominated the area of mutual funds investment performance for the next twenty-five years. Sharpe[30] in his study calculated the reward-to-volatility ratio and found that the better performing funds tended to be those with the lower expenses. Furthermore, he showed that performance could be evaluated with a simple theoretically meaningful measure that considers both average return and risk. These results were very soon confirmed by the results of Jensen's research work.[22] He used the capital market line in order to calculate a performance measure (Jensen's alpha) for his data. Using this measure he concluded that the examined mutual funds were on average not able to predict security prices well enough to outperform a "buy-the market-and-hold" policy.

Lehmann and Modest[23] in their research work tried to ascertain whether conventional measures of abnormal mutual fund performance are sensitive to a benchmark chosen to measure normal performance. They employed the standard CAPM benchmarks and a variety of Arbitrage Pricing Theory (APT) benchmarks in order to give an answer to the previous question. Cumby and Clen[6] examined the performance of internationally diversified mutual funds. They used two performance measures, the Jensen measure and the positive weighting measure proposed by Grinblatt and Titman[14] and found that there is no evidence that funds provide investors with performance that surpasses that of a broad, international equity index over the examined period.

Brockett *et al.*[2] in their empirical analyses of mutual fund investment strategies used a chance constrained programming approach in order to maximize the possibility of the performance of a mutual fund portfolio to exceed the performance of the S&P 500 index formalizing risk and return relations. Grinblatt and Titman[15] examined the sensitivity of performance inferences to benchmark choice; they compared the Jensen measure with two new measures that were developed in order to overcome the timing-related biases of the Jensen measure, and finally they analyzed the relationship between the mutual fund performance and the funds attributes. They concluded that the measures generally yield similar inferences when using different benchmarks and the tests of fund performance that employ fund characteristics suggest that turnover is significantly positively related to the ability of fund managers to earn abnormal returns.

Chiang *et al.*[4] used an artificial neural network method in order to develop forecasting models for the prediction of end-of-year net asset values of mutual funds, taking into account historical economic information. They compared their forecasting results to those of traditional econometric techniques and concluded that neural networks significantly outperform regression models in situations with limited data availability. Murthi *et al.*[28] examined the efficiency of mutual fund industry by different investment objectives. They tried to overcome the limitations of traditional indices, proposing a new measure of performance that is calculated through the data envelopment analysis. O'Neal[29] in his research work tried to investigate whether the investors can receive diversification benefits from holding more than a single mutual fund in their portfolios. The results given by the simulation analysis that he conducted showed that the time-series diversification benefits are minimal but the expected dispersion in terminal-period wealth can be substantially reduced by holding multiple funds.

Indro et al.[21] used artificial neural networks in order to predict mutual fund performance. Precisely, they used the fund's five-year annualized return, the turnover of the fund's portfolio, the price-earnings ratio, the price-book ratio, the median market capitalization, the percentage of cash and the percentage of stock (in relation to the fund's portfolio) to predict the mutual fund performance, which is measured by the fund's risk-adjusted return. They used a multi-layer model and a nonlinear optimizer taking into account fund-specific historical operating characteristics in order to forecast mutual funds' risk adjusted return. They concluded that whether the neural network approach is superior to linear models for predicting mutual fund performance depends on the style of the fund. Morrey and Morrey[25] in their empirical analysis used two basic quadratic programming approaches in order to identify those funds that are strictly dominated, regardless of the weightings on the different time horizons examined, relative to their mean returns and risks. Furthermore, these approaches endogenously determine a custom-tailored benchmark portfolio to which each mutual fund's performance is compared.

Dalhquist et al.[7] studied the relation between fund performance and fund attributes in the Swedish market. They examined 130 equity mutual funds for the period 1993-97. According to their work, performance is measured as the constant term in a linear regression of fund returns on several benchmark assets, allowing for time-varying betas. They came up with the conclusion that good performance occurs among small equity funds, low fee funds, funds whose trading activity is high and in few cases funds with good past performance. Wermers[37] in his study performed a comprehensive analysis of mutual fund industry through a new database that allows an analysis of mutual funds in both the stock holdings level and the net return level from 1975 to 1994. He decomposed performance into several components to analyze the value of active fund managers. According to the results of the application of the performance decomposition methodology (characteristic selectivity and timing measures, average style measure, and execution costs) followed in this study, funds that hold stocks outperform the market, whereas their net returns underperform the market. Thus, funds include stocks to cover their costs. Finally, there is evidence that supports the value of active mutual fund management.

These results are important for managing the performance of a portfolio of mutual funds. Gruber[17] in his study identified the risk structure of mutual fund returns for 270 funds over the period 1985-1994 and for 274 funds over the period 1985-1995. Precisely, he used a four-index model employing the S&P Index, and publicly available size, growth and bond indexes in order to examine what influences generate mutual fund returns and develop a model for measuring performance. He used factor analysis and proved that a fifth index appears to be present. In the case where he tested a publicly available index of growth mutual fund performance he found out that it explains a large proportion of the residuals from a four-index model. Finally, the data suggested that cluster analysis could be best used as an added influence to the based model. On the other hand, adding an index based on the dividend yield value index to the base model with a Morningstar Growth Fund Index explained correlation in a better way.

Zopounidis and Pendaraki[39] presented an integrated multicriteria decision aid methodology for the portfolio selection and composition problem in the case of equity mutual funds over the period 1997-1999. The methodology used consists of two stages. In the first stage the mutual funds are ranked according to their performance through the PROMETHEE II method based on several different weighting scenarios, in order to select from the total set of mutual funds, the best performers. In the second stage of this methodology a goal programming formulation was used in order to solve the mutual funds portfolio composition problem specifying the proportion of each fund in the constructed portfolio. The proposed integrated approach constitutes a significant tool that can be used to provide answers to two vital questions: (a) which funds are the most suitable to invest, and (b) what portion of the available capital should be invested in each one of these funds.

The present study explores the performance of a non-parametric approach based on the UTADIS method, in developing mutual fund's performance models.

3. The UTADIS Multicriteria Decision Aid Method

The method used to classify the mutual funds in two groups in this study, is the UTADIS multicriteria decision aid method. The UTADIS method is aimed at developing an additive utility model for the classification of a set of alternatives in predefined homogeneous classes with minimum classification error.[10] In the considered case, the alternatives correspond to the mutual funds, whereas the classification involves two groups, i.e., the high performance funds and the low performance ones.

The method operates on the basis of a non-parametric ordinal regression-based framework that is similar to the one commonly used in traditional statistical and econometric classification techniques (e.g., discriminant analysis, logit, probit, etc.). Initially, using a training sample the classification model is developed. If the classification accuracy of the model in the training sample is satisfactory, then it can be used to any other sample for extrapolating purposes. The model development process is briefly outlined below (a detailed description can be found in Ref. 38).

Let the training sample consist of n mutual funds (objects) $a_1, a_2, \ldots, a_n$ described over a set of m evaluation criteria (variables) $g_1, g_2, \ldots, g_m$. The funds under consideration are classified into q ordered classes $C_1, C_2, \ldots, C_q$ (C_k is preferred to C_{k+1}, k=1, 2, ..., $q-1$). The additive utility model, which is developed through the UTADIS method, has the following form:

$$U(a) = \sum_{i=1}^{m} u_i[g_i(a)] \text{ }^{a}$$

where $U(a)$ is the global utility of a fund a and $u_i[g_i(a)]$ is the marginal utility of the fund on the evaluation criterion g_i. To classify the funds, it is necessary to estimate the utility thresholds $u_1, u_2, \ldots, u_{q-1}$ (threshold u_k distinguishes the classes C_k and C_{k+1}, $\forall k \leq q-1$). Comparing the global utilities of a fund a with the utility thresholds, the classification is achieved through the following classification rules:

$$\left.\begin{array}{l} U(a) \geq u_1 \Rightarrow a \in C_1 \\ u_2 \leq U(a) < u_1 \Rightarrow a \in C_2 \\ \ldots\ldots\ldots\ldots\ldots\ldots\ldots \\ u_k \leq U(a) < u_{k-1} \Rightarrow a \in C_k \\ \ldots\ldots\ldots\ldots\ldots\ldots\ldots \\ U(a) < u_{q-1} \Rightarrow a \in C_q \end{array}\right\}$$

Estimations of the global utility model (additive utility function) and utility thresholds are accomplished through solution of the following linear program:

[a]This form implies that the marginal utility functions $u_i[g_i(a)]$ are not normalized between 0 and 1. In the case where the marginal utility functions of each criterion are normalized, then the utility function can be equivalently written as $U(a) = \sum_{i=1}^{m} p_i u_i[g_i(a)]$, where p_i represents the weight of criterion i.

$$\text{Minimize}$$
$$F = \sum_{a \in C_1} \sigma^+(a) + ... + \sum_{a \in C_k} [\sigma^+(a) + \sigma^-(a)] + ... + \sum_{a \in C_q} \sigma^-(a)$$
$$\text{s.t.}$$
$$\sum_{i=1}^{m} u_i[g_i(a)] - u_1 + \sigma^+(a) \geq 0, \forall a \in C$$
$$\left.\begin{aligned} \sum_{i=1}^{m} u_i[g_i(a)] - u_{k-1} - \sigma^-(a) \leq -\delta \\ \sum_{i=1}^{m} u_i[g_i(a)] - u_k + \sigma^+(a) \geq 0 \end{aligned}\right\}, \forall a \in C_k$$
$$\sum_{i=1}^{m} u_i[g_i(a)] - u_{q-1} - \sigma^-(a) \leq -\delta, \forall a \in C_q$$

$$\sum_{i=1}^{m} \sum_{j=1}^{a_i - 1} w_{ij} = 1$$

$$u_{k-1} - u_k \geq s,\ k = 2, 3, ..., q-1$$
$$w_{ij} \geq 0,\ \sigma^+(a) \geq 0,\ \sigma^-(a) \geq 0,$$

where a_i is the number of subintervals $[g_i^j, g_i^{j+1}]$ into which the range of values of criterion g_i is divided, $w_{ij} = u_i(g_i^{j+1}) - u_i(g_i^j)$ is the difference between the marginal utilities of two successive values g_i^j and g_i^{j+1} of criterion g_i ($w_{ij} \geq 0$), δ is a threshold used to ensure that $U(a) < u_{k-1}$, $\forall a \in C_k$, $2 \leq k \leq q-1$ ($\delta > 0$), s is a threshold used to ensure that $u_{k-1} > u_k$ ($s > \delta > 0$), and $\sigma^+(a)$ and $\sigma^-(a)$ are the classification errors (overestimation and underestimation errors, respectively).

After the solution F^* of this linear program has been obtained, a post–optimality stage is performed to identify, if possible, other optimal or near optimal solutions, which could provide a more consistent representation of the decision maker's preferences. These correspond to error values lower than $F^* + k(F^*)$, where $k(F^*)$ is a small fraction of F^*. Through post–optimality analysis, a range is determined for both the marginal utilities and the utility thresholds, within which there is an optimal or near–optimal solution. In this way, the robustness of the developed classification model is examined.[38,39]

The UTADIS method has been applied to several fields of financial management including bankruptcy prediction, credit risk assessment, country risk evaluation, credit cards assessment, portfolio selection and management.[9,10]

4. Application to Mutual Funds' Performance Assessment

4.1. *Data Set Description*

4.1.1. *Sample*

The sample used in this application is provided from the Alpha Trust Mutual Fund Company S.A. (A.E.D.A.K.) and consists of daily data of all domestic equity mutual funds over the period 1999-2001. Precisely, daily returns for all domestic equity mutual funds are examined for the 3-years period (1999-2001; 752 observations). At the end of 2001, the sample consisted of 72 domestic equity mutual funds. The number of mutual funds in the sample is not fixed through out the three-year period, examined. This occurs mainly because of the varying starting point of their operation. From the total set of mutual funds 33 are selected, which are the ones operating during the entire examined period. Further information is derived from the Athens Stock Exchange and the Bank of Greece, regarding the return of the market portfolio and the return of the three-month treasury bill respectively.

The starting year 1999 of the examined period has been characterized as the year of equity mutual funds. During the whole year, equity mutual funds presented high returns, in contrast to the subsequent two years, 2000 and 2001. In the examined period (1999-2001) the mean return of equity mutual funds ranged between –16,65% to 67,45%, while the percentage change of net asset value ranged between –22,72% to 2840,32%. The variation in these percentages among different mutual funds occur due to the excessive growth that some funds presented in their net asset value and the inflows by investors to these mutual funds.

The mutual funds under consideration are categorized into two groups according to their performance in the first semester of the year 2002:

(a) Group 1: High performance funds $[R_{pt} > R_{Mt} + 20\% R_{Mt}]$, and
(b) Group 2: Low performance funds $[R_{pt} < R_{Mt} + 20\% R_{Mt}]$,

where R_{pt} = return of mutual fund in 2002, and R_{Mt} = return of market portfolio in 2002.

4.1.2. *Evaluation Criteria*

The criteria that are used to evaluate mutual fund performance in the three years of the analysis are: (1) Return in the 3-years period, (2) Mean Return, (3) Standard Deviation of Returns, (4) Coefficient of Variation, (5)

Percentage change of net asset value in the 3-years period, (6) Geometric Mean of excess Return over Benchmark, (7) Value at Risk, (8) Sharpe Index, (9) Modigliani measure, (10) Information ratio, (11) beta coefficient (β), (12) Treynor Index, (13) Jensen's alpha (α) coefficient, (14) Treynor & Mazuy's α coefficient, (15) Treynor & Mazuy's γ coefficient, (16) Henriksson & Merton's α coefficient, (17) Henriksson & Merton's γ coefficient, and (18) Treynor and Black appraisal ratio. All these variables refer to different performance and risk measures and are briefly described below.

The return on a mutual fund investment includes both income (in form of dividends or interest payments) and capital gains or losses (the increase or decrease in the value of security). The return is calculated net of management fees and other expenses charged to the fund. Thus, a funds' return in the period t is expressed as follows:

$$R_{pt} = \frac{NAV_t + DIST - NAV_{t-1}}{NAV_{t-1}}$$

where NAV_t = net asset value per unit of the mutual fund in the period t, NAV_{t-1} = net asset value per unit of the mutual fund in the period $t-1$, and $DIST_t$ = dividend of the mutual fund in the period t.

The basic measure of variability is the standard deviation, also known as the volatility. For a mutual fund the standard deviation is used to measure the variability of daily returns presenting the total risk of the fund.

An alternative measure of risk refers to the coefficient of variation. The coefficient of variation measures the risk per unit of return achieved, and takes positive or negative values and values higher or lower than unity. The utility of this coefficient refers to the comparison of total risk among mutual funds.

The computation of the arithmetic average of daily returns for a period of time is not the same as the daily rate of return that would have produced the total cumulative return during the examined period. The latter is equivalent to the geometric mean of daily returns, calculated as follows:

$$R_{gt} = \sqrt[N]{\Pi(1 + R_t)}$$

where R_{gt} is the geometric mean for the period of N days. Investors are not interested in the returns of a mutual fund in isolation but in comparison to some alternative investment free of risk. Thus, another simple measure of return of a mutual fund refers to the geometric mean of excess return over a benchmark such as the return of the three months treasury bill (risk free interest rate). The excess return of a fund is referred as the fund's return minus the risk-free rate. The geometric mean of fund's excess return over a benchmark shows how well the manager of a fund was able to pick stocks. For example, a geometric mean of fund's excess return over the benchmark equal to 6% means that the fund was able to beat its benchmark by 6% in the examined period.

Another well-known measure of risk is Value at Risk (VAR). The popularity of VAR was much enhanced by the 1993 study by the Group of Thirty, *Derivatives: Practices and Principles*, which strongly recommended VAR analysis for derivatives trading. The VAR measure gives an answer in the question " How much can the value of a portfolio decline with given probability in a given time period?". The calculation of VAR is based on certain assumptions about the statistical distribution of the fund's return. Precisely, in order VAR to be calculated the assumption that returns follow normal distribution is done. The VAR measure is defined as follows: VAR in period t = Mean Return in period t - 1.96 Standard Deviation of Mean Return in period t. The power of VAR models refer to the construction of a measure of risk for a portfolio not from its own past volatility but from the volatilities of risk factors affecting the portfolio as it is constituted today. It is a measure highly correlated with volatility because it is proportional to standard deviation.

The traditional total performance measures, Sharpe index (1966), and Treynor index (1965) are used to measure the expected return of a fund per unit of risk. These measures are defined as follows:

$$\text{Sharpe index} = (R_{pt} - R_{ft})/\sigma_{pt},$$
$$\text{Treynor index} = (R_{pt} - R_{ft})/\beta_p,$$

where R_{pt} = return of mutual fund in period t, R_{ft} = return of Treasury bill (risk free interest rate) in period t, σ_{pt} = standard deviation of mutual fund return (total risk of mutual fund) in period t, and β_p = systematic risk of mutual fund.

The Sharpe index or alternatively the reward-to-variability ratio is a useful measure of performance. Precisely, the Sharpe index is calculated by dividing the fund's average excess return by its standard deviation. In other words, the numerator shows the reward provided by the investor for bearing risk, while the denominator shows the amount of risk actually bear. It is obvious that this ratio is the reward per unit of variability. Furthermore, the Sharpe index represents a relevant measure of mutual fund performance for investors who are not well diversified and, therefore, are concerned with their total risk exposure when evaluating mutual fund performance. The Sharpe performance measure reflects both the differences in returns to each fund and the level of mutual fund diversification.

The Treynor index is obtained by simply substituting variability (the change in the rate of return on a fund associated with 1% change in the rate of return on, say, the market portfolio) by volatility in the formula of the Sharpe index. Thus, the Treynor index is similar to the Sharpe index except that the performance is measured as the risk premium per unit of systematic (β_p) and not of total risk (σ_{pt}). Precisely, the Treynor index is calculated by dividing the fund's average excess return by the β_p coefficient.

The evaluation of mutual funds with those two indices (Sharpe & Treynor) shows that a mutual fund with higher performance per unit of risk is the best managed fund, while a mutual fund with lower performance per unit of risk is the worst managed fund.

Modigliani and Modigliani[24] proposed an alternative measure of risk-adjusted performance that an average investor can easily understand. This measure is defined as follows:

$$\text{Modigliani measure} = (R_{pt}/\sigma_{pt}) \times \sigma_{It},$$

where R_{pt} = fund's average excess return in period t, σ_{pt} = standard deviation of fund's excess return in period t, and σ_{It} = standard deviation of index excess return in period t.

The fund with the highest Modigliani measure presents the highest return for any level of risk. According to this measure, every portfolio is adjusted to the level of risk in its unmanaged benchmark, and then measures the performance of this risk-equivalent portfolio, comparing portfolios on the same scale. Ranking portfolios by this measure yields a score expressed in basis points. The main drawback of this measure refers to, as the Sharpe ratio, its limited practical use by investors who are not in a position to use leverage in their mutual fund investments.

Another performance measure that is derived from comparing a fund to its benchmark is called information ratio and is calculated as follows:

$$\text{Information ratio} = \frac{R_{pt} - R_{Mt}}{STDV(R_{pt} - R_{Mt})},$$

where R_{Mt} = return of market portfolio (benchmark return) in period t, and $STDV$ = standard deviation of the difference between the return of the mutual fund and the return of the market portfolio in period t.

This performance measure is an alternative version of the Sharpe ratio, where instead of dividing the fund's return in excess of the risk-free rate by its standard deviation, the ratio of the fund's return in excess of the return on the benchmark index to its standard deviation is considered. It should be mentioned that the rankings of funds through the information ratio will generally differ from the ones obtained through the Sharpe ratio, and its relevance is not obvious to an investor.

The beta (β) coefficient is a measure of fund risk in relation to the market risk. It is called systematic risk and the asset-pricing model implies that is crucial in determining the prices of risky assets. For the calculation of beta (β) coefficient the well-known capital asset pricing model is used:

$$R_{pt} = \alpha_p + \beta_p R_{Mt} + \varepsilon_p,$$

where α_p = coefficient that measures the return of a fund when the market is constant, β_p = estimated risk parameter (systematic risk), and ε_p = error term (independent normally distributed random variable with $E(\varepsilon_p) = 0$), that represents the impact of non systematic factors that are independent from the market fluctuations.

The Jensen alpha[22] measure is the intercept in a regression of the time series of fund excess returns against the time series of excess returns on the benchmark. Both the Treynor index and the Jensen alpha assume that investors are well diversified and, therefore, they are only taking into account systematic risk when evaluating fund performance. The Jensen alpha measure is given by the regression of the following model:

$$(R_{pt} - R_{ft}) = \alpha_p + \beta_p (R_{Mt} - R_{ft}) + \varepsilon_p,$$

where α_p = Jensen alpha measure.

The coefficient α_p will be positive if the manager has any forecasting ability and zero if he has no forecasting ability. On the other hand, we can rule out a negative coefficient α_p by perversing forecasting ability.

The Treynor and Mazuy model[35] measures both market timing and security selection abilities of funds' managers. Treynor and Mazuy add a quadratic term to Jensen equation to test for market timing skills. This model is defined as follows:

$$(R_{pt} - R_{ft}) = \alpha_p + \beta_p (R_{Mt} - R_{ft}) + \gamma_p (R_{Mt} - R_{ft})^2 + \varepsilon_p,$$

where α_p = intercept term (estimated selectivity performance parameter), β_p = estimated risk parameter, and γ_p = second slope coefficient (estimated market-timing performance parameter).

The market timing and the security selection performance of mutual funds are also examined through the Henriksson and Merton model.[18] This model is defined as follows:

$$(R_{pt} - R_{ft}) = \alpha_p + \beta_p(R_{Mt} - R_{ft}) + \gamma_p Z_{Mt} + \varepsilon_p,$$

where $Z_{Mt} = max[0, (R_{Mt} - R_{ft})]$.

In both Treynor–Mazuy and Henriksson–Merton models, the evaluation of the performance of portfolio manager is shown through the two estimated parameters α_p and γ_p. Precisely, the parameter α_p shows the stock selection ability of the portfolio manager, the parameter β_p shows the fund's systematic risk while the parameter γ_p shows the market-timing ability of the portfolio manager. Positive values of these parameters show the forecasting ability of the portfolio manager, while negative values show the forecasting inability of the portfolio manager. Values of these parameters close to zero or zero show that the portfolio manager has no forecasting ability at all.

Another measure that ranks managers of mutual funds according to their forecasting abilities involves the Treynor and Black appraisal ratio,[36] defined as follows:

$$\text{Treynor and Black appraisal ratio} = \frac{\alpha_p}{s_p},$$

where α_p = Jensen alpha coefficient, and s_p = standard deviation of the error term in the regression used to obtain the alpha coefficient. The results obtained from the Treynor and Black appraisal ratio require a number of assumptions before they are valid. These assumptions refer to: no ability to forecast the market, multivariate normal returns, exponential utility as the criterion for investment for all managers, and the tradability of all assets for all managers.

4.1.3. *Statistical Analysis*

The incorporation in the analysis of all the above evaluation criteria would result in the development of an unrealistic mutual fund assessment model with limited practical value. To overcome this problem, a factor analysis is performed to select the most relevant criteria, which best describe the performance of mutual funds. Of course, it could be possible to override factor analysis if a mutual fund expert was available to determine the most significant indicators.

In this case study, factor analysis is performed using all the available data on the study of the three years period. The application of factor analysis resulted in the development of four factors that account for 88,5% of the total variance in the data. The selection of the criteria is performed on the basis of their factor loadings. Initially, fourteen criteria are selected, having factor loadings higher than 0,8 (in absolute terms). Precisely, according to the first factor eight criteria are selected, and based on the other three factors, two criteria are selected each time. From each one of these four factors the most important criteria are selected according to their statistical significance (one criterion for each factor). Thus, on the basis of the factor analysis results and the statistical significance of the considered criteria, the following four evaluation criteria are finally selected: (a) Return in the 3-years period, (b) beta (β) coefficient, (c) Henriksson & Merton's γ coefficient, and (d) Treynor & Black appraisal ratio.

The significance in the differences between the group means for all the examined criteria is investigated through a one-way ANOVA test. The results presented in Table 1 indicate that most criteria (13 out of 18 criteria) present statistically significant differences between the groups at the 5% and 10% significant levels. Regarding the selected criteria, the Return in the 3-years period and the Treynor & Black appraisal ratio are statistical significant at the 5% level, while beta (β) coefficient and the Henriksson & Merton's γ coefficient are statistical significant at the 10% level.

5. Presentation of the Results

In order to investigate the performance of the UTADIS method and compare it with the linear discriminant analysis, several validation tests are conducted using the cross-validation approach.[33] Cross-validation is a widely used approach to evaluate the generalizing and predictive performance of classification and regression models. In general, during k-fold cross-validation the complete sample A consisting of n observations (mutual funds), is randomly split into k mutually exclusive sub-samples (folds) A_1, A_2, ..., A_k of approximately equal size $d(d \approx n/k)$. The UTADIS method is applied k times to develop and test an additive utility model: each time (t=1, 2, ..., k) the model is developed on A, *excluding* A_t, and validated using the holdout sample A_t.

In this study a leave-one-out (n-fold) cross-validation approach is used to estimate the performance of the UTADIS method. In each replication of the leave-one-out cross-validation approach the reference set (training sample) consists of 32 mutual funds, whereas the validation (holdout) sample consists of one fund. The UTADIS is applied to the reference set to develop an additive utility classification model, which is then tested on the excluded mutual fund.

On the basis of the above methodology, Table 2 summarizes some statistics on the significance of each criterion in the discrimination between high and low performance mutual funds according to the models developed through the UTADIS method. The results clearly indicate that two criteria, the Treynor & Black appraisal ratio and the beta coefficient (β) are the major factors, distinguishing the two groups of mutual funds, whose total weight exceeds 85%. In particular, the analysis showed that the funds risk in relation to the market risk and the forecasting ability of funds managers have very important role in the evaluation of the performance of mutual funds. This is consistent with the results of the work of other researchers.[11,12,3,8]

Table 3 summarizes the average classification results for the leave-one-out cross validation experiment obtained using the UTADIS method. For comparative purposes the results of linear discriminant analysis (LDA) are also reported. The elements "High performance-High performance" and "Low performance-Low performance" represent average classification accuracy for each of the two groups, while all the other elements correspond to average classification errors.

The obtained results indicate that UTADIS outperforms the linear discriminant analysis in both the training and the validation samples. Precisely, in the training sample, the overall classification accuracy of the UTADIS method is 80,52% while for the LDA method is 77,98%. Of course, higher model fit in the training sample does not ensure higher generalizing ability, which is the ultimate objective in decision models, developed through regression–based techniques. In that respect, the results on the validation tests are of particular interest towards the evaluation of the predictability of UTADIS and the other statistical methods. The comparison of the methods according to the validation sample results indicates that in terms of the overall classification accuracy, UTADIS performs better than LDA. In particular, in the validation sample, the overall classification accuracy of the UTADIS method is 78,33% while for the LDA method is 69,44%. Moreover, the average classification errors in the UTADIS method are lower than the ones in the LDA method for both the "low performance" group and the "high performance" group. The case of misclassification of the "low performance" group of funds may result to capital losses for the investor.

On the contrary, the case of misclassification of the "high performance" group may lead to opportunity costs.

6. Concluding Remarks and Future Perspectives

The performance of mutual funds has become an increasingly important issue among portfolio managers and investors. The aim of this study is to propose a methodological framework for evaluating a number of mutual funds (alternatives) based upon mutual funds' characteristics regarding their relative returns and risks. In order to achieve this goal we used a sample of 33 Greek domestic equity funds of high and low performance.

We used factor analysis to select the evaluation criteria and a MCDA classification technique (UTADIS) to explore the possibility of developing models that identify factors associated with the performance of the funds and classify the funds into two performance groups. The advantages of the UTADIS method refer to the development of powerful classification models through a computational tractable procedure and to real-time results and extrapolation ability. The results were compared to a well-known multivariate statistical technique (discriminat analysis). Four criteria were selected as mutual funds evaluation criteria. The criteria selected refer to the funds' return, their risk in relation to the market risk and the forecasting ability (market timing and stock selection) of funds managers.

A cross-validation procedure was employed to evaluate the predictive performance of the developed models in order to have an, as much as possible, unbiased estimation of the two methods employed. The results of these models suggest that there is a potential in detecting high performance mutual funds through the analysis of different performance and risk measures.

The proposed approach constitutes a significant tool that can be used from managers of financial institutions and institutional investors in order to provide evaluation of the performance of mutual funds in the future. Further examination of the proposed methodological framework in other performance assessment problems and comparative studies among other methods to identify their relative strengths and weakness is also very interesting to be conducted.

References

1. Association of Greek Institutional Investor in http://www.agii.gr.
2. P.L. Brockett, A. Charnes, and W.W. Cooper. Chance constrained programming approach to empirical analyses of mutual fund investment strategies. *Decision Sciences*, **23**, 385-403 (1992).

3. M. Carhart. On persistence in mutual fund performance. *The Journal of Finance*, LII (1), 57-82, (March) (1997).
4. W.C. Chiang, T.L. Urban, and G.W. Baldridge. A neural network approach to mutual fund net asset value forecasting. *Omega, Int. J. Mgmt Sci.*, **24**, 205-215 (1996).
5. G. Colson and M. Zeleny. Uncertain prospects ranking and portfolio analysis under the condition of partial information. in : *Mathematical Systems in Economics*, Verlag Anton Hain, ed. 44, (Maisenheim) (1979).
6. R.E. Cumby and J.D. Glen. Evaluating the performance of international mutual funds. *The Journal of Finance*, XLV, 497-521 (1990).
7. M. Dalhquist, S. Engstrom, and P. Soderlind. Performance and characteristics of Swedish mutual funds. *Journal of Financial and Quantitative Analysis*, **35** (3), (September), 409-423 (2000).
8. K. Daniel, M. Grinblatt, S. Titman, and R. Wermers. Measuring mutual fundperformance with characteristic-based benchmarks. *Journal of Finance*, **52** (3), 1035-1058 (1997).
9. M. Doumpos and C. Zopounidis. The use of the preference disaggregation analysis in the assessment of financial risks. *Fuzzy Economic Review*, **3** (1), 39–57 (1998).
10. M. Doumpos and C. Zopounidis. *Multicriteria Decision Aid Classification Methods*, Kluwer Academic Publishers, Dordrecht (2002).
11. E.J. Elton, M.J. Gruber, and C.R. Blake. The persistence of risk-adjusted mutual fund performance. *Journal of Business*, **69** (2), 133-157 (1996).
12. W. Ferson and R. Schadt. Measuring fund strategy and performance in changing economic conditions. *Journal of Finance*, **51**, 425-461 (1996).
13. I. Friend, F. Brown, E. Herman, and D. Vickers. A study of Mutual Funds. *U.S. Securities and Exchange Commission* (1962).
14. M. Grinblatt and S. Titman. Portfolio performance evaluation: Old issues and new insights. *Review of Financial Studies*, **2**, 393-421 (1989).
15. M. Grinblatt and S. Titman. A study of monthly fund returns and performance evaluation techniques. *Journal of Financial and Quantitative Analysis*, **29**, 419-443 (1994).
16. Group of Thirty *Derivatives: Practices and Principles*, Washington, DC (1993).
17. M. J. Gruber. Identifying the risk structure of mutual fund returns. *European Financial Management*, **7** (2), 147-159 (2001).
18. R. Henriksson and R. Merton. On market timing and investment performance. *Journal of Business*, **54** (4), 513-534 (1981).
19. Ch. Hurson and C. Zopounidis. On the use of Multicriteria decision aid methods for portfolio selection. *Journal of Euro-Asian Management*, 1/2, 69-94 (1995).
20. Ch. Hurson and C. Zopounidis., *Gestion de portfeuille et Analyse Multicritère*, *Economica*, Paris (1997).
21. D.C. Indro, C.X. Jiang, B.E. Patuwo, and G.P. Zhang. Predicting mutual fund performance using artificial neural networks. *Omega*, **27**, 373-380 (1999).

22. C.M. Jensen. The Performance of Mutual Funds in the Period 1945-1964. *Journal of Finance*, 23, 389-416 (1968).
23. B.N. Lehmann and D.M. Modest. Mutual Fund Performance Evaluation: A Comparison of Benchmarks and Benchmark Comparisons. *The Journal of Finance*, XLII (2), (June), 233-265 (1987).
24. F. Modigliani and L. Modigliani. Risk-adjusted performance. *Journal of Portfolio Management*", **23** (2) (Winter), 45-54 (1997).
25. M.R. Morey and R.C. Morey. Mutual fund performance appraisals: A multi-horizon perspective with endogenous benchmarking. *Omega, Int. J. Mgmt Sci.*, 27, 241-258 (1999).
26. Morningstar at http://www.morningstar.com.
27. Moody's investor service at http://www.moodys.com.
28. B.P.S. Murthi, Y.K. Choi, and P. Desai. Efficiency of mutual funds and portfolio performance measurement: A non-parametric approach, *European Journal of Operational Research,* **98**, 408-418 (1997).
29. E.S. O' Neal. How many mutual funds constitute a diversified mutual fund portfolio. *Financial Analysts Journal*, (March/April), 37-46 (1997).
30. W.F. Sharpe. Mutual Fund Performance. *Journal of Business*, **39**, 119-138 (1966).
31. W.F. Sharpe. Morningstar's risk adjusted ratings. *Financial Analysts Journal*, (July/August), 21-23 (1998).
32. Standard & Poor's investor service at http://www.moodys.com.
33. M. Stone. Cross-validation choice and assessment of statistical predictions. *Journal of the Royal Statistical Society*, **36**, 111-147 (1974).
34. J.L. Treynor. How to rate management of investment funds. *Harvard Business Review*, **43**, 63-75 (1965).
35. J.L. Treynor and K.K. Mazuy. Can mutual funds outguess the market. *Harvard Business Review*, 131-136 (1966).
36. J. Treynor and F.Black. How to use security analysis to improve portfolio selection. *Journal of Business*, **46**, 66-68 (1973).
37. R. Wermers. Mutual fund performance: An empirical decomposition into stock-picking talent, style, transactions costs, and expenses. *The Journal of Finance*, LV (4), (August), 1655-1703 (2000).
38. C. Zopounidis and M. Doumpos. A multicriteria decision aid methodology for sorting decision problems: The case of financial distress. *Computational Economics*, **14** (3), 197-218 (1999).
39. C. Zopounidis and K. Pendaraki. An integrated approach on the evaluation of equity mutual funds' performance. *European Journal of Business and Economic Management*, in press (2002).

Table 1. One-Way ANOVA results

Criteria		Sum of Squares	fd	Mean Square	F
x_1	Between Groups	1765,437	1	1765,437	0,031*
	Within Groups	10750,312	31	346,784	
	Total	12515,749	32		
x_2	Between Groups	128,681	1	1284,681	0,048*
	Within Groups	9368,233	31	302,201	
	Total	10652,914	32		
x_3	Between Groups	0,437	1	0,437	0,861
	Within Groups	434,395	31	14,013	
	Total	434,832	32		
x_4	Between Groups	3760,071	1	3760,071	0,312
	Within Groups	110551,250	31	3566,169	
	Total	114311,320	32		
x_5	Between Groups	886812,230	1	886812,233	0,197
	Within Groups	15786056,000	31	509227,627	
	Total	16672869,000	32		
x_6	Between Groups	1616,804	1	1616,804	0,044*
	Within Groups	11317,780	31	365,090	
	Total	12934,584	32		
x_7	Between Groups	1379,238	1	1379,238	0,047*
	Within Groups	9986,593	31	322,148	
	Total	11365,831	32		
x_8	Between Groups	0,429	1	0,429	0,063**
	Within Groups	3,571	31	0,115	
	Total	4,000	32		
x_9	Between Groups	1419,409	1	1419,409	0,063**
	Within Groups	11813,760	31	381,089	
	Total	13233,169	32		
x_{10}	Between Groups	3,637	1	3,637	0,058**
	Within Groups	29,122	31	0,939	
	Total	32,758	32		
x_{11}	Between Groups	0,009	1	0,009	0,088**
	Within Groups	0,092	31	0,003	
	Total	0,101	32		
x_{12}	Between Groups	1193,967	1	1193,967	0,108
	Within Groups	13520,808	31	436,155	
	Total	14714,775	32		
x_{13}	Between Groups	0,002	1	0,002	0,143
	Within Groups	0,024	31	0,001	
	Total	0,026	32		
x_{14}	Between Groups	0,003	1	0,003	0,059**
	Within Groups	0,028	31	0,001	
	Total	0,031	32		
x_{15}	Between Groups	0,000	1	0,000	0,087**
	Within Groups	0,001	31	0,000	
	Total	0,001	32		
x_{16}	Between Groups	0,011	1	0,011	0,024*
	Within Groups	0,059	31	0,002	
	Total	0,070	32		
x_{17}	Between Groups	0,008	1	0,008	0,069**
	Within Groups	0,066	31	0,002	
	Total	0,074	32		
x_{18}	Between Groups	0,009	1	0,009	0,043*
	Within Groups	0,063	31	0,002	
	Total	0,072	32		

* Significant at the 5% level.

** Significant at the 10% level.

Table 2. Statistics on the weights of the evaluation criteria according to the UTADIS method (leave-one-out cross validation results

Criteria	**Average weight**	**St. error**
Annual Return	12,37%	5,03%
beta (β) coefficient	29,87%	7,53%
Henriksson & Merton's γ coefficient	0,06%	0,01%
Treynor & Black Appraisal ratio	57,70%	8,00%

Table 3. Average (validation) classification results (leave-one-out cross validation)

TRAINING SAMPLE				
		High Performance	Low Performance	Overall accuracy
UTADIS	High Performance	75,50%	24,50%	80,52% (0,45)
	Low Performance	14,46%	85,54%	
LDA	High Performance	72,48%	27,52%	77,98% (0,36)
	Low Performance	16,52%	83,48%	
VALIDATION SAMPLE				
		High Performance	Low Performance	Overall accuracy
UTADIS	High Performance	73,33%	26,67%	78,33% (7,12)
	Low Performance	16,67%	83,33%	
LDA	High Performance	66,67%	33,33%	69,44% (8,00)
	Low Performance	27,78%	72,22%	

Note: Parentheses indicate the standard error of overall accuracy.

CHAPTER 6

STACKED GENERALIZATION FRAMEWORK FOR THE PREDICTION OF CORPORATE ACQUISITIONS

E. Tartari

Mediterranean Agronomic Institute of Chania
Dept. of Economics, Marketing and Finance
73100 Chania, Greece

M. Doumpos

Technical University of Crete
Dept. of Production Engineering and Management
Financial Engineering Laboratory
University Campus
73100 Chania, Greece

G. Baourakis

Mediterranean Agronomic Institute of Chania
Dept. of Economics, Marketing and Finance
73100 Chania, Greece

C. Zopounidis

Technical University of Crete
Dept. of Production Engineering and Management
Financial Engineering Laboratory
University Campus
73100 Chania, Greece

Over the past decade the number of corporate acquisitions has increased rapidly worldwide. This has mainly been due to strategic reasons, since acquisitions play a prominent role in corporate growth. The prediction of acquisitions is of major interest to stockholders, investors, creditors and generally to anyone who has established a relationship with the acquired and non-acquired firm. Most of the previous studies on the prediction of corporate acquisitions have focused on the selection of an appropriate methodology to develop a predictive model and the comparison with other techniques to investigate the relative efficiency of the methods. On the contrary, this study proposes the combination of different methods in a stacked generalization context. Stacked generalization is a general framework for combining different classification models into an aggregate estimate that is expected to perform better than the individual models. This approach is employed to combine models for predicting corporate acquisitions which are developed through different methods into a combined model. Four methods are considered, namely linear discriminant analysis, probabilistic neural networks, the rough set theory and the UTADIS multicriteria decision aid method. An application of the proposed stacked generalization approach is presented involving a sample of 96 UK firms.

Keywords: Stacked generalization, classification, corporate acquisitions.

1. Introduction

During the period 1998-2001 more than 3000 acquisitions/mergers of UK firms were reported from the National Statistics Office, London, with an expenditure value of £371.58 billion. The increased employment brought on this method of corporate growth has generated a number of studies explaining certain segments of the merger movement. Attempts have been made to explain why firms merge, how firms merge, and how mergers have affected subsequent performance of firms.

Stevens[41] considers acquisitions as an investment alternative similar to other large capital budgeting decisions, which compete for limited funds. Therefore, the decision to acquire a firm should be consistent with shareholder wealth maximization criteria, thus financial characteristics play a role in the total decision making process. For this reason the analysis of financial characteristics of the acquired firms has been the subject of many studies. Generally, acquisitions can be considered as investment projects that often require significant funds and entail major risks.

The study of financial characteristics of the acquired firms has been the object of a decade of studies trying to determine the financial characteristics for the discrimination of the acquired firms from the non-acquired ones. These studies may be classified by country: United States,[35,41,23,46,14,29,6,51,34] Canada,[7,26,32] United Kingdom,[45,43,24,4,5] France,[8] Australia,[11,31] New Zeland,[3] and Greece.[51,37] The main evaluation methods used in the above studies were discriminant analysis,[7,32,45] logit analysis,[14] probit analysis,[23] and a combination of the above mentioned methods (factor and discriminant analysis,[4,41] factor, discriminant and logit analysis[50]).

Most of these works tried to identify financial characteristics for discriminating between acquired and non-acquired firms. They found that acquired firms suffer from the characteristics of having lower P/E ratios, lower dividend payout ratio, low growth in equity, and are considered to be smaller firms and more inefficient in comparison to non-acquired firms.

Several of the proposed approaches adopt a classification perspective. Classification refers the assignment of a set of objects into predefined groups. Over the past decades several methodologies for the construction of efficient classification models have been proposed from a variety of quantitative disciplines. However, there has been theoretical evidence (no free lunch theorem) showing that there is no method that is consistently better than any other method in terms of its classification performance.[48] This implies that while specific applications and data sets may suggest the use of a specific method, on average, it should be expected that all methods should perform almost equally well. In a sense, any method provides a piece of useful information for the problem under consideration. However, for a variety of reasons (data availability, time and cost limitations), the training sample cannot be exhaustive and fully comprehensive enough to cover all aspects of the problem. Thus, the developed models become sample–based and possibly unstable.

The above issues have motivated the development of algorithm–independent approaches that exploit the instability inherent in classification models and the differences between methods to improve classification performance. Stacked generalization is such an approach. Stacked generalization[47] is a general framework for combining classification models developed by a classification method or a set of classification methods. The general idea of stacked generalization is to develop a set of base models from the available data and then combine them at a higher level by a meta–model that provides the final classification . Given that the group assignments of the base models are independent and that all the base models perform better than chance, the combined model will perform better than any of the base models.

The following research proposes the combination of different methods in a stacked generalization context. In particular, the focus in this chapter was not on the comparison of different methods, but instead on their combination in order to obtain improved predictions for corporate acquisition. The considered methods originate from different quantitative disciplines and include the linear discriminant analysis, a probabilistic neural network,[39] the rough set theory[30] and the UTADIS multicriteria decision aid method (UTilités Additives DIScriminantes).[16] The performance of this approach was explored using data from annual reports of 96 UK public firms listed in the London Stock Exchange. The obtained results are quite encouraging towards the efficiency of the stacked generalization framework in predicting corporate acquisitions, since the combined model performs consistently better than all the methods in both applications and throughout all the years of analysis.

The rest of the chapter is organized as follows. The next section is devoted to the main features of stacked generalization model and empirical methods used in the analysis. Section 3 focuses on presenting the application study, describing the data and the variables used. The obtained results of the empirical study are described in Section 4. Finally, Section 5 summarizes the main findings of this chapter and discusses some issues for future research.

2. Methodology

2.1. *Stacked Generalization Approach*

Stacked generalization has been proposed by Wolpert[47] as an algorithm–independent approach for combining classification and regression models developed by an appropriate algorithm (i.e., classification or regression method) or a set of algorithms. Generally stated a classification problem involves the assignment of objects into a set C of predefined groups $C=\{C_1, C_2, \ldots, C_q\}$. Each object is described by a set of attributes $x_1, x_2, \ldots, x_n$. Thus each object can be considered as a vector of the form $\mathbf{x}_i=(x_{i1}, x_{i2}, \ldots, x_{in})$, where x_{ij} is the description of object $\mathbf{x}_i$ on attribute x_j (henceforth $\mathbf{x}$ will be used to denote the attribute vector). Essentially, the objective in a classification problem is to identify an unknown function $f(\mathbf{x})$ that assigns each object into one of the predefined groups. The function f can be real–valued in which case a numerical score is assigned to each object and the classification decision is made through the use of a classification rule. Alternatively, f can also directly produce a classification recommendation instead of a numerical score (this is the case of rule-based models and decision trees).[10]

Similarly to regression analysis the construction of the classification function f is performed through a training sample T consisting of m pairs $(\mathbf{x}_1, c_1)$, $(\mathbf{x}_2, c_2)$, $\ldots$, $(\mathbf{x}_m, c_m)$, where $c_i \in C$ denotes the group assignment for object $\mathbf{x}_i$. Given such a training sample, the specification of the function f can be performed in many different ways using several well–known methods. The expected performance of a classification method in providing correct estimations for the classification of the objects (expected error rate) is affected by three factors:

(1) The noise that is inherent in the data. This noise cannot be eliminated and consequently it defines the lower bound for the expected error rate.
(2) The squared bias of the error rate over all possible training samples of a given size.
(3) The variance of the classification estimations over all possible training samples of a given size.

The stacked generalization framework attempts to reduce the squared bias component of the expected error rate. Conceptually, stacked generalization can be considered similar to cross–validation.[42] Cross–validation is a widely–used resampling technique for the estimation of the error rate of classification models. Cross–validation is also often used for the comparison of classification methods and the selection of classification models. In this case, the model with the lower average cross–validation error rate is selected as the most appropriate one; this is a "winner takes all" strategy.

Stacked generalization seeks to extend this naive strategy to a more sophisticated one through the development of a more intelligent approach for combining the different classification models. These models can be developed either through a single classification method or through different methods. The latter (combination of different methods) is the most commonly used way for the implementation of stacked generalization strategies. The general steps followed in the stacked generalization framework for developing a combined classification model considering a set of w methods are the following (Figure 1).

(1) Using a resampling technique, p partitions of the training sample T into sub–samples T_{s1} and T_{s2} (s=1, 2, ..., p) are formed. Originally, Wolpert[47] suggested leave-one-out cross validation as the resampling technique, but other approaches are also applicable, such as k–fold cross validation[42] or bootstrapping.[19]
(2) For each partition s=1, 2, ..., p, the sub–sample T_{s1} is used to develop a classification model f_{ls} (base model) using method l (l=1, 2, ..., w). Each model is then employed to decide upon the classification of the objects belonging into the validation sub–sample T_{s2}.
(3) After all the p partitions have been considered, the group assignments

for the objects included in every validation sub–sample T_{s2} are used to form a new training sample for the development of a meta–model that combines the results of all base models at a higher level. The meta–model can be developed by any of the w considered methods.

Once the meta-model has been developed through the above procedure, it can be easily used to perform the classification of any new object (Figure 2). In particular, when a new object is considered, all the methods which are combined in the stacked generalization framework, are employed to obtain a classification assignment for the object. The classification of the object by a method l is determined on the basis of a model F_l developed by the method using the initial training sample T. The different group assignments c_l (l=1, 2, ..., w) determined by the models F_1, F_2, ..., F_w developed by all the w methods, are then combined by the developed meta–model to obtain the final classification decision.

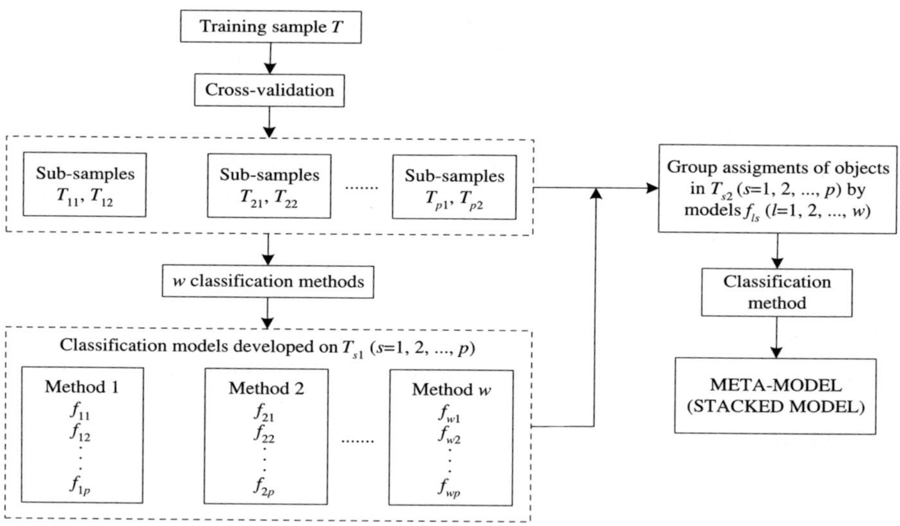

Fig. 1. Development of a stacked generalization model combining multiple methods

2.2. *Methods*

The successful implementation of the stacked generalization framework for the prediction of corporate acquisition depends on the methods that are combined. Obviously, if all the methods provide the same group assign-

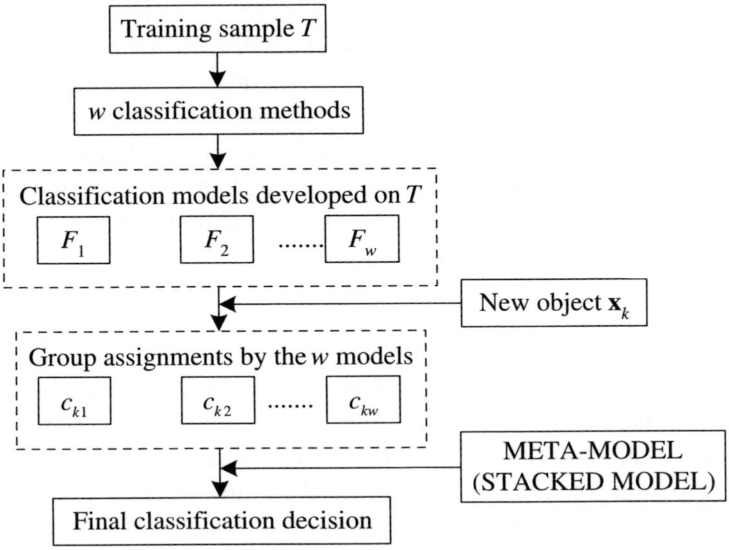

Fig. 2. Use of a stacked generalization model for the classification of new objects

ments, then any combined model will also lead to the same results as the ones of the methods considered. The classification performance of the methods is of limited interest in this context, i.e., one is not interested in combing highly accurate methods, but methods that are able to consider different aspects of the problem and the data used. Of course, it is difficult to ascertain which methods meet this requirement. However, it is expected that the consideration of different types of methods (e.g., methods which are not simple variations of one another) should be beneficial in the stacked generalization framework.[47] On the basis of this reasoning, in this study four classification methods are considered, namely linear discriminant analysis, probabilistic neural networks, the rough set theory and the UTADIS multicriteria decision aid method. These four methods originate from different quantitative disciplines (statistics, neural networks, rule induction, multicriteria analysis), they are based on different modelling forms (discriminant functions, networks, decision rules, utility functions) and they employ different model development techniques for the specification of a classification model. These existing differences between the four methods used in the analysis are expected to lead to the development of divergent classification models that are able to cover different aspects of the corporate acquisition problem and the data used for developing appropriate models. At this point it should be noted that several experiments were also made with the consideration of additional classification methods, such as logistic regression, artificial neural networks and the MHDIS multicriteria decision aid method (**M**ulti–group **H**ierarchical **DIS**crimination).[54] Nevertheless, the results obtained from the combination of a richer set of methods were not found to be better than the results obtained from combining the four aforementioned methods. The following sub–sections briefly outline the four methods used in the proposed stacked generalization framework.

2.2.1. *Linear Discriminant Analysis*

Discriminant analysis proposed by Fisher[20] can be viewed as the first approach to consider classification problems in a multidimensional context. Discriminant analysis is a multivariate statistical technique, which leads to the development of a set of discriminant functions so that the ratio of among–group to within–group variance is maximized, assuming that the variables follow a multivariate normal distribution. Assuming that the variance–covariance matrices across all groups are equal, then the developed discriminant functions are linear (linear discriminant analysis – LDA). For dichotomous classification problems, the developed linear discriminant function has the following form:

$$f(\mathbf{x}) = b_0 + b_1 x_1 + b_2 x_2 + \ldots + b_n x_n \tag{1}$$

where the constant term b_0 and the vector $\mathbf{b}$ of discriminant coefficients $\mathbf{b}=(b_1, b_2, \ldots, b_n)^T$ are estimated on the basis of the common within–groups variance–covariance matrix Σ and the vectors μ_1 and μ_2 corresponding to the attributes' averages for the objects belonging in the two groups C_1 and C_2, respectively:

$$\mathbf{b} = \mathbf{\Sigma}^{-1}(\mu_1 - \mu_2) \quad \textit{and} \quad b_0 = -\frac{1}{2}(\mu_1 - \mu_2)^{\mathrm{T}}\mathbf{b} \tag{2}$$

Assuming that the *a-priori* group membership probabilities are equal and that the misclassification costs are also equal, an object $\mathbf{x}_i$ will be classified in group C_1 if $f(\mathbf{x}_i) \geq 0$, and in group C_2 otherwise.

Despite its restrictive statistical assumptions regarding the multivariate normality of the attributes and the homogeneity of the group variance–covariance matrices, LDA has been the most extensively used methodology for developing classification models for several decades. Even today, the method is often used in comparative studies as a benchmark for evaluating the performance of new classification techniques. Furthermore, LDA has been extensively used in financial classification problems, including credit risk assessment, bankruptcy prediction, country risk evaluation, prediction of mergers and acquisitions, *etc.*[2,1]

2.2.2. *Probabilistic Neural Networks*

Probabilistic neural networks (PNN) have initially been developed as a density estimation technique for classification problems (Parzen window method).[18] Organized in a neural network structure,[39] they constitute a classification methodology that combines the computational power and flexibility of artificial neural networks, while managing to retain simplicity and transparency. PNNs can be realized as a network of three layers (Figure 3). The input layer includes n nodes, each corresponding to one attribute. The inputs of the network are fully connected with the m nodes of the pattern layer, where m is the number of objects in the training sample. Each pattern node k (k=1, 2, ..., m) is associated with a weight vector $\mathbf{w}_k=(x_{k1}, x_{k2}, \ldots, x_{kn})$. The input $\mathbf{x}_i$ to a pattern node k together with the associated weight vector $\mathbf{w}_k$ is passed to an activation function that produces the output of the pattern node k. The most usual form of the activation function is the exponential one (σ is a smoothing parameter):

$$Output_k = \exp\left(-\frac{\|\mathbf{w}_k - \mathbf{x}_i\|^2}{2\sigma^2}\right) \tag{3}$$

The outputs of the pattern nodes are passed to the summation layer. The summation layer consists of q nodes each corresponding to one of the q predefined groups $C_1, C_2, \ldots, C_q$. Each pattern node is connected only to the summation node that corresponds to the group where the object assigned to the pattern node belongs (recall that each pattern node represents an object of the training sample). The summation nodes simply sum the output of the pattern nodes to which they are connected with. Conceptually, this summation provides q numerical scores $g_h(\mathbf{x}_i)$, h=1, 2, ..., q, to each object $\mathbf{x}_i$, representing the similarity of the object $\mathbf{x}_i$ to group C_h. The object is classified to the group to which it is most similar.

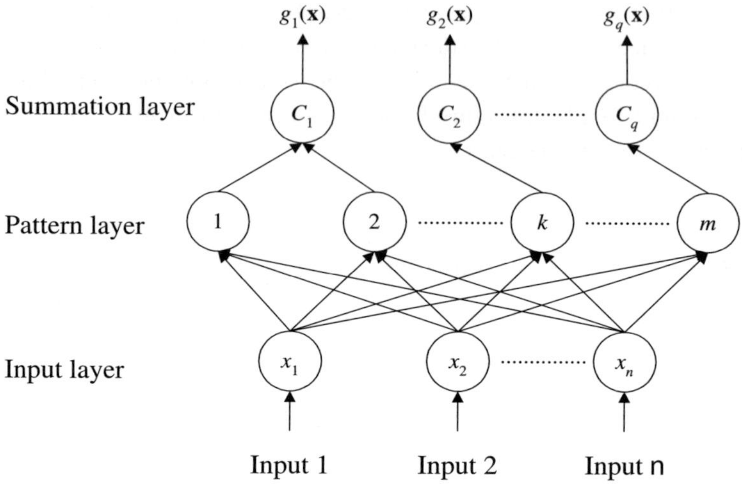

Fig. 3. Architecture of a probabilistic neural network

2.2.3. *Rough Set Theory*

Pawlak[30] introduced the rough set theory as a tool to describe dependencies between attributes, to evaluate the significance of attributes and to deal with inconsistent data. As an approach to handle imperfect data (uncertainty and vagueness), it complements other theories that deal with data uncertainty, such as probability theory, evidence theory, fuzzy set theory, etc. The rough set philosophy is founded on the assumption that with every object some information (data, knowledge) is associated. This information involves two types of attributes: condition and decision attributes. Condition attributes are those used to describe the characteristics of the objects, whereas the decision attributes define a partition of the objects into groups according to the condition attributes.

Objects that have the same description in terms of condition attributes are considered to be indiscernible. The indiscernibility relation constitutes the main mathematical basis of the rough set theory. Any set of all indiscernible objects is called an elementary set and forms a basic granule of knowledge about the universe. Any set of objects being a union of several elementary sets is referred to as crisp (precise). Otherwise the set is rough (imprecise, vague). Consequently, each rough set has a boundary–line consisting of cases (objects) which cannot be classified with certainty as members of the set or of its complement. Therefore, a pair of crisp sets, called the lower and the upper approximation can represent a rough set.

The lower approximation consists of all objects that certainly belong to the set and the upper approximation contains objects that possibly belong to the set. The ratio of the cardinality of the lower approximation of a rough set to the cardinality of its upper approximation defines the accuracy of approximating the rough set. Given this accuracy, the first major capability that the rough set theory provides is to reduce the available information so as to retain only what is absolutely necessary for the description and classification of the objects. This is achieved by discovering subsets of the attributes' set, which provide the same accuracy of classification as the whole attributes' set. Such subsets of attributes are called *reducts*. Generally, the number of reducts is greater than one. In such case the intersection of all reducts is called the *core*. The core is the collection of the most relevant attributes, which cannot be excluded from the analysis without reducing the quality of the obtained description (classification). The decision maker can examine all obtained reducts and proceed to the further analysis of the considered problem according to the reduct that best describes reality. Heuristic procedures can also be used to identify an appropriate reduct.[36]

The subsequent steps of the analysis involve the development of a set of rules for the classification of the objects into the classes where they actually belong. The rules developed through the rough set approach have the following form:

IF conjunction of elementary conditions

THEN disjunction of elementary decisions

The developed rules can be consistent if they include only one decision in their conclusion part, or approximate if their conclusion involves a disjunction of elementary decisions. Approximate rules are consequences of an approximate description of the considered groups in terms of blocks of objects (granules) indiscernible by condition attributes. Such a situation indicates that using the available knowledge, one is unable to decide whether some objects belong to a given group or not. The development of decision rules can be performed through different rule–induction algorithms.[40,38] In this study, the MODLEM algorithm is employed.[22]

The rough set theory has found several applications in financial decision making problems, including the prediction of corporate mergers and acquisitions.[37] A comprehensive up-to-date review on the application of rough sets in economic and financial prediction can be found in Ref. 44.

2.2.4. *The UTADIS method*

The UTADIS method originates from the preference disaggregation approach of multicriteria decision aid.[16] The preference disaggregation approach refers to the analysis (disaggregation) of the global preferences (judgment policy) of the decision maker in order to identify the criteria (attribute) aggregation model that underlies the preference result (ranking or classification). Similarly to the multiattribute utility theory,[25] preference disaggregation analysis uses common utility decomposition forms to model the decision maker's preferences. Nevertheless, instead of employing a direct procedure for estimating the global utility model, as in the multiattribute utility theory, preference disaggregation analysis uses regression–based techniques (indirect estimation procedure).

Given a training sample, the objective of the model development process in the UTADIS method is to develop a criteria aggregation model that performs well in discriminating among objects belonging to different groups. The developed criteria aggregation model has the form of an additive utility function:

$$U(\mathbf{x}) = \sum_{j=1}^{n} p_j u_j(x_j) \tag{4}$$

This utility function characterizes all the objects and assigns a score to each of them. This score (global utility) measures the overall performance of each object along all criteria (attributes), in a utility/value scale between 0 and 1 (the higher the global utility the higher the performance of an object). The global utilities are calculated considering both the criteria weights p_j and the performance of the objects on the evaluation criteria (attributes). The criteria weights sum up to 1 and they indicate the significance of each criterion in the developed classification model. On the other hand, the marginal utility functions $u_j(x_j)$ are used to consider the partial performance of each object on a criterion x_j. The marginal utilities are functions of the criteria's scale and they range between 0 and 1. Similarly to the global utilities, the higher the marginal utility of an object $\mathbf{x}_i$ on criterion x_j, the higher the performance of the object on the criterion. Both the criteria weights and marginal utility functions are specified as outputs of the model development process.

On the basis of this functional representation form, the classification of any object $\mathbf{x}_i$ in the q predefined groups is performed through the introduction of q–1 cut–off points called utility thresholds $u_1, u_2, \ldots, u_{q-1}$

$(u_1 > u_2 > \ldots > u_{q-1} > 0)$ in the global utility scale:

$$\left.\begin{array}{l} U(\mathbf{x}_j) \geq u_1 \Rightarrow \mathbf{x}_j \in C_1 \\ u_2 \leq U(\mathbf{x}_j) < u_1 \Rightarrow \mathbf{x}_j \in C_2 \\ \cdots\cdots\cdots\cdots\cdots\cdots\cdots\cdots \\ U(\mathbf{x}_j) < u_{q-1} \Rightarrow \mathbf{x}_j \in C_q \end{array}\right\} \qquad (5)$$

The goal of the model development process is to specify all the parameters of the model, including the marginal utilities, the criteria weights and the utility thresholds such that the use of the above classification rules for the objects of the training sample leads to a minimum classification error. A linear programming problem is solved for this purpose, followed by a post–optimality stage to investigate the robustness of the obtained optimal solution. The details of the model development process employed in the UTADIS method are described in the book by Doumpos and Zopounidis.[16]

3. Description of the Case Study

3.1. *Sample Data*

The estimation sample consists of data for public firms, listed in the London Stock Exchange (LSE), that were subject to a takeover bid in the year 2001 in the UK. The sample is composed of 96 firms, from which two sets of samples are obtained, each containing an equal number of acquired and non-acquired firms, with 48 firms in each group. For each acquired firm, their annual reports for the three years (1998-2000) preceding their takeover was collected. For the same years, corresponding data are also collected for the sample of non-acquired firms. These are equivalent data for UK listed firms that are as similar as possible to the takeover targets in terms of their principal business activity, asset size, sales volume, and the number of employees. In this study, only industrial/commercial firms are considered. The inclusion of financial firms in the sample was rejected because it would add tremendous heterogeneity to the sample. Table 1 presents the synthesis of the considered sample in terms of the industry sector of the selected firms.

3.2. *Variables*

Once the sample groups are determined, the financial statements of these firms (balance sheet, profit and loss account, and income statement), provided by Hemscott Data Services (`www.hemscott.net`), are collected and a group of financial ratios is calculated. Twenty-three financial ratios for each

Table 1. Industrial sectors represented in the sample

Sector	Type of company	
	Acquired	Non-acquired
Aerospace & Defence	1	1
Chemicals	1	1
Construction & Building Materials	6	6
Distributors	2	2
Engineering & Machinery	1	1
Food Producers & Processors	3	3
General Retailers	4	4
Household Goods & Textile	4	4
Investment Companies	2	2
Leisure Entertainment & Hotels	4	4
Media	1	1
Oil, integrated	1	1
Pharmaceuticals	3	3
Real Estate	4	4
Software & Computer services	2	2
Financial Services	3	3
Telecommunications	1	1
Transportation	3	3
Water	2	2
Total	48	48

company, three years prior to the acquisitions, were computed using the financial model base on the FINCLAS system,[53] as presented in Table 2. The selected ratios measure financial qualities such as profitability, liquidity, solvency and managerial performance. They are chosen after considering a) data availability, b) existing empirical studies in the area, c) their popularity in the financial literature,[12,15,27,28,41]) and d) their appearance in the literature as predictors of corporate takeovers.

3.3. *Factor Analysis*

Before developing any predictive model, multivariate factor analysis was performed in order to identify any multicollinearity among the financial ratios in our study. Indeed, when the variables are highly collinear, the weights in the resulting model are highly unstable, the model tends to be highly sample sensitive, and the interpretation becomes very difficult. Altman[1] observed the high multicollinearity in the ratios from which he derived a bankruptcy discriminant model; for this reason, he emphasized the need for choosing the financial ratios very carefully. The factor analysis technique was applied in the prediction of corporate acquisitions by several

Table 2. Financial ratios used as predictive evaluation criteria

Class	Number	Financial ratios
I. Profitability	1	EBIT/ Total Assets
	2	Net Income/ Net Worth
	3	Net Income/ Sales
	4	Sales/ Net Fixed Assets
	5	Sales/ Total Assets
	6	Gross Profit/Total Assets
	7	Net Income/ Net Fixed Assets
	8	Cash Flow/ Total Assets
	9	Cash Flow/ Net Fix Assets
II. Liquidity and Solvency	10	Net Income/ Working Capital
	11	Net Worth/ Total Liabilities
	12	Total Liabilities/ Total Assets
	13	Long-term Liabilities/ Long-term Liab. + Net Worth
	14	Current Assets/ Current Liabilities
	15	Quick Assets/ Current Liabilities
	16	Cash/ Current Liabilities
	17	Current Liabilities/ Net Worth
	18	Working Capital/ Working Capital Required
	19	Total Liabilities/Working Capital
	20	Working Capital / Total Assets
III. Managerial performance	21	Interest Expenses/ Sales
	22	Accounts Receivable/ Sales*365
	23	Inventories/Current Assets

authors.[41,26,4]

Our factor analysis was performed using the SAS package with varimax rotation. The criterion for selecting the number of factors was minimum eigenvector greater than 1, which is a common selection rule in practice. The original group of ratios is factored into nine distinct and orthogonal dimensions, with each dimension being a linear combination of the original twenty-three ratios. The nine extracted factors explain more than 77.48 % of the total variance in the sample. Table 3 presents the factor loadings for the considered financial ratios in the extracted factors.

The factor loadings were employed to select a limited set of financial ratios. In particular it was decided to include in the analysis the ratios with the highest loadings in each factor. Therefore nine financial ratios were finally selected (Table 4). Each of the selected ratios corresponds to one of the extracted factors and has the highest loading in this factor.

According to the results included in Table 4 these factors include ratios from all three classes of financial ratios: profitability (ratios 6 and 15), solvency (ratios 10, 11, 14, 16, 17 and 19) and managerial performance (23).

Table 3. Significant financial ratios (factor loadings)

	Factors								
Ratios	1	2	3	4	5	6	7	8	9
1		0,702							
2		0,670							
3									
4									
5				0,808					
6				0,830					
7					0,984				
8									
9			0,849						
10			0,877						
11	0,941								
12	0,907								
13	0,892								
14								0,824	
15		0,762							
16						0,941			
17									0,887
18						0,903			
19					0,987				
20	0,775								
21									
22									
23							0,854		

Note: Empty entries correspond to factor loading lower than 0.65 (in absolute terms)

Table 4. Financial ratios selected through factor analysis

11	Current Assets/ Current Liabilities
15	Net Income/ Net Fixed Assets
10	Long-term Liabilities/ Long-term Liab. +Net Worth
6	Gross Profit/Total Assets
19	Total Liabilities/Working Capital
16	Cash Flow/ Total Assets
23	Inventories/Current Assets
14	Current Liabilities/ Net Worth
17	Working Capital/ Working Capital Required

4. Results

The methods outlined in Section 2 were applied in the corporate acquisitions data set described in subsection 3.1. The most recent year is used as the training sample, whereas the data for the other two years are used to test the generalizing performance of the proposed stacked generalization approach in providing accurate early warning signals for predicting acquisition. Therefore, the training sample T consists of 96 firms (m=96).

On the basis of this training sample an 8–fold cross validation approach is employed to build the base models using the four considered methods. The 8-fold cross validation partitions the training sample into eight disjoint sub–samples $T_1, T_2, ,\ldots, T_8$,each of which consists of $m/8$ firms. Each method is applied eight times (replications). In each replication s, a predictive classification model f_{ls} is developed on T excluding T_s using the method l (l=1–LDA, 2–PNN, 3–rough sets, 4–UTADIS). This model is then used to classify the firms in T_s. The group assignments for the firms in each T_s are then used as a training sample for the development of a stacked generalization model predicting corporate acquisitions.

The development of the stacked model is performed using the UTADIS method (i.e., a model is developed using the UTADIS method that combines –at the meta–level– the group assignment of all the four methods considered in the analysis).

Table 5 presents the classification results for the data set. Initially the results obtained using each of the four methods in the conventional way are presented (i.e., each method is employed to construct a predictive classification model using the data for the most recent year as the training sample, and then this model is used to classify the firms in the other two years). Furthermore, the results of the stacked model that combines the four methods are also reported.

In Table 5, the results are analyzed in terms of the type I and II errors as well as in terms of the overall error. The type I error refers to the classification of an acquired corporate as a non-acquired one. Similarly, the type II error refers to the classification of a non-acquired corporate as an acquired one. The overall error is estimated as the average of these two error types.

The most interesting finding derived from the results of Table 5 is that the stacked model performs at least as well (in terms of the overall error rate) as any of the four methods upon which it is based, throughout all years of the analysis. In particular, in 1999 (two years prior to acquisition) the best model developed by the individual methods used in the analysis is the PNN method. The types I and II error rates of this model are respectively 29.17% and 54.17%, and the overall error model is 41.67%. The rough set method provides the same type I error, but a slightly type II and overall error. Thus, for the year 1999 the PNN method performs better. The stacked model that combines all four methods provides a higher type I error, but its type II error rate is significantly reduced to 45.33% (8.33%

Table 5. Classification results (error rates in %)

Methods	Error type	2000	1999	1998
LDA	Type I	43.75	56.25	56.25
	Type II	39.58	43.75	35.42
	Overall	**41.67**	**50.00**	**45.83**
PNN	Type I	0.00	29.17	33.33
	Type II	0.00	54.17	41.67
	Overall	**0.00**	**41.67**	**37.50**
Rough sets	Type I	0.00	29.17	29.17
	Type II	0.00	56.25	54.17
	Overall	**0.00**	**42.71**	**41.67**
UTADIS	Type I	27.08	37.50	45.83
	Type II	29.17	47.92	43.75
	Overall	**28.13**	**42.71**	**44.79**
Stacked model	Type I	22.92	33.33	39.58
	Type II	0.00	45.83	33.33
	Overall	**11.46**	**39.58**	**36.46**

less than the corresponding type II error of the PNN model), thus leading to a reduction of the overall error rate down to 39.58%.

Similarly to 1999, in 1998 (three years prior to acquisition) the PNN model provides the lowest overall error rate compared to the other methods, while its type I error (33.33%) is higher than the type I error of the rough set model (29.17%) and its type II error (41.67%) is higher than the type II error of the LDA (35.42%). The overall error rates for the rough sets, the UTADIS method and LDA are all well above 40%. On the other hand the overall error rate of the stacked generalization model is 36.46%, which lower compared to the error rates of the four considered methods.

Overall, in both 1999 and 1998 the stacked generalization model performs consistently better than the four methods in terms of the overall error rate and the type II error. On the other hand, its type I error rate is higher compared to the type I errors of the rough set model and the PNN.

Finally, it is worth noting that, generally, the error rates for all the models in this study developed for predicting corporate acquisitions are rather high, at least for the years 1999 and 1998 (the years preceding acquisition). This is party due to the similarities of the financial characteristics of acquired and non-acquired firms and the consideration of different (non-homogeneous) business sectors in the sample. However, similar results have been obtained in other studies involving the development of acquisition prediction models,[7,4,37,50]) thus highlighting the importance of considering non-financial (strategic) factors that often lead to acquisitions. Such information, however, is usually not available to the public and thus it is difficult to consider in research studies such as the one presented in this chapter.

5. Conclusion

The development of models for predicting corporate acquisitions is often considered as a classification problem. Over the past decades several methodologies for the construction of efficient classification models have been proposed from a variety of quantitative disciplines. Most of the existing studies on the use of classification methods for predicting corporate acquisitions have relied on the identification of the most appropriate method based on comparisons with other existing approaches.

This study followed a different line of research path. In particular, the focus in this chapter was not on the comparison of different methods, but instead on their combination in order to obtain improved predictions for corporate acquisition. For this purpose a stacked generalization framework was employed to combine different prediction estimates for corporate acquisitions obtained from models developed through a set of four classification methods (LDA, PNN, rough sets, and UTADIS method). The performance of this approach was explored using data from the annual reports of 96 UK public firms listed in the London Stock Exchange. The obtained results clearly indicate that a stacked generalization approach for the combination of different methods may contribute to the development of corporate acquisition models that are more reliable than the methods that are combined. Therefore, the increased computational effort required to implement the stacked generalization approach is compensated by an increased performance.

Of course, the reliability of the stacked generalization approach depends on the methods that are combined. Further research is required on this issue to investigate the similarities and the differences in the group assignments made by models developed by different classification methods. Such a study will provide useful guidelines on the methods which could be considered in a combined context. Furthermore, it would be interesting to investigate different approaches to perform the combination of the methods. Finally, it would be interesting to consider alternative model ensemble approaches which have recently been proposed in the classification literature, such as new bagging and boosting algorithms.[9,21]

References

1. E.I. Altman. *Corporate Financial distress and Bankruptcy.* John Wiley and sons, New York (1993).

2. E.I. Altman, R. Avery, R. Eisenbeis, and J.Stinkey. *Application of Classification Techniques in Business, Banking and Finance.* JAI Press, Greenwich (1981).
3. M.I. Amery and D.M. Emanuel. Takeover announcement and shareholder returns: New Zealand evidence. *Pacific Accounting Review*, **1**, 42-58 (1988).
4. P. Barnes. The Prediction of Takeover Targets in the U.K by means of Multiple Discriminant Analysis. *Journal of Business Finance and Accounting* **17**:73-84 (1990).
5. P. Barnes. Predicting UK Takeover Targets: Some Methodological Issues and an Empirical Study. *Review of Quantitative and Accounting*, **12**(3) 283-301 (1999).
6. J.W. Bartley and M. Calvin. The Relevance of Inflation Adjusted Accounting Data to the Prediction of Corporate Takeovers. *Journal of Business Finance and Accounting* **17**, 53-72 (1990).
7. A. Belkaoui. Financial ratios as predictors of Canadian takeovers. *Journal of Business Finance and Accounting* **5**:93-107 (1978).
8. J.F. Boulier, and S.Demay. Predicting takeover targets in the French Stock Exchange. *Recherche et de L'Innovation* (1993).
9. L. Breiman. Bagging Predictors. *Machine Learning*, Vol. 24(2), 123-140 (1996).
10. L. Breiman, J. Friedman, R. Olshen, and C.J. Stone. *Classification and Regression Trees*, Chapman and Hall, New York (1984).
11. A.D. Castagna and Z.P. Matolczy. Financial ratios as predictors of company acquisitions. *Journal of the Securities Institute of Australia* 6-10 (1976).
12. J.K. Courtis. Modelling a financial ratios categoric framework. *Journal of Business Finance and Accounting*, **5**, 4, 371–386 (1978).
13. H. Dahl, A. Meeraus, and S.A. Zenios. Some Financial Optimization Models: I Risk Management. In: *Financial Optimization*, S.A. Zenios, ed. Cambridge University Press, Cambridge, 3–36, (1993).
14. J. Dietrich and E. Sorenson. An application of logit analysis to prediction of mergers targets. *Journal of Business Research* **12**, 393-412 (1984).
15. A.I. Dimitras, S.H. Zanakis, and C. Zopounidis. A survey of business failures with an emphasis on prediction methods and industrial applications. *European Journal of Operational Research*, **90**, 487-513 (1996).
16. M. Doumpos, and C. Zopounidis. *Multicriteria Decision Aid Classification Methods.* Kluwer Academic Publishers, Dordrecht, (2002).
17. M. Doumpos, K. Kosmidou, G. Baourakis, and C. Zopounidis. Credit Risk Assessment Using a Multicriteria Hierarchical Discrimination Approach: A Comparative Analysis. *European Journal of Operational Research*, Vol. **138**(2), 392–412 (2002).
18. R.O. Duda, P.E. Hart, and D.G. Stork. *Pattern Classification* (2nd Edition), John Wiley, New York (2001).
19. B. Efron. Estimating the Error Rate of a Prediction Rule: Improvement of Cross–validation. *Journal of the American Statistical Association.* Vol. **78**, 316–330 (1983).
20. R.A. Fisher. The Use of Multiple Measurements in Taxonomic Problems.

Annals of Eugenics, Vol. **7**, 79–188, (1936).
21. Y. Freund. Boosting a Weak Learning Algorithm by Majority. *Information and Computation*, Vol **121**(2), 256-285, (1995).
22. J.W. Grzymala–Busse and J. Stefanowski. Three Discretization Methods for Rule Induction. *International Journal of Intelligent Systems*, Vol. **26**, 29–38 (2001).
23. R.S. Harris, J.F. Stewart, and W.T. Carleton. Financial characteristics of acquired firms. In: *Mergers and Acquisitions Current Problems in Perspective*, M. Keenan and L. White (eds), D.C. Health and Co., Lexington, 223-241 (1982).
24. A. Hughes. The impact of Merger: A Survey of Empirical Evidence for the UK. In: *Mergers and Mergers Policy*, J.A. Farburn and J.A. Kay (eds), Oxford University Press (1989).
25. R.L. Keeney, and H. Raiffa. *Decisions with Multiple Objectives: Preferences and Value Trade–offs.* Cambridge University Press, Cambridge (1993).
26. D. Kira and D. Morin. Prediction of takeover targets of Canadian firms. *Proccedings of the 6-th International Symposium on Data Analysis* (1993).
27. C.F. Lee. *Financial Analysis and Planning: Theory and Application*, Addison-Wesley, Reading, MA (1985).
28. R. Libby. Accounting ratios and the prediction of failure: Some behavioral evidence. *Journal of Accounting Review,* **13**, 1, 150-161 (1975).
29. K. G. Palepu. Predicting Takeover Targets: A Methodological and Empirical Analysis. *Journal of Accounting and Economics*, **8**, 3-35 (1986).
30. Z. Pawlak. Rough Sets. *International Journal of Information and Computer Sciences*, Vol. **11**, 341–356 (1982).
31. R.G. Powell. Modelling takeover likelihood. *Journal of Business, Finance and Accounting* **24**, 1009-30 (1997).
32. U.P. Rege. Accounting ratios to locate takeovers targets. *Journal of Business Finance and Accounting* **11**,301-311 (1984).
33. B.D. Ripley. *Pattern Recognition and Neural Networks.* Cambridge University Press, Cambridge (1996).
34. P.S. Rose. Characteristics of merging banks in the United States: Theory, empirical results, and implications for public policy. *Review of Business and Economic Research* **24**, 1-19 (1998).
35. M. Simkowitz and R. J. Monroe. Investment Characteristics of Conglomerate Targets: A discriminant Analysis. *Southern Journal of Business* (November 1971).
36. R. Slowinski and C. Zopounidis. Application of the Rough Set Approach to Evaluation of Bankruptcy Risk. *International Journal of Intelligent Systems in Accounting, Finance and Management*, Vol. **4**, 27–41 (1995).
37. R. Slowinski C. Zopounidis, and A.I. Dimitras. Prediction of company acquisition in Greece by means of the rough set approach. *European Journal of Operational Research* **100**, 1-15 (1997).
38. R. Slowinski, J. Stefanowski, S. Greco, and B. Matarazzo. Rough Sets Processing of Inconsistent Information. *Control and Cybernetics*, Vol. **29**(1), 379–404 (2000).

39. D.F. Specht. Probabilistic Neural Networks. *Neural Networks*, Vol. **3**, 109–118 (1990).
40. J. Stefanowski. On Rough Set Based Approaches to Induction of Decision Rules. In: *Rough Sets in Knowledge Discovery*, L. Polkowski and A. Skowron (eds.), Physica–Verlag, Heidelberg, 500–529 (1998).
41. D.L. Stevens. Financial Characteristics of Merged Firms: A Multivariate Analysis. *Journal of Financial and Quantitative Analysis*, pp. 149-158 (March, 1973).
42. M. Stone. Cross-validation Choice and Assessment of Statistical Predictions. *Journal of the Royal Statistical Society B*, Vol. **36**, 111–147 (1974).
43. R. Taffler. The Z-Score approach to measuring company solvency. *The Accountant's Magazine*, **87**:921, 91-96 (1983).
44. F.E.H. Tay and L. Shen. Economic and financial prediction using rough sets model. *European Journal of Operational Research*, **141**, 641-659 (2002).
45. J. Tzoannos and J.M. Samuels. Mergers and takeovers: The financial characteristics of companies involved. *Journal of Business Finance* **4**, 5-16 (1972).
46. T.R. Wansley and W.R. Lane. A financial profile of merged firms. *Rev Business and Economic Research*, **19**, 87–98 (1983).
47. D.H. Wolpert. Stacked Generalization. *Neural Networks*, Vol. **5**, 241–259 (1992).
48. D.H. Wolpert and W.G. Macready. No Free Lunch Theorems for Search. *Technical Report SFI–TR–95–02–010*, Santa Fe Institute (available at: http://citeseer.nj.nec.com/ wolpert95no.html) (1995).
49. L.A. Zadeh. Fuzzy Sets. *Information and Control*, Vol. **8**, 338–353 (1965).
50. S.H. Zanakis, and C. Zopounidis. Prediction of Greek company takeovers via multivariate analysis of financial ratios. *Journal of the Operational Research Society*, **47**, 678-687 (1997).
51. S.H. Zanakis and G. Walter. Dicriminant Characterisitics of U.S banks acquired with or without federal assistance. *European Journal of Operational Research*, **77**, 440-465 (1994).
52. C. Zopounidis. *Operational Tools in the Management of Financial Risks.* Kluwer Academic Publishers, Dordrecht (1998).
53. C. Zopounidis and M. Doumpos. Developing a Multicriteria Decision Support System for Financial Classification Problems: The FINCLAS System. *Optimization Methods and Software*, Vol. **8**, 277–304 (1998).
54. C. Zopounidis and M. Doumpos. Building Additive Utilities for Multi-group Hierarchical Discrimination: The M.H.DIS Method. *Optimization Methods and Software*, Vol. **14**(2), 219–240 (2000).
55. C. Zopounidis and M. Doumpos. Multicriteria Classification and Sorting Methods: A Literature Review. *European Journal of Operational Research*, Vol. **138**(2), 229–246 (2002).

CHAPTER 7

SINGLE AIRPORT GROUND HOLDING PROBLEM - BENEFITS OF MODELING UNCERTAINTY AND RISK

K. Taafe

Industrial and Systems Engineering Department
University of Florida, PO Box 116595, Gainesville, FL 32611
E-mail: taafe@ufl.edu

Air travel has become a primary mode of travel for many individuals and, with increased demand for flights, airports are quickly reaching their capacities. Individual airline schedules are usually constructed to provide flexibility and convenience to the passenger, but they must also adhere to the capacity of each airport. Under poor weather conditions, which usually are called irregular operations, an airport's arrival capacity, or airport acceptance rate (AAR), can be severely affected. Under these circumstances, we look for ground holding policies which hold certain aircraft at their originating or upline stations in order to reduce the more expensive airspace delays incurred when an aircraft circles, waiting for available arrival runway capacity.

The Ground Holding Problem (GHP) has been researched actively over the past 10 years, including work on both static and dynamic, and single and multiple airport, versions of the problem. Much of this work has presented efficient methods for solving each version of the GHP. We provide a foundation for this research by presenting the underlying motivation for evaluating the GHP using stochastic programming as opposed to using a deterministic or expected value approach. Also, the majority of past work has centered on using an objective that minimizes "total delay costs." We will introduce risk-aversion objectives to quantify penalties for deviating from expected runway capacities.

1. Introduction

Over the past 20 years, business and leisure air travel have consistently increased in popularity. With more and more passengers wanting to travel, airlines and airports have continued to expand to meet passengers' needs. And now, many airports are at or near capacity with few options for expansion. As more airports approach their capacity, the air travel industry is witnessing higher average delays. While some delays result from an airline's operations (ground servicing, flight crews, late baggage, etc.), a majority of the severe delays are weather related. During bad weather, the Federal Aviation Administration (FAA) imposes restrictions on the number of aircraft an airport can accept in an hour. In technical terms, the airport will be instructed to operate under one of three flight rule policies: VFR (Visual Flight Rules), IFR1 (Instrument Flight Rules 1), or IFR2 (Instrument Flight Rules 2 – more restrictive than IFR1). An airport operates under VFR during good weather or normal conditions. As the weather conditions deteriorate, the FAA may restrict airport capacity by requiring an airport to operate under IFR1 or IFR2. In the most extreme cases, an airport will temporarily close until the poor weather conditions subside.

As a result of these rules, the airport and airlines must decide what to do with all of the aircraft wanting to arrive at an airport experiencing bad weather. The aircraft can be allowed to take off and approach the airport, resulting in some air delays while flight controllers sequence these arriving aircraft. Alternatively, the aircraft can be held at their originating stations, incurring what is called a ground holding delay. Finding the desired balance between ground delays and air delays under severe weather conditions that achieves the lowest cost is the focus of this paper.

The airport acceptance rate (AAR) plays an important role in determining ground holding policies at airports across the nation. Since all airports have finite capacity, there is a continuing effort to maximize capacity utilization, while avoiding unwanted ground and air delays, which impact fuel costs, crew and passenger disruptions, and other intangible costs. We use a stochastic programming approach, which takes into consideration the random nature of the events that put a ground holding plan into place. As these plans cannot predict the future, there may be unnecessary ground holds at originating or upline stations, resulting in unused capacity at the airport in question if the actual delays (weather-related or not) are not realized.

Research has been conducted on the ground holding problem for single and multiple airports. In addition, both static and dynamic versions of the problem exist. The static version assumes that the capacity scenarios are defined once at the beginning of the ground holding period under evaluation. In contrast, a dynamic version has been studied in which capacity scenarios are updated as the day progresses. In other words, there will exist k sets of capacity scenarios for time period k, defined in periods 1, 2,...,k, respectively. In this paper, we focus on the static, stochastic single airport version of the problem. For additional background on the static, single-airport problem, see Ball *et al.*,[2] Hoffman and Ball,[4] Richetta and Odoni,[6] and Rifkin.[7] For research on the dynamic or multiple airport problems, please see Vranas, Bertsimas, and Odoni,[10] Vranas, Bertsimas, and Odoni,[11] and Navazio and Romanin-Jacur.[5] In Section 2, we present the Rifkin[7] stochastic formulation of the ground holding problem, along with some solution properties. We adopt this model formulation and develop new findings, which are presented in Sections 3 and 4. First, Section 3 illustrates the benefit of studying a stochastic as opposed to a deterministic ground holding problem. A justification for the use of a stochastic model is presented through a series of computational experiments performed with various input data. Benchmark measures, such as the value of the stochastic solution and expected value of perfect information (see Birge and Louveaux[3]), are used for this justification. In Section 4, we consider the effect of introducing risk-averse constraints in the model. In other words, by restricting the size of the worst-case delays, how is the overall expected delay affected? This is not the same as a maximum delay model, which would place a strict upper bound on worst-case delays. Finally, we summarize the findings of this report and discuss directions for future work.

2. Static Stochastic Ground Holding Problem

2.1. *Problem Definition and Formulation*

We define a potential ground holding period to consist of a finite number, T, of 15-minute increments or periods. Suppose the arrival schedule contains Fflights during the time horizon under evaluation. These individual flights are scheduled to arrive during the day such that we can represent flights in the same time periods as:

D_t = Number of arrivals initially scheduled to arrive in period t

While the airport may have a nominal arrival capacity of Xaircraft per period, the estimates based on the poor weather conditions will produce Q possible capacity scenarios within any interval. For each capacity scenario q, there is a probability p_q of that scenario actually occurring. For each time period and scenario, let M_{qt} be the arrival capacity for scenario $q$during period t. Let c_g denote the unit cost of incurring a ground delay in period t. Assume that ground delays do not increase in cost beyond the first time period for any aircraft. Similarly, define an air delay cost, c_a, as the unit cost of incurring an air delay for one period. We will use these parameters to examine model performance for different ground/air delay ratios in Section 3. We next define the following decision variables:

- A_t= Number of aircraft allowed to depart from an upline station and arrive "into the airspace" of the capacitated airport during period t
- W_{qt}= Number of aircraft experiencing air delay during period t under scenario q
- G_j= Number of aircraft incurring ground delays in period j. This is the difference between the total number of expected arrivals $(\sum_{t=1}^{j} A_t)$ through period j and the actual number of arrivals $(\sum_{t=1}^{j} D_t)$ through period j.

As stated previously, we focus on the static, stochastic single airport version of the problem. Based on the problem first presented in Richetta and Odoni[6] and later revised in Rifkin,[7] the formulation is as follows:

[SSGHP] Static Stochastic Ground Holding Problem (General Case)

$$\text{minimize} \quad c_g \sum_{t=1}^{T} G_t + c_a \sum_{q=1}^{Q} \sum_{t=1}^{T} p_q W_{qt} \tag{1}$$

subject to:

Scheduled Arrival Times:

$$\sum_{t=1}^{j} A_t \le \sum_{t=1}^{j} D_t, \quad j = 1, \ldots, T \tag{2}$$

$$\sum_{t=1}^{T+1} A_t = \sum_{t=1}^{T+1} D_t$$

Arrival Period Capacities:

$$A_t + W_{q,t-1} - W_{qt} \le M_{qt}, \quad t = 1, \ldots, T, \quad q = 1, \ldots, Q, \tag{3}$$

Initial Period Air Delays:

$$W_{q0} = 0 \quad q = 1, \ldots, Q, \tag{4}$$

Ground Delays:

$$G_j + \sum_{t=1}^{j} A_t = \sum_{t=1}^{j} D_t \quad j = 1, \ldots, T, \tag{5}$$

Integrality:

$$A_t \in Z^+, W_{qt} \in Z^+, G_t \in Z^+ \quad t = 1, \ldots, T, q = 1, \ldots, Q. \tag{6}$$

The objective function minimizes total expected delay cost, accounting for both ground and air delays. Constraint set (2) shows that no aircraft will arrive earlier than its planned arrival period. The equality constraint that includes a summation to period $T+1$ requires that we account for all aircraft that couldn't land by period t. Therefore, any aircraft not landing by T will land in period $T+1$. When examining an entire planning day, this is fairly realistic since even the busiest airports will reduce their operations to 10–20% of capacity late in the evening. When we only want to consider a portion of a planning day, then we need to realize that there may not be enough capacity in period $T+1$ to land all aircraft. So some additional delay will be present. Constraint set (3) requires that, for a given time period, all en-route aircraft, including those on-time and those already experiencing air delay, will either land or experience an air delay until the next time period. Constraint set (4) assumes that there are no aircraft currently experiencing delays and waiting to land, prior to the beginning of the ground holding policy. Constraint set (5) assigns ground delays to those aircraft not landing in their originally desired time periods. Thus, one can see that aircraft can be assigned ground delay, air delay, or both, when airport capacity is restricted.

2.2. *Solution Properties*

It has been shown[6,7] that the constraint matrix of [SSGHP] is totally unimodular, and it follows that the linear programming relaxation of [SSGHP] is guaranteed to yield an integer solution. Thus, constraint set (6) can be replaced with the nonnegativity constraint set:

Nonnegativity:

$$A_t, W_{qt}, G_t \geq 0, \quad t = 1, \ldots, T, \quad q = t, \ldots, Q \tag{7}$$

This property will not hold when we introduce the risk aversion measures in Section 4, so we cannot remove the integrality constraints from all of our models in this analysis.

While the original problem itself is not very large, it is always desirable to reduce the problem to a linear program. The number of integer decision variables is $O(T + T * Q + T) = O(QT)$, where Q is the number of scenarios and T is the number of 15-minute periods. As will be shown, the experiments presented in this report are based on a 22-scenario, 24-period problem, which implies the evaluation covers a six-hour timeframe. This would translate to 576 integer variables. However, a 50-scenario model would contain 1248 integer variables, and the benefit of not requiring the integer restriction will increase.

3. Motivation for Stochastic Programming Approach

3.1. *Arrival Demand and Runway Capacity Data*

When a ground holding policy is being considered, the expected arrivals to the airport will be affected. So it is important to know the arrival stream. We consider a six-hour timeframe. Based on actual data from a major airport, an estimated arrival demand in 15-minute increments was obtained. Figure 1 presents this arrival demand data. Each chart shows typical arrival patterns for an airport with hub operations in the U.S. This is represented in the cyclical demand for arrivals throughout the period under consideration. Typically, arrivals will dominate the traffic pattern of an airport for approximately one hour, followed by an hour of traffic dominated by departures. The length of these cycles will depend on the airport and airlines operating at the airport. Test 1 and Test 2 use the same underlying arrival pattern, with Test 2 having 25% more arrivals per period. At the beginning of the six-hour period, there is a low arrival demand, indicating that the ground holding policy is being put into place during a departure peak. Test 3 and 4 also use the same distribution as Tests 1 and 2, with a "time shift" to recognize that a ground holding policy is just as likely to begin during a peak arrival flow.

Even though the FAA may impose only one of three polices (VFR, IFR1, IFR2), the actual weather and flight sequencing will further affect an airport's capacity. So, although there may only be three official AARs under a given runway configuration, many more capacity cases will be seen. Consider first the possibility that no inclement weather materializes. We denote this case as Capacity Scenario 0, or CP0. We then include reduced capacity scenarios in sets of three. For each capacity-reduced set, the first scenario represents a particular weather-induced capacity restriction. The second and third scenarios in the set reduce the capacity in each period by an additional 15% and 30%, respectively. Under these scenarios, there will exist no periods in which the nominal or VFR capacity is realized.

Figure 2 presents several "bad weather" scenarios and their effect on realized runway capacity. We only show the first scenario from each set of

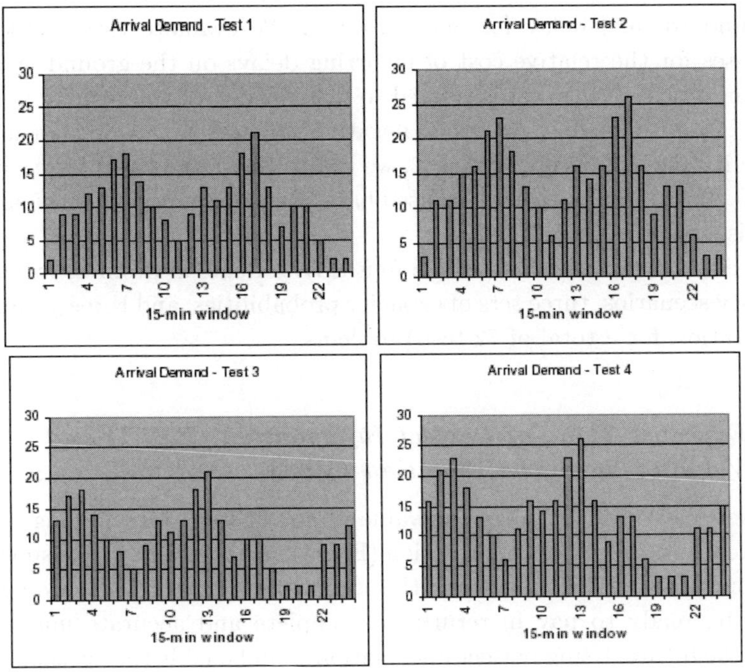

Fig. 1. Aircraft Arrival Demand at Capacitated Airport

three created. In some extreme cases, the bad weather may appear twice within one six-hour period, and this is considered in CP16 – CP21. The probabilities associated with these severely affected capacity scenarios are relatively small. We created seven capacity-reduced sets, for a total of 22 capacity scenarios (including the full capacity scenario, CP0) in our stochastic problem. A second set of "bad weather" scenarios (not presented) was also used in the computational experiments.

With all of the arrival capacity scenarios, we have assigned reasonable probabilities. This, of course, is where much of the difficulty of using stochastic programming is seen. Air traffic controllers, FAA, and airline personnel do not have historical data to provide them with such capacity scenarios and probabilities. So, until this information becomes more readily available, we must make some assumptions about how to obtain such data. Since there would be some debate as to the appropriate probabilities to assign to each scenario, we test three sets of probabilities.

Recall that [SSGHP] uses c_g (or c_a) to denote the unit cost of incurring a ground (or air) delay for one period, t. We evaluate three reasonable estimates for the relative cost of incurring delays on the ground or in the air. Since most of the cost is related to fuel, air delays will usually be much higher. But there may be other negative impacts of holding an aircraft before it takes off. Keeping $c_g = 2$, we create three test cases for $c_a = 3$, 5, and 10. These test cases are also used based on prior experiments conducted and discussed in Refs. 2, 6, 7.

In all, the experiments include four arrival demand profiles, two sets of capacity scenarios, three sets of capacity probabilities, and three ground/air delay ratios, for a total of 72 test problems.

3.2. *Expected Value of Perfect Information (EVPI) and Value of Stochastic Solution (VSS)*

Two key measures to gauge the value of stochastic programming are the expected value of perfect information (EVPI) and the value of the stochastic solution (VSS). EVPI measures the maximum amount a decision maker would be ready to pay in return for complete and accurate information about the future. Using perfect information would enable the decision maker to devise a superior ground holding policy based on knowing what weather conditions to expect. It is not hard to see that obtaining perfect information is not likely. But we can quantify its value to see the importance of having accurate weather forecasts. VSS measures the cost of ignoring uncertainty in making a planning decision. First, the deterministic problem, i.e., the problem that replaces all random variables by their expected values, is solved. Plugging this solution back into the original probabilistic model, we now find the "expected value" solution cost.

This value is compared to the value obtained by solving the stochastic program, resulting in the VSS. Applying this to the GHP, we obtain a solution to the deterministic problem, which provides a set of recommended ground holds per period based on the expected reduction of arrival capacity per period. We then resolve the stochastic problem using these recommended ground holds as fixed amounts. The difference between the original stochastic solution and the solution using this pre-defined ground holding policy is the VSS. Both the EVPI and VSS measures are typically presented in terms of either unit cost or percent. (We have chosen to show EVPI and VSS as percent values.) For a more thorough explanation, refer to Birge and Louveaux.[3]

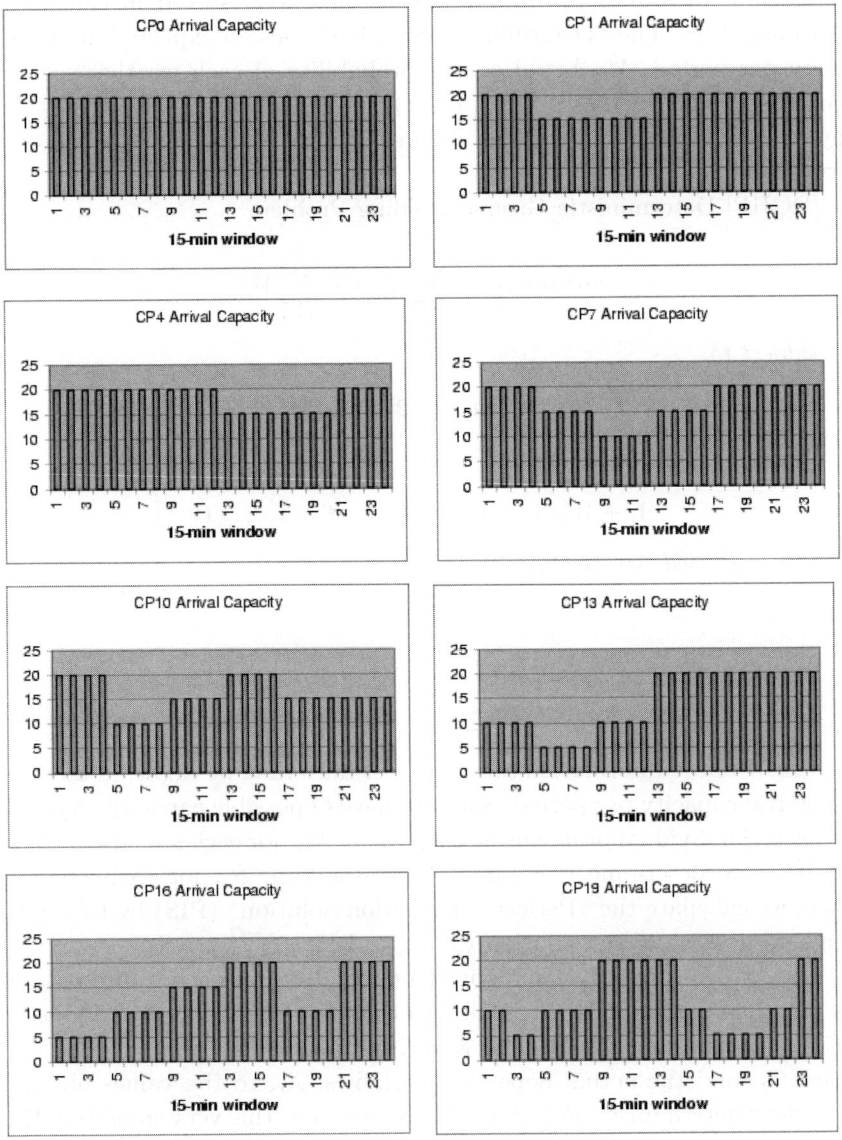

Fig. 2. Weather-Induced Arrival Capacity Scenarios (Note: Only the first scenario in each set of three is shown.)

We first introduce the four problems that were solved in calculating EVPI and VSS. The "Deterministic Solution" uses an expected arrival capacity per period, $\bar{M}_t$, based on the probability of each weather scenario occurring. Denoting $\bar{M}_t = \sum_{q=1}^{Q} p_q M_{qt}$, we can rewrite [SSGHP] without any scenarios and, thus, without any uncertainty. We present the following formulation:

[DGHP] Deterministic Ground Holding Problem

$$minimize \;\; c_g \sum_{t=1}^{T} G_t + c_a \sum_{t=1}^{T} W_t \tag{8}$$

subject to:

Constraints (2) and (5),

Arrival Period Capacities:

$$A_t + W_{t-1} - W_t \le \bar{M}_t, \qquad t = 1, \ldots, T, \tag{9}$$

Initial Period Air Delays:

$$W_0 = 0 \tag{10}$$

Nonnegativity:

$$A_t, W_t, G_t \ge 0, \qquad t = 1, \ldots, T \tag{11}$$

The "Perfect Information Solution" assumes that we know, in advance, the arrival capacity per period. Since we have Q possible capacity scenarios, we solve Q individual problems, setting $\bar{M}_t = M_{qt}$ for each scenario q. Using [DGHP], we determine a minimum cost solution, S_q, for each scenario. Then, we calculate the "Perfect Information Solution" (PIS) by taking the weighted average of the solution values, or PIS $= \sum_{q=1}^{Q} p_q S_q$.

The "Stochastic Solution," our recommended approach, represents the results of solving [SSGHP]. Finally, to calculate the "Expected Value Solution," we will use [SSGHP]. However, we first set the ground delay variables, G_t, and the actual departure variables, A_t, to the values obtained with the "Deterministic Solution." When we solve this version of [SSGHP], we are actually supplying a fixed ground holding plan and observing the additional air delays that result from not taking the randomness of each weather scenario, q, into account explicitly.

Runs were performed across all of the combinations of demand profiles, capacity profiles, probability sets, and ground/air delay ratios. In order to arrive at some summary statistics, the two arrival capacities and three sets of probabilities were grouped together. Thus, each summary test case is an average of six runs. Denote each run's ground/air delay ratio as G2A#, where G2 represents a unit cost of 2 for incurring ground delay and A# represents a unit cost of # for incurring air delay. Each summary test case is then unique based on its arrival demand profile and its ground/air delay ratio (G2A#). Table 1 summarizes the results over these groups of test cases:

Table 1. Overall EVPI and VSS Statistics (Minimize Total Expected Delay Cost Model)

Summary Test Case	Deterministic Solution	Perfect Info. Solution	Stochastic Solution	Expected Value Solution	EVPI	VSS
ArrDem1, G2A3	199	562	821	862	46%	4.9%
ArrDem1, G2A5	199	562	1246	1304	122%	4.6%
ArrDem1, G2A10	199	562	1874	2409	235%	28.3%
ArrDem2, G2A3	1034	1459	2033	2154	39%	6.0%
ArrDem2, G2A5	1034	1459	2728	2901	87%	6.4%
ArrDem2, G2A10	1034	1459	3484	4767	140%	36.8%
ArrDem3, G2A3	130	602	897	930	49%	3.6%
ArrDem3, G2A5	130	602	1387	1464	131%	5.4%
ArrDem3, G2A10	130	602	2094	2798	250%	33.3%
ArrDem4, G2A3	1035	1497	2145	2340	43%	9.1%
ArrDem4, G2A5	1035	1497	3007	3210	101%	6.8%
ArrDem4, G2A10	1035	1497	3906	5384	162%	37.8%

Note: Delays are represented in terms of cost units. They are not scaled to represent a monetary figure.

Both the deterministic and perfect information solutions do not change within a particular arrival demand profile. This indicates that all delays are being taken as ground holds. Since the ground delay cost is less than the air delay cost in each test case, the model will always assign ground delays first, assuming that future arrival capacity information is available. Since deterministic information is not usually available, introducing uncertainty through stochastic programming results in solutions with much higher total delay costs. Arrival demand profiles 2 and 4 both increase the amount of traffic arriving to the capacitated airport. This is clearly shown through the large increase in delays, even in the deterministic case.

For the G2A3 cases, the value of the stochastic solution (VSS) is at least 3.6%, which can be quite important given the magnitude in the cost per delay unit. And, as we move to the G2A10 cases, VSS is greater than 28%. For example, in the G2A10 case under Arrival Demand Profile 4, the expected value solution gives a value of 5384, and the expected savings by using a stochastic solution would be 1478. This indicates that, if air delays are expected to be more than five times as costly as ground delays, then evaluating the ground holding policy using stochastic programming is essential. Similarly, with EVPI values ranging from 40% to 250%, it is quite evident that obtaining higher quality weather forecasts would be very beneficial. The EVPI measure can be used as a justification for investing in improved weather forecasting techniques.

4. Risk Aversion Measures

4.1. *Conditional Value at Risk (CVaR) Model*

The solution to the SSGHP model is determined by minimizing total expected delay cost over the entire set of scenarios presented. However, there still may be instances where, in certain weather scenarios, the delay incurred as a result of a particular ground holding strategy is much longer than the delay incurred under any other scenario. In this situation, we may want to find solutions that attempt to minimize the spread of delays across all scenarios, or to minimize the extent to which extremely poor outcomes exist. This can be done through the addition of risk aversion measures. Such measures allow us to place a relative importance on other factors of the problem besides total delay. The Value-at-Risk (VaR) measure has been extensively studied in the financial literature. More recently, researchers have discovered that another measure, Conditional Value-at-Risk (CVaR), proves very useful in identifying the most critical or extreme delays from the distribution of potential outcomes, and in reducing the impact that these outcomes have on the overall objective function. For a more detailed description of CVaR and some applications, see Rockafellar and Uryasev.[8,9]

CVaR can be introduced in more than one form for the GHP, depending on the concerns of the airlines, the air traffic controllers, and the FAA. We can define a new objective that focuses on the risk measure, or we can add the risk measure in the form of risk aversion constraints (see Section 4.3 for alternate CVaR models). In this section, we present a new formulation that attempts to minimize the expected value of a percentile of the worst-case delays, i.e., we place the CVaR measure in the objective function.

In order to set up the CVaR model, additional variables and parameters are required. Let α represent the significance level for the total delay cost distribution across all scenarios, and let ζ be a decision variable denoting the Value-at-Risk for the model based on the α-percentile of delay costs. In other words, inα % of the scenarios, the outcome will not exceedζ. Then, CVaR is a weighted measure of ζ and the delays exceeding ζ, which are known to be the worst-case delays. Next, we introduce τ_q to represent the "tail" delay for scenario q. We define "tail" delay as the amount by which total delay cost in a scenario exceeds ζ, which can be represented mathematically as$\tau_q = \text{MAX}\{TD_q - \zeta, 0\}$, where $TD_q = c_g \sum_{t=1}^{T} G_t + c_a \sum_{t=1}^{T} W_{qt}$. The risk aversion problem is now formulated:

[GHP–CVaR] Ground Holding Problem (Conditional Value-at-Risk)

$$minimize \;\; \zeta + (1-\alpha)^{-1} \sum_{q=1}^{Q} p_q \tau_q \tag{12}$$

subject to:

Scheduled Arrival Times:

$$\sum_{t=1}^{j} A_t \leq \sum_{t=1}^{j} D_t \quad j = 1, \ldots, T, \tag{2}$$

$$\sum_{t=1}^{T+1} A_t = \sum_{t=1}^{T+1} D_t$$

Arrival Period Capacities:

$$A_t + W_{q,t-1} - W_{qt} \leq M_{qt}, \qquad t = 1, \ldots, T, \;\; q = 1, \ldots, Q, \tag{3}$$

Initial Period Air Delays:

$$W_{q0} = 0, \qquad q = 1, \ldots, Q, \tag{4}$$

Ground Delays:

$$G_j + \sum_{t=1}^{j} A_t = \sum_{t=1}^{j} D_t, \qquad j = 1, \ldots, T, \tag{5}$$

Worst-Case (Tail) Delays:

$$\tau_q \geq c_g \sum_{t=1}^{T} G_t + c_a \sum_{t=1}^{T} W_{qt} - \zeta q = 1, \ldots, Q, \tag{13}$$

Nonnegativity:

$$\zeta, \tau_q \geq 0, \qquad q = 1, \ldots, Q, \tag{14}$$

Integrality:

$$A_t \in Z^+, W_{qt} \in Z^+, G_t \in Z^+, \qquad t = 1, \ldots, T, \;\; q = 1, \ldots, Q. \tag{6}$$

This model will actually have an objective function value equal to α-CVaR. In order to compare this solution to the solution provided by [SSGHP], we must still calculate total expected delay cost. Total delay cost, as well as maximum scenario delay cost, can be determined after [GHP–CVaR] is solved.

Operating under this holding policy, we can address the risk involved with incurring extremely long ground and air delays. This may sacrifice good performance under another capacity scenario since the bad weather is not realized under all scenarios. In other words, a capacity scenario that would result in little or no delay may now experience a greater delay based on the holding policy's attempt to reduce the delay under a more severely constrained capacity case. So these differences would need to be dealt with on a case-by-case basis, and we present some alternative models to accommodate the goals of different decision makers in Section 4.3.

4.2. *Minimize Total Delay Cost Model vs. Minimize Conditional Value-at-Risk Model*

The CVaR model requires the additional input of a significance level, and $\alpha = 0.9$ is chosen for the analysis. Table 2 presents a comparison of the delay statistics for the Minimize Total Delay Cost Model (SSGHP) and the Minimize Conditional VaR model (GHP–CVaR).

Table 2. Overall Model Comparisons

	Minimize Total Expected Delay			*Minimize Conditional VaR*		
Summary Test Case	Total Delay Cost	α-VaR	α-CVaR	Total Delay Cost	α-VaR	α-CVaR
ArrDem1, G2A3	821	1661	2628	1767	1875	2087
ArrDem1, G2A5	1246	2337	3943	2059	2222	2471
ArrDem1, G2A10	1874	3075	4856	2757	2855	2934
ArrDem2, G2A3	2033	3667	4777	3426	3505	3717
ArrDem2, G2A5	2728	4450	6299	3743	3852	4101
ArrDem2, G2A10	3484	4809	6755	4411	4485	4564
ArrDem3, G2A3	897	2109	3078	1981	2139	2378
ArrDem3, G2A5	1387	3087	4703	2371	2531	2749
ArrDem3, G2A10	2094	3209	5624	2960	3166	3225
ArrDem4, G2A3	2145	4315	5520	3878	3953	4192
ArrDem4, G2A5	3007	5544	7552	4263	4345	4563
ArrDem4, G2A10	3906	5092	7624	4869	4980	5039

Note: Delays are represented in terms of cost units. They are not scaled to represent a monetary figure.

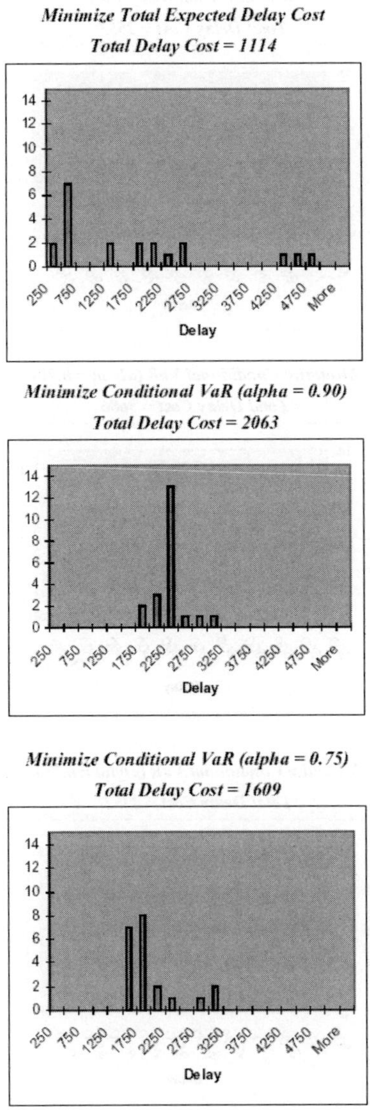

Fig. 3. Total Delay Output for Arrival Demand Level 1

Results of the model comparisons show that in the CVaR model, total delay will be increased in order to reduce the worst-case delays across all test cases. This supports the explanation stated in Section 4.1 in describing

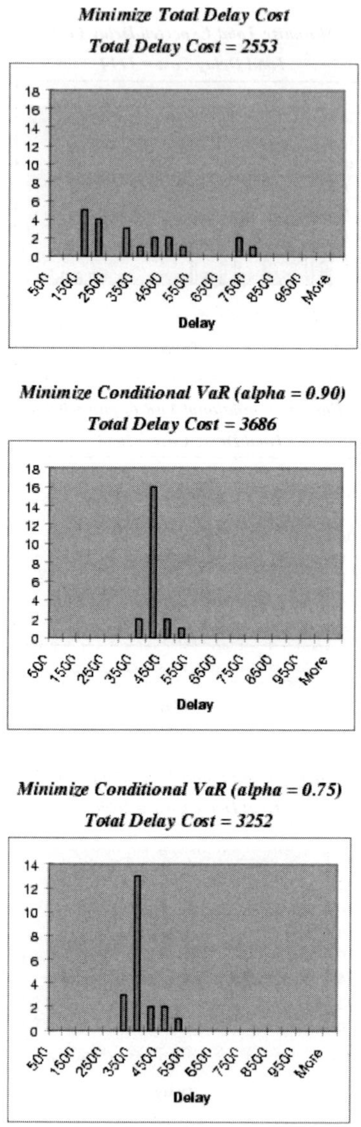

Fig. 4. Total Delay Output for Arrival Demand Level 2

the use of risk aversion. By examining the results more closely, we note some interesting findings.

Observe the difference in values for α-VaR and α-CVaR when no risk is modeled. This illustrates the importance of considering the average of

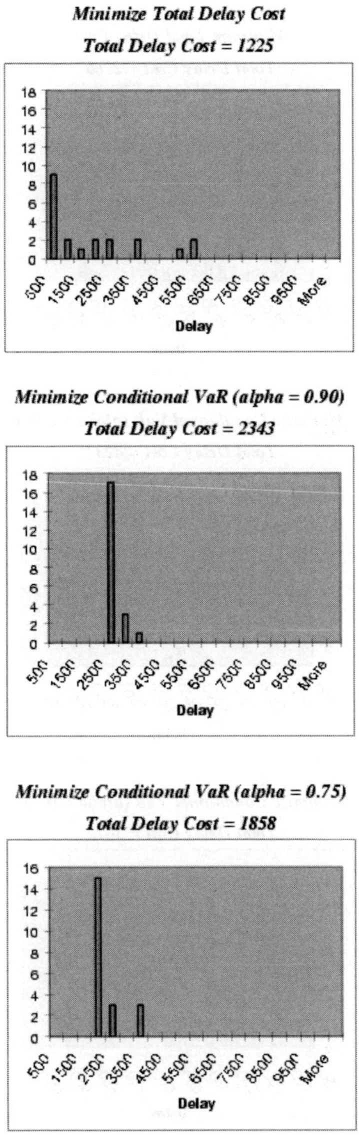

Fig. 5. Total Delay Output for Arrival Demand Level 3

worst-case delay costs when you choose to model risk. VaR tends to overlook the differences in delays beyond the critical value, and it may not be able to reduce the worst-case delay costs as effectively. When minimizing α-CVaR

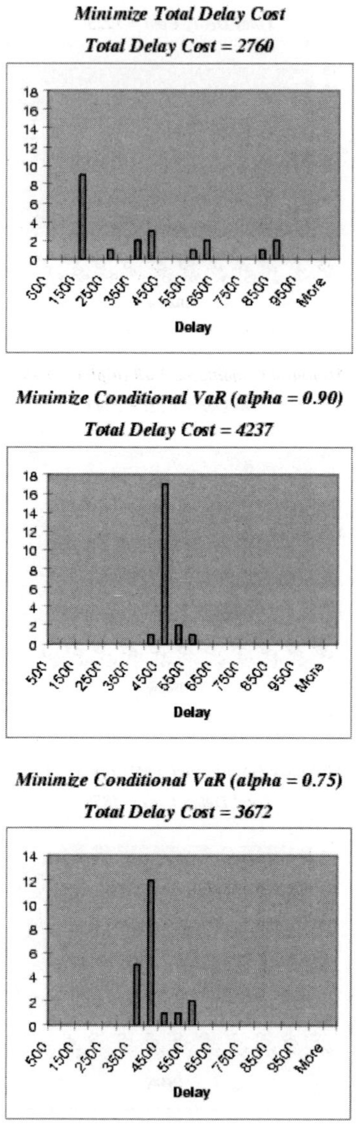

Fig. 6. Total Delay Output for Arrival Demand Level 4

in the second model, notice that the α-VaR and α-CVaR values are much closer.

Also, the percentage increase in total expected delay cost between the two models is more drastic for smaller air delay costs. But as the air delay cost rises, the total delay cost incurred when minimizing α-CVaR is not severely affected. Consider the following results observed for the Arrival Demand 4 profile. The G2A3 case experiences an increase in average delay cost of 80%, while the G2A10 case experiences only a 25% increase. The magnitude of air delay costs will significantly impact the effectiveness of using risk constraints. Recall from Table 1 the example that was previously described. In the G2A10 case the expected value solution gives a value of 5384, and the expected savings without risk constraints would be 1478. Now, by minimizing the 10% worst-case delays (using the GHP–CVaR model), the expected savings reduces to 515. But we also have reduced the worst-case delays from 7624 to 5039.

Since the CVaR analysis up to this point only considers using $\alpha = 0.9$, it is worthwhile to show how CVaR can shape the distribution of outcomes at another significance level. Consider $\alpha = 0.75$, which implies that the 25%-largest delay costs are minimized. Figure 3 shows the actual delay cost outcomes with and without CVaR, and at both levels of significance. The data is for a specific test case only (ArrDem 1, G2A5, with arrival capacity 1 and probability set 1). Depending on what the decision makers are willing to accept, different delay outcomes can be achieved. This is one of the underlying powers of introducing risk aversion such as minimizing α-CVaR. Additional figures presenting the results for the other three arrival demand profiles are presented in Appendix A.

4.3. *Alternate Risk Aversion Models*

Depending on the input from each group involved in constructing a ground holding policy, there will be conflicting desires to reduce total expected delay and to reduce the worst-case outcomes. For these purposes, we can actually choose among several risk aversion models. If your sole desire were to reduce the total expected delay cost, you would not require the use of risk aversion. But if you want to reduce maximum delay costs, you might use the GHP–CVaR model. For other cases, which will account for most collaborative efforts, some combination of these models will be chosen.

A logical extension to the GHP–CVaR model would therefore be the inclusion of total delay costs in the objective function. We are now concerned with reducing both the total delay costs and the worst-case delay costs. Formulating this extension, we have:

[GHP–CVaR1] Alternate GHP Risk Formulation 1

$$minimize \quad c_g \sum_{t=1}^{T} G_t + c_a \sum_{q=1}^{Q} \sum_{t=1}^{T} p_q W_{q,t} + \zeta + (1-\alpha)^{-1} \sum_{q=1}^{Q} p_q \tau_q \tag{15}$$

subject to:

Constraints (2)–(6), (13)–(14)

We introduce a second alternate formulation that imposes a restriction on allowable losses. We use the original objective function from [SSGHP], minimizing the expected total delay cost, while satisfying a constraint requiring the percentile of worst-case delays to be no more than some parameter, v.

[GHP–CVaR2] – Alternate GHP Risk Formulation 2

$$minimize \quad c_g \sum_{t=1}^{T} G_t + c_a \sum_{q=1}^{Q} \sum_{t=1}^{T} p_q W_{qt} \tag{1}$$

subject to:

Constraints (2)–(6), (13) – (14)

Worst-Case Delay Bound:

$$\zeta + (1-\alpha)^{-1} \sum_{q=1}^{Q} p_q \tau_q \leq v \tag{16}$$

By placing an upper bound on a loss function, as in (16), it approaches a maximum loss constraint. But some scenarios can actually exceed this parameter value, as long as the weighted average of losses within the percentile remains below v. We provide an illustration of the effect of using each model to set the ground holding policy.

Notice that SSGHP provides the lowest expected total delay cost, based on considering the likelihood of each weather scenario actually occurring. On the other hand, GHP–CVaR produces the best value of α-CVaR and the lowest maximum delay cost in any scenario. The tradeoff is that the more likely scenarios will now encounter increased ground holdings. Combining these two objectives with GHP–CVaR1, we gain a substantial amount of the benefit of the previous two models, with total expected delay cost at 3746 and maximum scenario delay cost at 5696. And we can even fine-tune

Table 3. Performance Comparison of Alternate Risk Models

Model	Total Expected Delay Cost	α-VaR	α-CVaR	Maximum Delay Cost
SSGHP	3336	4890	7354	8570
GHP–CVaR	4325	4446	4521	5076
GHP–CVaR1	3746	4136	4703	5696
GHP–CVaR2 ($v = 5000$)	3566	4068	5000	5898
GHP–CVaR2 ($v = 6000$)	3383	4332	6000	6882

Note: Results are based on the data set (ArrDem 2, G2A10, Arrival Capacity 1, Probability Set 1)

our objective further through the use of the CVaR constraint. As α-CVaR is increased, we approach our original SSGHP model.

In addition to the above risk models, Rifkin[7] briefly presents the Maximum Air Delay Model (MADM). MADM can be thought of as a maximum-loss constraint for any scenario, and if such a number exists, this could be added to any of the above formulations. What MADM fails to address is the continuing effort to minimize total delays. A max-loss constraint can be added to any of the formulations presented in this paper, allowing the user additional insight into a particular airport's ground holding policies. As with the parameter v, setting the maximum loss too tight may prevent the model from finding a feasible solution.

There is no one answer when deciding which problem formulation to use. Each will shape the resulting total delay in different ways, and thus it is dependent on the groups making the decisions in determining the amount of acceptable delay.

5. Conclusions and Future Work

Modeling the ground holding problem as a stochastic problem is most certainly beneficial. Even under cases when delay costs are low and uniform, the value of the stochastic solution is significant. Additionally, introducing risk aversion allows a decision maker to offer several potential outcomes based on various worst-case delay scenarios.

There are several issues that were not addressed in this report and that are areas for future research. By modeling the problem at the individual flight detail, we may be able to gain more accuracy in determining the true capacity and realized arrival flows into an airport. Researchers can determine if the additional detail will bring about enough benefit to merit the undertaking of working with a more complex model. Once the model is at the flight level, arrival sequencing, banking, and other arrival/departure disruptions can also be modeled. Considering the originating stations of the aircraft could also be worthwhile. A multiple airport model would be able to provide more realistic information on the decisions and actions of each individual departing aircraft en route to the capacitated airport under study.

Finally, Collaborative Decision Making [CDM], described in Ball, *et al.*,[1] has been an area of focus recently. It allows airlines to be involved in the decisions on which aircraft will be delayed during a ground holding plan. This is likely to achieve a reduction in overall costs to individual airlines by allowing "more critical" aircraft to take off at their scheduled departure times and not incur ground holding delays. CDM may be more difficult to model, but it is important to include this fundamental interactive approach in order to represent or simulate the actual environment.

References

1. M. O. Ball, R. Hoffman, C. Chen, and T. Vossen, "Collaborative Decision Making in Air Traffic Management: Current and Future Research Directions," Technical Report T.R. 2000-3, NEXTOR, University of Maryland (2000).
2. M. O. Ball, R. Hoffman, A. Odoni, and R. Rifkin, "The Static Stochastic Ground Holding Problem with Aggregate Demands," Technical Report T.R. 99-1, NEXTOR, University of Maryland and Massachusetts Institute of Technology (1999).
3. J. Birge and F. Louveaux, *Introduction to Stochastic Programming*, Springer Series in Operations Research (1997).
4. R. Hoffman and M. Ball, "A Comparison of Formulations for the Single-Airport Ground Holding Problem with Banking Constraints," Technical Report T.R. 98-44, NEXTOR, University of Maryland (1998).
5. L. Navazio and G. Romanin-Jacur, "The Multiple Connections Multi-Airport Ground Holding Problem: Models and Algorithms," *Transportation Science* 32, No. 3, 268-276 (1998).
6. O. Richetta and A. R. Odoni, "Solving Optimally the Static Ground-Holding Policy Problem in Air Traffic Control," *Transportation Science* 27, No. 3, 228-238 (1993).
7. R. Rifkin, *The Static Stochastic Ground Holding Problem.* Master's Thesis, Massachusetts Institute of Technology (1998).
8. R. T. Rockafellar and S. Uryasev, "Conditional Value-at-Risk for General Loss Distributions," Research Report 2001-5, ISE Department, University Of Florida (2001).
9. R. T. Rockafellar and S. Uryasev, "Optimization of Conditional Value-at-Risk," *The Journal of Risk*, Vol. 2, No. 3, 21-41 (2000).
10. P. Vranas, D. Bertsimas, and A. Odoni, "Dynamic Ground-Holding Poli-

cies for a Network of Airports," *Transportation Science* 28, No. 4, 275-291 (1994).

11. P. Vranas, D. Bertsimas, and A. Odoni, "The Multi-Airport Ground-Holding Problem in Air Traffic Control," *Operations Research* 42, No. 2, 249-261 (1994).

CHAPTER 8

MEASURING PRODUCTION EFFICIENCY IN THE GREEK FOOD INDUSTRY

A. Karakitsiou

DSS Laboratory
Department of Industrial and Production Engineering
Technical University of Crete, Greece
E-mail: karakitsiou@ergasya.tuc.gr

A. Mavrommati

DSS Laboratory
Department of Industrial and Production Engineering
Technical University of Crete, Greece
E-mail: athanasia@ergasya.tuc.gr

A. Migdalas

DSS Laboratory
Department of Industrial and Production Engineering
Technical University of Crete, Greece
E-mail: migdalas@ergasya.tuc.gr

Despite of the high rate of economic growth which the Greek Food industry has achieved in last decades it is clear that those benefits have not been distributed evenly. The Greek Food Sector still remains concentrated in centralized regions. The primary purpose of this study is to measure the technical efficiency in Greek Food Industry. Unlike past studies this study decomposes technical efficiency into its pure technical efficiency and scale efficiency, using a non-parametric linear programming approach called Data Envelopment Analysis. The DEA is a modeling methodology for deriving the relative efficiency of units where there are multiple incommensurate inputs and outputs. The evidence suggests that at least at the aggregate level, the major part of technical efficiency in Greek food sector is due to scale inefficiencies, i.e., a consequence of wrong size election. Therefore, there are significant possibilities for the

Greek food sector to increase its competitiveness by changing size.

Keywords: Technical efficiency, pure technical efficiency, scale efficiency, data envelopment analysis, linear programming, Greek food sector.

1. Introduction

The Greek food sector plays a significant role in the country's development. First, it defines the country's level of self-sufficiency in foodstuffs and consequently the level of the country's foreign dependence. Second, it seriously influence the nation's balance of payments in the framework of international division of labor. However, the level of its development and competitiveness is not satisfactory. Its main weakness are small size firms, the lack of modern technology, and poor management.

The continuous growth of manufacturing costs has driven governments to encourage industries to increase their efficiency. Manufacturing operations are affected nowadays by increased demand for quality products and continuous technology investments. Such issues include debate concerning the trade-offs between objective of efficiency, effectiveness and equity in the way resources are allocated in the food sector. Improving the efficiency in the manufacturing operations requires, first of all, to define the term efficiency in order to determine the complexity of activities taking place within industry production function. The technical efficiency of food industry production concerns the extent to which maximum output is produced for given levels of resource or minimum input is employed for a given level of production.

In the literature of productivity and efficiency analysis, much discussion has focused on economic efficiency measures, as well as on the economic justification of technical efficiency measures. Following the seminal paper by Farell,[9] economic efficiency can be decomposed into two components: allocative efficiency and technical efficiency. The technical component requires quantitative volume data of inputs and outputs only, while associated prices or cost shares are also necessary for measuring allocative efficiency. Debreu[6] and Farell[9] have already expressed their concern about the ability to measure prices accurately enough to make good use of economic efficiency measurement. For example, accounting data can give a poor approximation for economic prices (i.e. marginal opportunity costs), because of debatable valuation and depreciation schemes. Several authors cite this concern as a motivation for emphasizing technical efficiency measurement. Consequently, many studies in the more application-oriented side of operations research, including the seminal articles on data envelopment analysis (DEA) by *Charnes et al.*, [5] Banger *et al.*,[2] Førsund *et al.*,[10] Ali *et al.*,[1] Charnes *et al.*,[4] Dyson and Thanassoulis,[7] assess efficiency solely in terms of technical efficiency.

Interestingly, the radial (i.e. Debreu-Farrell) technical efficiency measure provides a theoretical upper bound for economic efficiency, as already noticed by Ref. 6. However, this measure need not be a particularly good approximation for economic efficiency, as it does not utilize any price information whatsoever. In numerous empirical studies, at least some rough information on the economic prices is available from theory or practical knowledge of the industry under evaluation. One source of price information is prior knowledge on the quality or risk of the different inputs and outputs. For example, primary inputs are typically more expensive than secondary inputs. Therefore, the unit of labor input of key personnel (e.g. detectives, dentists, university professors, surgeons, teachers) is more expensive than that of assisting staff (e.g. secretaries, research assistants, cleaners, janitors). As capital inputs are concerned, the unit cost of equity capital exceeds that of debt, because equity involves more risk for the capital suppliers than debt does. Consequently, there exist both need and opportunities to include incomplete price information in efficiency analysis, so as to improve approximation of economic efficiency concepts.

In this study we measure the relative level of technical efficiency of the Greek food manufacturing sector in Greece in 1999,(decomposed into its pure technical and scale components) at aggregate level. To estimate efficiency scores cross DEA method is applied to cross section data comprising of nine region in Greece for the year 1999. This study is to our knowledge the first application of DEA in order to measure technical efficiency and its components at the aggregate level in Greek food sector. This enables more detailed understanding of the nature of technical efficiency of the sector.

The outline of the chapter is as follows: In Section 2 we introduce the notions of technical efficiency and clearly identify the importance of efficiency. In Section 3 the analytical theoretical concept of DEA method. Section 4 is devoted to data and their source description. The last two sections cover the empirical findings of this study, and conclusion and suggestion for further research.

2. Technical Efficiency

Technical inefficiency reflects the failure of some firms to obtain the maximum feasible output given the amount of inputs used. Its measurement is crucial to quantify the importance of poor performances in a productive activity. Unfortunately, measurement is not enough. In order to improve technical efficiency (TE), firms should be able to identify the sources of miss-performances and the alternatives available to make better use of their resources. Therefore, the question to be answered is "how can a firm become efficient in practice"? The answer to this question depends on the sources of inefficiency.

Some studies consider technical inefficiency as the result of a lack of motivation or effort, as suggested by Leibenstein (1966). Thus, the question of efficiency improvement is assessed within the framework of principal-agent contractual theory. In this line, Bogetoft (1994) suggests that efficiency improvements may be achieved introducing an appropriate incentive scheme to induce the desired (efficient) effort level from the agent. A different approach considers technical inefficiency as the result of a lack of knowledge or managerial ability.[9] Under this view, efficiency improvements may be achieved through learning processes, as is the case of management programs. Thus, the main difference between the two approaches is the assumption made about the motivation of the productive agents.

In recent years, a number of studies on the theoretical and empirical measurement of technical efficiency has been generated by researchers, and two different notions of technical efficiency have emerged in the economic literature. The first notion due to Koopmans,[11] defines a producer as technically efficient if a decrease in any input requires an increase of at least one other input. This definition is closely related to the Pareto efficiency concept, and its great intuitive appeal has led to its adoption by several authors, in particular by Färe and Lovell.[13] The second notion introduced by Debreu[6] and Farell,[9] is based on radial measures of technical efficiency. In the input case, the Debreu and Farrell index measures the minimum amount that a vector can be shrunk along a ray while holding output levels constant. This efficiency index is constructed around a technical component that involves equiproportionate modification of inputs, and this has received a growing interest during the last few years. Following Charnes *et al.*,[5] several empirical papers have implemented the Debreu and Farrell measure. In particular, describing the production set as a piece-wise linear technology, it can be computed by linear programming.

The production technology transforming inputs into outputs, for $j = 1, \ldots, J$, firms can be modeled by an input mapping L.[8] More specifically, let y denote an output vector in $\mathcal{R}_+^m$, and x an input vector in $\mathcal{R}_+^n$. Then, $L(y)$ is the set of all input vectors x which yield at least the output vector y, that is, $L(y)$ is a subset of $\mathcal{R}_+^n$. Then the isoquant and the efficient subset of all $L(y)$ are defined as follows:

$$\text{Isoq } L(y) = \begin{cases} \{x|x \in L(y), \lambda x \notin L(y)\} & \text{for } \lambda < 1 \\ 0 & \text{for } y = 0 \end{cases}$$

$$\text{Eff } L(y) = \begin{cases} \{x|x \in L(y), \bar{x} \le x \Rightarrow \bar{x} \notin L(y)\} & \text{for } y \in \mathcal{R}_+^m \\ 0 & \text{for } y = 0. \end{cases}$$

Figure 1 illustrates the concept of a piece-wise linear convex isoquant in a production process, where in two inputs X_1 and X_2 are used to produce output Y. $L(y)$ consists of input vectors in or inside $SBCS$. $IsoqL(y)$ consists of the set of input vectors on $SBCS$. Firms B and C are, the most efficient ones (since they utilize the least input combination to produce the same level of output), and hence they are used to establish the piece-wise linear convex isoquant frontier. Firm A is technically inefficient and wants to move as close to the frontier SS as possible. As such, point F on the frontier SS would become the target for firm A, and the distance AF can be treated as its technical inefficiency. The ratio of AF/AO to represent technical inefficiency. This ratio score of efficiency will result in a value from the range of 0 to 1. The higher score indicates a higher technical efficiency.

There are two different approaches to measure technical efficiency: parametric and non-parametric production frontier.[14] The parametric approach requires assumptions about the functional form of the production frontier. It uses statistical estimation to estimate the coefficients of the production function as well as the technical efficiency.[12] Since the parametric production frontier is assumed to be the "true" frontier, the scores of technical efficiency obtained are regarded as absolute technical efficiency. A potential disadvantage of the parametric production frontier is the possible miss-specification of a functional form for the production process.

Non-Parametric production frontier on the other hand, are based on mathematical programming and do not make assumptions about the functional form. The data points in the set are compared for efficiency. The most efficient observation are utilized to construct the piece-wise linear convex non-parametric frontier. As a result, non-parametric production frontiers

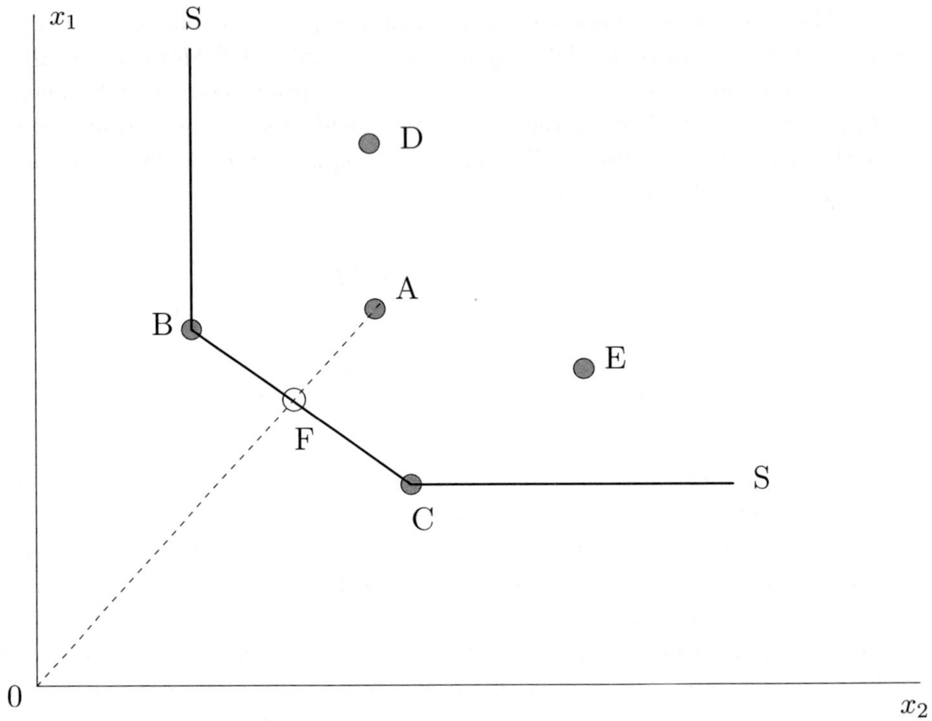

Fig. 1. Piece-wise linear convex isoquant SS and technical efficiency

are employed to measure relative technical efficiency among the observations.

3. Research Methodology

The methodology of data envelopment analysis, initially introduced by Charnes *et al.*,[5] is a mathematical programming technique used to evaluate the relative efficiency of homogeneous units. This efficiency evaluation derives from analyzing empirical observations obtained from decision-making units (DMUs), to define productive units which are characterized by common multiple outputs and common designated inputs.

DEA was originally developed for use in non-profit organizations but the fields of applications have increased. Unlike the classic econometric approaches that require a pre-specification of a parametric function and several implicit or explicit assumptions about the production function, DEA requires only an assumption of convexity of the production possibility set and uses only empirical data to determine the unknown best practice frontier.

DEA can be a powerful tool when used wisely. A few of the characteristics that make it powerful are:

- DEA can handle multiple input and multiple output models.
- It doesn't require an assumption of a functional form relating inputs to outputs.
- DMUs are directly compared against a peer or combination of peers.
- Inputs and outputs can have very different units.

The same characteristics that make DEA a powerful tool can also create problems. An analyst should keep these limitations in mind when choosing whether or not to use DEA.

- Since DEA is an extreme point technique, noise (even symmetrical noise with zero mean) such as measurement error can cause significant problems.
- DEA is good at estimating "relative" efficiency of a DMU but it converges very slowly to "absolute" efficiency. In other words, it can tell you how well you are doing compared to your peers but not compared to a "theoretical maximum."
- Since DEA is a non-parametric technique, statistical hypothesis tests are difficult and are the focus of ongoing research.
- Since a standard formulation of DEA creates a separate linear program for each DMU, large problems can be computationally intensive.

Under the assumption of non-parametric frontiers, the technical efficiency of one unit can be calculated by solving the following linear program:

$$\begin{array}{lll} P1 & \min \lambda_1 & \\ \text{s.t} & u & \leq zU \\ & \lambda_1 x & \geq zX \\ & z & \geq 0, \end{array} \qquad (1)$$

where λ_1 is the technical efficiency value, and because we have adopted an input orientation it can be interpreted as the proportion in which all inputs can be diminished in order to obtain an efficient performance. u is the vector of the m output obtained by the analyzed unit, U is the k $\times$ m matrix of outputs for the k units in the sample, x represents the values for the n inputs used by the studied unit, X is the k $\times$ n matrix of inputs for all units, and z is the vector of the intensity coefficients which determine convex combinations of observed input and output combination.

When λ_1 equals 1, the analyzed unit is on the isoquant and it is impossible to obtain its output with radial reduction of all its inputs. In this chapter, we define a company as technical efficiency in the Pareto sense, i.e., it is not possible to reduce just one input factor and keep the level of output constant.

So far we have considered constant returns to scale; we can nevertheless, relax this assumption to consider variable returns to scale. In this case, it is possible to calculate the efficiency of each unit, not only in relation to the whole sample, but also in relation to units of similar size. Efficiency under variable returns to scale is calculated then by solving the following problem:

$$\begin{array}{lllll} P2 & \min \lambda_2 & & & \\ \text{s.t.} & u & \leq & zU & \\ & \lambda_2 x & \geq & zX & \\ & \sum z_i & = & 1 & \\ & z & \geq & 0 & \end{array} \qquad (2)$$

Thus, a company is agreed to be purely technical efficient when its λ_2 equals 1 and is agreed as efficient under Pareto criteria.

The efficiency losses due to a wrong choice of the unit size, or scale inefficiency, is the ratio $\frac{\lambda_1}{\lambda_2}$. To determine if the scale inefficiency is due to increasing returns, Berg *et al.*[3] state that if the sum of the z coefficients in problem P1 is greater than 1, the analyzed unit is operating under diminishing return to scale, and if it is smaller than 1 the unit operates with increasing returns to scale.

The Farrell measure is radial; for a given company j, it determines the maximal amount by which the input x_j can be proportionally reduced while maintaining production of output y_j. Note that the Farrell measure does not require comparison of a given input vector to an input vector that belongs to the identified efficient subset. This is clear from the definitions of the Farrell measure, the isoquant Isoq L(y) and the efficient subset Eff L(y).

4. Input and Output Measures

For the empirical implementation of this study we separate Greece into 9 geographical regions, based on National Bureau of Statistics, which are the following:

- ◇ Attica
- ◇ Macedonia
- ◇ Crete
- ◇ Aegean Islands
- ◇ Thessaly
- ◇ Peloponnese
- ◇ Thrace
- ◇ Epirus & Ionian Islands
- ◇ Sterea Hellas

In addition the Greek food sector is separated into 13 three-digit industries. More specifically:

- ◇ Bakery products(B)
- ◇ Dairy products(D)
- ◇ Flour(FL)
- ◇ Legumes (LE)
- ◇ Meat products (M)
- ◇ Pasta products (P)
- ◇ Sugar products(SU)
- ◇ Canned food products (C)
- ◇ Seafood products(SF)
- ◇ Fruits and Vegetables(FV)
- ◇ Livestock (LI)
- ◇ Seed oils(SO)
- ◇ Snack products (SN)

We consider the two inputs (labor and capital) and one output case. Aggregate data from each of the nine regions of Greece and from each industry, are based on annual surveys conducted by ICAP (Financial Directories of Greek Companies). It should, however, be mentioned that our measures of output and inputs were constrained by available data at the regional level in Greece.

For our measure of output we use turnover. Output is aggregated into a single index of each sector to avoid any further complexity.

The measure of the quantity labor is based on annual accounting data for the number of employees from the ICAP. Since there is no objective way to account for the contribution of part-time versus full-time employees, just the 25% of part- time employees is used in the labor input measure. This, however, does not introduce a substantial bias in the labor quantity measure since part-time employees accounted for less that 0.7% of the work force in 1999. In addition we use gross fixed assets as proxy of capital stock.

Table 1. Technical Efficiency Results

	S1 B	S2 C	S3 D	S4 F	S5 FL	S6 EV	S7 ALE	S8 AI	S9 M	S10 CA	S11 P	S12 AN	S13 DU
Attica	1	0.81	0.7	1	1	0.98	1	0.6	0.77	1	1	0.31	1
Macedonia	1	1	1	0.8	0.47	0.97	1	0.51	0.93	0.43	0.57	0.2	1
Crete	0.79	0.29	0.62		0.67	1		1	0.45	0.32		1	0.77
Aegean Islands	0.79	0.61	0.65	0.66				0.53	1	0.82			0.61
Thessaly	1	0.94	0.63	0.37	0.81	0.55		0.6	0.81	0.6		0.43	0.81
Epirus & Ionian Islands	0.94	1	1	0.91	0.55	1	0.74	0.75	0.56	0.63		0.46	0.55
Thrace	0.73	0.88	0.48	0.11	0.68	0.43	0.89	0.37	0.66	0.26			0.4
Peloponnese	0.23	0.98	1	0.45				0.96	1	0.43			0.58
Sterea Hellas	0.79	0.58	0.69	0.62	1		0.74	0.46	0.79	0.38	0.61	0.23	0.96

Table 2. Purely Technical Efficiency Results

	S1	S2	S3	S4	S5	S6	S7	S8	S9	S10	S11	S12	S13
	B	C	D	F	FL	EV	ALE	AI	M	CA	P	AN	DU
Attica	1	1	1	1	1	1	1	0.78	1	1	1	1	1
Macedonia	1	1	1	1	0.54	1	1	1	0.96	0.44	0.74	0.57	1
Crete	0.97	0.34	0.70		0.68	1		1	0.45	0.34		1	1
Aegean Islands	1	1	1	1				1	1	1			0.93
Thessaly	1	0.96	0.66	0.72	1	0.87		1	0.96	0.63		0.87	0.90
Epirus & Ionian Islands	0.96	1	1	0.98	0.60	1	1	1	0.57	0.63		1	1
Thrace	0.93	0.98	0.49	0.17	0.81	0.48	0.98	0.81	0.71	0.29			0.58
Peloponnese	1	1	1	0.69				1	1	0.74			1
Sterea Hellas	0.86	0.61	0.71	0.77	1		1	0.49	0.84	0.44	1	0.61	1

Table 3. Scale Efficiency Results and Type of Return

	S1	S2	S3	S4	S5	S6	S7	S8	S9	S10	S11	S12	S13
	B	**CF**	**D**	**F**	**FL**	**EV**	**L**	**AI**	**M**	**CA**	**P**	**AN**	**DU**
Attica	1	0.81	0.70	1	1	0.98	1	0.77	0.77	1	1	0.31	1
Macedonia	1	1	1	0.80	0.88*	0.97	1	0.51	0.97*	0.98*	0.77*	0.35	1
Crete	0.82	0.87*	0.88*		0.98*	1		1	1	0.95*		1	0.77
Aegean Islands	0.79	0.61	0.65	0.66				0.53	1	0.82			0.65
Thessaly	1	0.98*	0.96*	0.52	0.81	0.64		0.60	0.85	f 0.96*		0.49	0.65
Epirus & Ionian Islands	0.98*	1	1	0.93	0.92*	1	0.74	0.75	0.98*	0.99*		0.46	0.55
Thrace	0.78	0.90	0.99*	0.62*	0.83*	0.89*	0.90	0.46	0.93*	0.89*			0.69*
Peloponnese	0.23	0.98	1	0.65				0.96	1	0.58			0.58
Sterea Hellas	0.93*	0.96*	0.98*	0.81*	1		0.74	0.93*	0.95*	0.86*	0.61	0.38	0.96

5. Empirical Results

Technical efficiency, purely technical efficiency and scale efficiency scores of each sector in each region were calculated by using the models (1) and (2). Results are shown respectively in Tables 1-3, where the latter also shows the types of return to scale of the inefficient sectors. It has also been assured that sectors reaching 1 in their efficiency indexes are also efficient under Pareto criteria.

A summary of the results show (Table 1) shows that the most industrialized region, Attica, show up on the efficiency frontier more frequently than any other region, in seven industries. It is followed by Macedonia, a region which also can be characterized as industrialized. The regions of Crete and Peloponnese show up on the efficiency frontier three time, and Aegean Islands, Epirus, and Sterea Hellas once. The most inefficient region appears to be Thrace, which does not show on the efficiency frontier. One possible explanation for the observed differences in efficiency is that latter four regions are recognizably poorer and technologically less developed.

Since for the measuring of the purely technical efficiency the hypothesis of constant returns to scale is relaxed, results for each region indicate the efficiency of that region by taking into account the effect of regions with the same size. From the results of table 2, Thrace gets the lowest purely technical efficiency figures, followed by Sterea Hellas, Thessaly, and Crete. Consequently, it can be seen that the situation of these four regions is difficult since the negative results of their purely technical efficiencies show the need to improve their internal organization to reach higher efficiency levels.

In addition our findings show that the technical inefficiency of Greek Food sector is due to inadequate dimension decisions. This situation seems clear in those sectors with purely technical efficiency of one and a technical efficiency less than one.

The analysis we made in this chapter enable us to make recommendations on whether the sectors should increase/decrease their size, depending on whether they operate under increasing/diminishing returns to scale. To explore the role scale effects can play, we calculate indices of efficiency of the region in each sector (Table 3) An index of one indicates scale efficiency, and an asterisk indicates decreasing returns to scale. In the majority of the case, highly industrialized regions on the frontier operate at inefficient scales. More specifically they are operating under increasing returns to scale. This implies that, from the economic policy view point, if production efficiency of Greek Food sector is to be improved, increasing firm size would be better than decreasing the size of firms.

6. Conclusion

Throughout this research, an input-oriented DEA model was used for estimating technical, pure technical and scales efficiency in Greek food sector. The results indicate that technical efficiency scores of some regions were low, especially those with low technological development. This implies that there is a significant scope to increase efficiency levels in Greek food sector. Our results also indicate that high industrialized areas exhibit higher level of efficiency.

Regarding purely technical efficiency, the values achieved by each area are rather high except for Thrace. Although in general this is to say that internal running and administration of Greek food sector is adequate, it is also true that management is to be improved in those cases where this efficiency is not 1.

In addition, the results of scales efficiency state clearly that a major part of the technical efficiency of Greek food sector is a consequence of scale inefficiency, in other words due to mistakes in size election. As a conclusion, the results recommend smaller sizes for those areas and sectors which exhibits diminishing returns of scale and larger for those which show increasing returns to scale.

The analysis in this chapter can be improved in a number of areas. As future research, allocative and overall efficiency indexes across the regions could be calculated in order to investigate the links between industrial location, concentration and economic efficiency of Greek food sector. Furthermore, a comparison of stochastic and DEA frontier analyses would be of great interest.

References

1. A. Ali, W. Cook and L. Seiford. Strict vs. weak ordinal relations for multipliers in data envelopment analysis. *Management Science*, **37**:733–738 (1991).
2. R. Banger, A. Charnes, and W. W. Cooper. Some models for estimating technical efficiency and scales inefficiencies in data envelopment analysis. *Management Science*, **30**(9):1078–1092 (1984).
3. S. A. Berg, F. Försund, and E. Jansen. Bank output measurement and the construction of best practice frontiers. Technical report, Norges Bank (1989).
4. A. Charnes, W. Cooper, A. Lewis, and L. Seiford. *The Measurement of Efficiency of Production*. Kluwer Academic Publishers, Norwell, MA (1994).

5. A. Charnes, W. Cooper, and E. Rhodes. Measuring the efficiency of decision-making units. *European Journal of Operations Research*, **3**:429–444 (1978).
6. G. Debreu. The coefficient of resource utilization. *Econometrica*, **19**:273–292 (1951).
7. R. Dyson and E. Thanassoulis. Reducing weight flexibility in data envelopment analysis. *Journal of Operations Research Society*, 39:563–576 (1988).
8. R. Färe, S. Grosskopf, and C. Lovell. The structure of technical efficiency. *Scandinavian Journal of Economic*, **85**:181–190 (1983).
9. M. Farell. The measurement of productive efficiency. *Journal of the Royal Statistical Society*, **120**(PART III):253–281 (1957).
10. F. Førsund, C. Lovel, and P. Schmidt. A survey of frontier production function and of their relationship to efficiency measurement. *Journal of Econometrics*, **13**:278–292 (1980).
11. T. Koopmans. Analysis of production as an efficient combination of activities. In *Activity Analysis of Production and Allocation*, T. Koopmans, ed., pp. 27–56 (1951).
12. C. Lovell. Production frontiers and productivity efficiency. In *The measurement of productive Efficiency: Techniques and Applications*, H. Fried, C. Lovell, and S. Schmidt, eds., pp. 84–110. Oxford University Press, New York (1993).
13. R. Färe and C. Lovell. Measuring the technical efficiency of production. *Journal of Economic Theory*, **19**:150–162 (1978).
14. P. Schmidt. Frontier production function. *Econometric Review*, **42**:289–328 (1986).

CHAPTER 9

BRAND MANAGEMENT IN THE FRUIT JUICE INDUSTRY

G. Baourakis

Mediterranean Agronomic Institute of Chania
Dept. of Economic Sciences, Management / Marketing
P.O.BOX 85, 73 100 Chania, Crete, Greece
E-mail: baouraki@maich.gr

G. Baltas

Athens University of Economics and Business
76 Patission Avenue, 10434 Athens, Greece
E-mail: gb@aueb.gr

This study considers the fruit juice market and investigates its preference and competitive structure by examining empirical survey data from Greece. Data on eight hundred individuals are collected through personal interviews and subsequently analyzed through multivariate methods in order to explore main product choice criteria and consumption patterns. The data reveal considerable category penetration and high individual consumption rates. Multi-criteria and multi-dimensional scaling analysis provides an insight into the structure of consumer preferences and brand competition in the examined market. The primary structure of the market lies in price and product form differences, which are also associated with specific characteristics of the examined brands.

Keywords: Brand management, marketing management, food and beverage marketing, multidimensional scaling, market research.

1. Introduction

General acceptance that the consumption of healthy products can reduce the risk of chronic diseases and improve health has raised the demand for natural products such as vegetables, organic produce and fruit juices. The fact that, fruit juices, in particular, have become an important part of the daily diet, can be attributed mostly to changes in consumer attitudes and preferences.

However, fruit juice consumption varies considerably across different countries. In the EU, Germany is the biggest consumer of fruit juices with the average German consuming 42 liters, followed by Austria where the annual per capita consumption is 34 liters, while the Netherlands with an annual 26.5 liters per capita consumption has moved from second to third place.[11,13]

In Greece, the fruit juice sector has exhibited a significant growth rate and per capita consumption has more than doubled over the last decade due to changes in lifestyle and the recognition of the health aspects of fruit juices.[3] The latest figures show that the rate of growth in domestic consumption has ranged between 3 and 4 percent on an annual basis for juices and between 2 and 3 percent for soft drinks.[7] In the early nineties many new companies started to penetrate the market increasing the existing competition, adopting new marketing strategies, leading to considerable proliferation of alternative brands, some of which are currently directed either to the tourist industry or the international market.[4]

Despite the differences among the various countries, the demand for juices is constantly increasing. This is particularly true for products made exclusively from squeezed fruits rich in liquids. As alluded to earlier, the most important drivers of demand are purity, nutritional properties and the health aspect.[1,2] One may here parenthetically note the steady ageing trend of the population, which contributes to the attention paid to health issues.[8]

The fruit juice market is divided into two main sub-categories, namely long-life juices, which dominate 70 percent of the market, and short-life juices, which comprise 30 percent of the market. Juices can also be divided into three main subcategories, based on natural juice content: a) 100 percent juices, b) nectars with a content of more than 50 percent, and c) fruit drinks with a content of more than 20 percent.[12] Short-life juices are mainly directed to the Athenian market, where volume sales account for 75 percent. Long-life juices have a more even distribution since Athenians consume approximately 40 percent of total production.

The increase in aggregate consumption can be attributed to changes in dietary habits, the introduction of improved products, attractive packaging, efficient distribution channels, intense advertising, and new production technologies that upgrade the quality of the final product.

However, little is currently known about consumer behaviour and brand competition in this market. The purpose of this empirical study is to explore the Greek juice market and determine the factors that influence consumer purchasing behaviour. To this end, a market survey was carried out in major Greek cities. Data on consumer preferences and perceptions with regard to fruit juice characteristics such as packaging, colour, price and taste, were collected through personal interviews with eight hundred randomly selected consumers. The questionnaire was highly structured and contained standard itemized scales.[15]

2. Consumption Patterns

The vast majority of the surveyed individuals (88.8 percent) were found to drink fruit juices, indicating that the product category has deep market penetration. As regards consumption rates, 24.5 percent of people drink more than one glass per day, 30.8 percent drink approximately one glass per day, 26.6 percent drink two to three glasses per week, 9.2 percent drink approximately one glass per week, and finally only 8.9 percent consume less than one glass per week.

The distribution of sample preferences over long-life, short-life and very short-life products was 60.3 percent, 20.9 percent, and 18.8 percent, respectively. This suggests that most people like the convenience of long-life products, although short-life products are fresher and perhaps of better nutritional quality.

The distribution of preferences over pack sizes of 1000, 330, and 500ml is 54.5, 17.4, and 12.8 percent, respectively. The remaining 15.2 percent of the sample has expressed indifference about the size. Again, most people were found to prefer the larger convenient size, although smaller sizes were considered more practical. Finally, most people drink fruit juices with breakfast (about 54 percent) or in the evening (approx. 59 percent).

3. Brand Preferences

3.1. *Consumer Attitudes*

The dominant brands in the juice sector were examined, namely Amita, Ivi, Life, Florina, Refresh, Frulite, and Creta Fresh. In addition, a fictitious brand named Fresh Juice was included for control purposes in order to evaluate respondents' awareness and overall attentiveness in the course of the survey.

Consumers were asked to assess the following basic criteria: packaging, color, price, taste and advertising. Table 1 indicates consumer preferences for the aforementioned brands in a ranking order according to an estimated score (ratio index). The ratio index was computed based on the ratings of a series of evaluation criteria with respect to their importance in decision-making. This offered the advantage of being able to highlight extreme cases and indicate the relatively important and unimportant attributes.[14,9] The last row in the table presents the overall preference order, according to the ratio index method.

Table 1. Consumer preferences in ranking order according to certain criteria

	BRAND NAME							
Criteria	Amita	Ivi	Creta Fresh	Fresh Juice	Life	Florina	Refresh	Frulite
Price	1st	3rd	7th	8th	5th	2nd	4th	6th
Taste	1st	4th	7th	8th	2nd	6th	3rd	5th
Advertisement	1st	5th	8th	7th	3rd	6th	4th	2nd
Packaging	1st	3rd	7th	8th	2nd	6th	5th	4th
Color	1st	3rd	7th	8th	2nd	6th	5th	4th
Overall	**1st**	**3rd**	**7th**	**8th**	**2nd**	**6th**	**4th**	**4th**

3.2. *Multidimensional Scaling Approach*

If one is interested in a concise and accessible interpretation of the market structure, the brand preference data can be used as input to multidimensional scaling models to produce brand maps that summarize graphically the structure of consumer demand in the examined product category.[6,5,10] More specifically, brand preferences can provide alternative indices of brand similarities, which are usually constructed from survey similarity data. Figure 1 presents an ALSCAL solution using the two estimated correlation matrices. The MDS map provides interesting insights into the competitive structure.

Figure 1 reveals that two important brand characteristics are associated with the two dimensions of the perceptual map. The vertical axis is associated with product life and the horizontal axis with product cost.

More specifically, Amita and Ivi, which dominate the long-life submarket, are located in the two upper quadrants. Frulite, Refresh and Life, which dominate the short-life submarket are placed in the two lower quadrants.

Therefore, product life is seen as an important attribute that differentiates the brands. In fact, product life determines not only the period in which the juice is suitable for consumption, but also other aspects of the product such as taste, production process, conservation, storage, and distribution requirements.

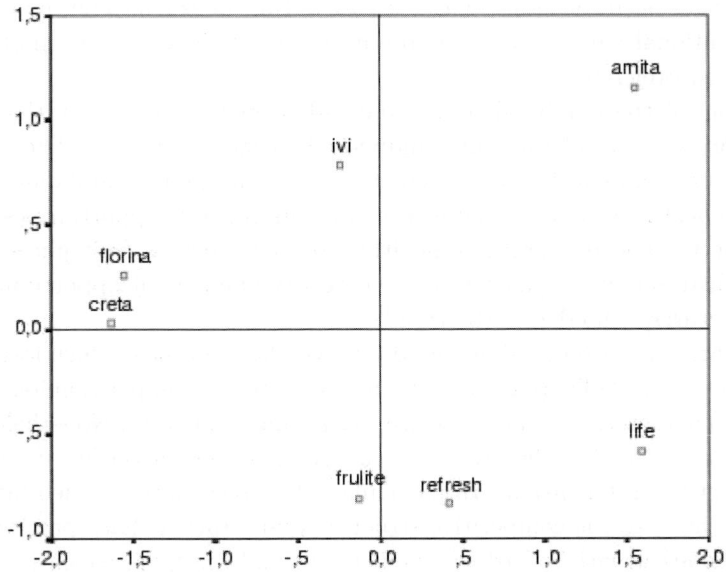

Fig. 1. MDS map of the competitive structure in the fruit juice category

For instance, the short-life brands should be kept refrigerated, while the long-life brands are less sensitive to temperature and storage conditions. However, the short-life brands taste more natural and fresh. Therefore, consumers face a tradeoff between convenience and quality.

While the vertical dimension is associated with product type, the horizontal axis is associated with product cost. Two relatively expensive brands (Amita and Life) are located in the right quadrants and two economy brands (Florina and Creta) are placed in the left quadrants. Florina and Creta brands are sold at considerably lower prices and are also associated with specific regions of the country, namely Florina and Crete.

The other three brands (Refresh, Frulite, and Ivi) are in the middle of the price range, although their projections on the horizontal axis do reflect price differences. For instance, Refresh is usually somewhat more expensive than Frulite.

4. Concluding Remarks

This study has been concerned with the case of the Greek fruit juice market. A large, national survey was carried-out to collect data on consumption patterns and preferences.

Analysis of the empirical data indicated deep market penetration of the category and considerably high individual consumption rates. The level of current demand can be attributed to changes in dietary habits and an increase in health awareness. These developments generate opportunities for the production and marketing of healthy products such as fruit juices. In the examined market, expansion of the category offers great opportunities to manufacturers who detect the trend.

The market is currently dominated by seven large brands, which formulate a rather typical oligopolistic structure. Inter-brand competition focuses on several characteristics such as price, packaging, and taste. Nonetheless, this study revealed that the primary structuring of the market lies in price and product form dimensions. In particular, the MDS map provided interesting insights into the competitive structure and a rather sharp partitioning of the market with regard to product cost and life-span, which is also indicative of other product properties, such as storage requirements and freshness.

References

1. G. Baltas. Nutrition labeling: issues and policies. *European Journal of Marketing* **35**, 708-721 (2001).
2. G. Baltas. The Effects of Nutrition Information on Consumer Choice. *Journal of Advertising Research* **41**, 57-63 (2001).
3. G. Baourakis, Y. Apostolakis, P. Drakos. Identification of market trends for Greek fruit juices. In C. Zopounidis, P. Pardalos, G. Baourakis, eds. *Fuzzy sets in management, economics and marketing*, World Scientific, 99-113 (2001).
4. G. Baourakis. The tourism industry in Crete: the identification of new market segments In C. Zopounidis, P. Pardalos, G. Baourakis, eds. *Fuzzy sets in management, economics and marketing*, World Scientific, 115-126, (2001).
5. D.J. Carrol, P. E. Green, and J. Kim. Preference mapping of conjoint-based profiles: an INDSCAL approach. *Journal of the Academy of Marketing Science* **14**, 273-281 (1989).
6. L. G. Cooper. A review of multidimensional scaling in marketing research. *Applied Psychological Measurement* **7**, 427-450, (1983).
7. ICAP, 1999. Greek financial directory.
8. A. Kouremenos and G. Avlonitis. The Changing Consumer in Greece. *International Journal of Research in Marketing* **12**, 435–448 (1995).

9. R. Lehmann. *Market research analysis.* Third edition, Irwin: Homewood, IL (1989).
10. G. Lilien and A. Rangaswamy. *Marketing engineering: computer-assisted marketing analysis and planning.* Addison-Wesley: Reading, MA (1998).
11. Panorama, 1996. Fruit and vegetable processing and conserving. Panorama of EU industry, EUROSTAT.
12. Z. Psallas. An x-ray of a "cool" market. *Industrial Review* **62**, 596-599 (1996).
13. USDA, 1999. www.fas.usda/gain files/199912/25556602.
14. R. Wiers. *Marketing research.* Second edition. Prentice-Hall International, Englewood Cliffs, NJ (1988).
15. W. G. Zikmund. *Exploring Marketing Research.* Dryden Press, Fort Worth, TX (1997).

CHAPTER 10

CRITICAL SUCCESS FACTORS OF BUSINESS TO BUSINESS (B2B) E-COMMERCE SOLUTIONS TO SUPPLY CHAIN MANAGEMENT

I. P. Vlachos

Agricultural University of Athens

The purpose of this chapter is to examine the critical success factors (CSFs) which are relevant in the Supply Chain Management.

Critical Success Factors can be defined as those few areas in which satisfactory results sustain competitive performance for the organization. The CSF approach represents an established top-down methodology for corporate strategic planning. This approach identifies a handful of factors that can be controllable by and informative to top management in order to formulate or adjust strategic decisions. In this chapter, CSFs are identified from the various business strategies adopted. Because the quest for competitive advantage from CSFs is the essence of the business level, as opposed to that of the corporate level, the business strategy is then the focus of attention.

Recent advances in the field of computer networks and telecommunications have increased the significance of electronic commerce. Electronic commerce is the ability to perform business transactions involving the exchange of goods and services between two or more parties using electronic tools and techniques. Companies across many industries are seeking to negotiate lower prices, broaden their supplier bases, and streamline procurement processes using e-commerce. The rapid diffusion of the Internet offers huge potential in building communities of interests, forging alliances, and creating technology-intense economies of scales.

Business-to-business e-commerce (B2B) is the largest portion of transactions performed online, including Electronic Data Interchange (EDI). Approximately 90-95% of the total e-commerce revenues are attributable to B2B. Business-to-Business E-commerce evolved from traditional EDI, which is one-to-one technology, to a diversity of business models. EDI has been a standard utility for Supply Chain Management.

Supply chain management aims at optimizing the overall activities of firms working together to manage and co-ordinate the whole chain. A Supply Chain is considered as a single entity. SCM aims at reducing the sub-optimization which results from the conflicting objectives of different functions. It is assumed that firms have a common understanding and management of their relationships as well as they recognize the need for those relationships to provide some form of mutual benefit to each party. SCM requires integration of independent systems and is of strategic importance in addition to being of operational importance.

This study reviews the literature on SCM and develops a framework for examining the effect of B2B adoption. Three research streams of B2B solutions (innovation adoption, organizational behavior, and critical mass) are reviewed and a conceptual framework. Then, the critical success factors of B2B solutions are identified and classified into two levels, the corporate level and the supply level that incorporate two critical areas: the value of B2B solutions and their limitations. Key factors are (i) strategy "co-operate to compete", (ii) win-win strategy, (iii) commitment to customer service, and (iv) common applications. The study is concluded with suggestions and recommendations for further research.

Keywords: Supply chain management, critical success factors, Business-to-Business (B2B) E-commerce, innovation adoption, organizational behavior, critical mass.

1. Introduction

Recent advances in the field of computer networks and telecommunications have increased the significance of electronic commerce. Electronic commerce is the ability to perform business transactions involving the exchange of goods and services between two or more parties using electronic tools and techniques. Companies across many industries are seeking to negotiate lower prices, broaden their supplier bases, and streamline procurement processes using e-commerce. The rapid diffusion of the Internet offers huge potential in building communities of interests, forging alliances, and creating technology-intense economies of scales. The use of Business-to-Business (B2B) e-commerce in supply chain management (SCM) presents new opportunities for further cost savings and gaining competitive advantage. However, B2B applications for supply chain management are still in an embryonic stage. There is a lack of working models and conceptual frameworks examining those technologies. This chapter addresses the need for a new approach in order to understand B2B adoption in supply chain. It examines the critical success factors of B2B solutions and in doing so it sheds lights into important aspects of SCM.

2. The Critical Success Factors Approach

The concept of critical success factors (CSF) was first defined by Rochart[21] as "the limited number of areas in which results, if they are satisfactory, will ensure successful competitive performance for the organization". Rochart indicated that CSF focus attention on areas where "things must go right", thus its usefulness is greater for applied managerial problems with little or none theoretical support. Boynton and Zmud [3] also defined CSF as the "few things that must go well to ensure success for a manager or an organization"'. They recognized the CSF approach as an appropriate planning instrument.

Among various studies that have used the CSF approach, Leidecker and Bruno[13] identified that critical success factors should be less than six in a successful firm. Furthermore, Guimaraes[8] attempted to rank CSFs based on their relative importance. Martin[14] argued that computers can facilitate the CSFs approach when the objective is to arrive at an effectively business strategy planning. Crag and Grant[5] used the CSF approach to identify significant competitive resources and their contexts. Kay *et al.*[11] identified several CSFs applicable to insurance agency sales in high performance and low performance groups.

3. Supply Chain Management

Supply Chain Management (SCM) is concerned with the linkages in the chain from primary producer to final consumer with the incentive of reducing the transaction costs incurred within. It seeks to break down barriers between each of the units so as to achieve higher levels of service and substantial cost savings. Mentzer *et al.*[16] conducted a meticulous literature review on supply chain management (SCM) and defined it as "the systematic, strategic coordination of the traditional business functions and the tactics across these business functions within a particular company and across businesses within the supply chain for the purposes of improving the long-term performance of the individual companies and the supply chain as a whole".

It is more and more evident that the present business environment is becoming highly competitive. A way of reacting to the intensified competition is through cooperation. Supply chain management is based on cooperation between the supply partners in order to co-ordinate the physical distribution of goods and to manage the related flows of information and capital. The goal of supply chain management is to reduce costs and generate gains for every participating supply partners. It requires enterprises to co-operate with their trading partners to achieve an integrated supply chain. Figure 1 depicts the types of supply chain.

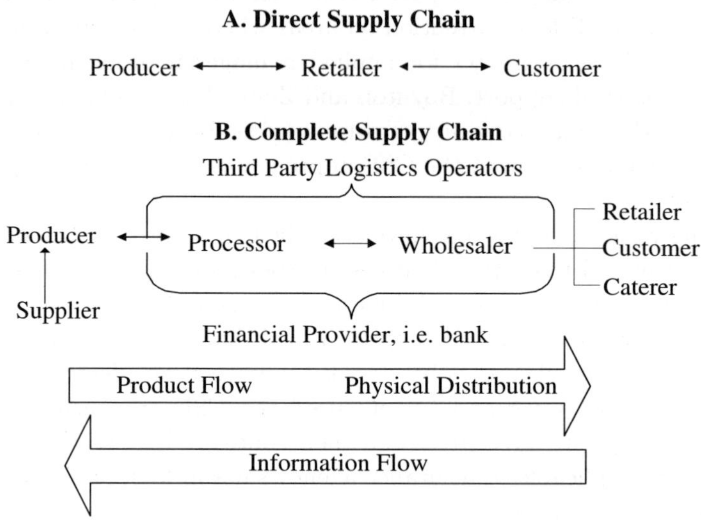

Fig. 1. Examples of Supply Chains.

Trust between partners appears a significant factor in supply chain management. Myoung *et al.*[17] argued that the successful implementation of SCM means that all participants in production, distribution, and consuming could trust each other in order to gain mutual benefits by sharing information. In this way, partners are involved in win-win relations, which are considered the cornerstone for long-term co-operation. Partners that perceive SCM results in mutual gains are most likely to implement common investments in technology. Vorst *et al.*[26] found that the availability of real-time information systems (i.e. Electronic Data Interchange-EDI) was a requirement for obtaining efficient and effective supply chains. Those systems require commitment by all trading partners in order to function at the peak of their operational capacity.

3.1. *Supply Chain Management Activities*

SCM imposes supply partners have to get involved into new activities:

- **Integrated behavior**

Bowersox and Closs[2] argue that enterprises need to incorporate customers and suppliers in their business behavior. In fact, Bowersox and Closs defined Supply Chain Management as this extension of integrated behaviors that is, to consider customers and suppliers an integrated part of the business.

- **Mutual sharing of information**
  Supply partners need to share data and information in order to achieve true coordination of product and information flows. As partners become to work closer and closer to each other, information sharing becomes more a tactical operation than strategic choice. Sharing of strategic and tactical information such as inventory levels, forecasts, sales promotion strategies, marketing strategies reduces uncertainty between supply partners, facilitates planning and monitoring processes and enhances supply performance.[4,16]
- **Mutual sharing of risks, and rewards**.
  Enterprises that take the initiate to work together should be prepared to share both benefits and losses and to share risks and rewards. This action should be a formal agreement between partners in order to help cooperation bloom and facilitate long range planning.
- **Cooperation**
  Co-operation takes place at all business levels (strategy, operational, tactical) and involves cross-functional coordination across supply partners. Cooperation initially focus on cost reductions through joint planning and control of activities and in the long run it extends on strategy issues such as new product development and product portfolio decisions.
- **Customer Policy Integration**
  All partners need to share the same goal and focus on serving customers. This requires the same level of management commitment to customer service and compatible cultures for achieving this goal.
- **Integration of processes**
  SCM requires all partners to integrate their processes from sourcing to manufacturing, and distribution.[16] This is similar to internal integration of processes in which an enterprise integrate fragment operations, staged inventories and segregate functions in order to reduce costs and enhance its performance. The extension of scope of integration with supply partners is an important activity of SCM.

- **Partners to build and maintain long term relationships**
  Probably the key to successful SCM is supply partners to forge long term relationships. Cooper *et al.* [4] argue that the number of partners should be kept small to make cooperation work. For example, strategic alliances with few key supply partners are considered to create customer value.
- **SCM and the Forrester Effect**
  Traditionally, the way of communicating demand for products or services across a supply chain was the following: a customer of each stage (Figure 1) keeps his internal data hidden from his suppliers, regarding, for example, sales patterns, stock levels, stock rules, and planned deliveries. This phenomenon, in which orders to the supplier tend to have larger variance than sales to the buyer and the distortion propagates upstream in an amplified form is called the **Forrester Effect**.[23] Forrester[7] showed that the effect is a consequence of **industrial dynamics** or time varying behaviors of industrial organizations and the lack of correct feedback control systems. Figure 2 shows an example of the **Forrester Effect** repercussions in the vehicle production and associated industries for the period 1961-1991.[15] The rational of the Bullwhip Effect is attributed to the non-integrated, autonomous behavior of supply partners. For instance, processors and retailers incur excess materials costs or material shortages due to poor product forecasting; additional expenses created by excess capacity, inefficient utilization and overtime; and mostly excess warehousing expenses due to high stock levels.[26,12]

4. B2B E-Commerce Solutions

E-commerce has received a plethora of operational definitions, which supports the observation that this is an area of business in continuous change.[25] Electronic commerce (e-commerce) can literally refer to any use of electronic technology relevant to a commercial activity. E-commerce includes a number of functions such as buying and selling of information, products, and services via computer networks. In USA, the National Telecommunications and Infrastructure Administration (NTIA) declared e-commerce has the following core functions:

- Bring products to market (e.g. research and development via

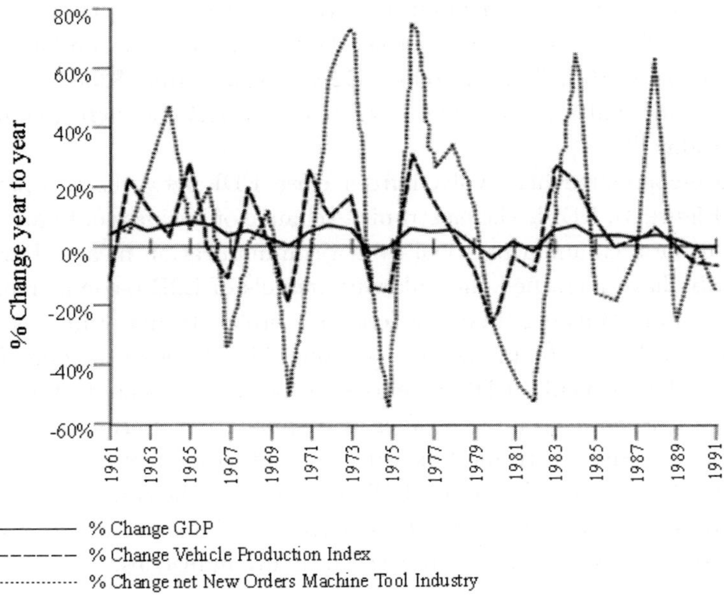

Fig. 2. The Forrester Effect (Supply Chain Bullwhip).

telecommunications).

- Match buyers with sellers (e.g. electronic brokers, or electronic funds transfer).
- Communicate with government in pursuit of commerce (e.g. electronic tax filings).
- Deliver electronic goods and services (e.g. information about electronic goods).

Business-to-Business e-commerce (B2B) is the largest portion of transactions performed online, including Electronic Data Interchange (EDI). Approximately 90-95% of the total e-commerce revenues are attributable to B2B. Business-to-business procurement activities amount to approximately $5 trillion annually worldwide and growth is expected to continue at a fast pace. Estimations of the potential growth of B2B e-commerce are attributed to the fact that businesses in every industry are replacing paper-based systems with a suitable type of electronic communication. For example, shippers in transportation industry replace phone and fax with Internet when communicating with customers. In addition to tangible cost savings, shippers perceive intangible benefits from better real-time tracking and delivery information. Estimations indicate that the US business will conduct \$2 trillion by 2003 and \$6 trillion by 2005 in B2B purchases from \$336 billion now. Internet trade will represent about 42 % of all B2B commerce, compared to 3 % today.[10]

B2B e-commerce has evolved from close EDI networks to open networks (Figure 3). EDI is the electronic exchange of business data and information using a common protocol over a communication means. Barnes & Claycomb[1] have identified the following models of B2B e-commerce: 'One Seller to Many Buyers', 'Many sellers to a broker to many buyers', 'One seller to one broker to many buyers', and 'Many Sellers to One Buyer' (Table 1). Traditionally, EDI systems have been one-to-one technology: A large organization, e.g., a big retailer or manufacturer, performed substantial work to create electronic link with its trading partners. A big retailer often forced its suppliers to adopt EDI systems with the threat of discontinuing paper-based procurements. This pattern of diffusion, which is known as 'hub and spokes', has been observed in many industries.[9]

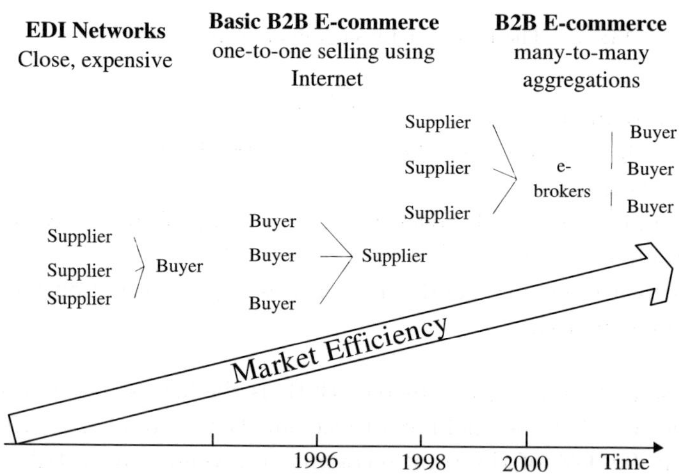

Fig. 3. Business-to-Business E-commerce Evolution

B2B e-commerce is considered the evolution of EDI systems. There are two major limitations of EDI systems that current B2B technologies seem to have made substantial progress to overcome them. First, EDI systems have usually been developed over a dedicated Value-Added-Network, which

Table 1. Business-to-Business E-Commerce Models

Models	**Description – Applications.**
One Seller to Many Buyers	Lack of online intermediaries strengthens business-to-business relationships. Focus on Customer satisfaction and Retention.
Many sellers to a broker to many buyers.	An e-broker is an intermediary which is also called content aggregator, 'hub' or 'portal'.
One seller to one broker to many buyers	It resembles an online auction. Applied to highly differentiated or perishable products and services that can be marketed to disparate buyers and sellers with varying perceptions of product value.
Many Sellers to One Buyer	It is an extension of pre-existing EDI models based on Internet and Web Technologies

is far more expensive than the Internet. This is a major shortcoming of EDI systems as the factor mostly associated with the explosion in Internet-based B2B is economics. Second, EDI transactions need to be codified in advance. This makes difficult any modification in EDI transactions as companies need to considerably redesign their information systems i.e. when a new invoice has to be exchanged electronically. On the contrary, B2B are developed on flexible designs which do not tie up companies in a specific technology to conduct their business operations.

In the past few years, the supply chain concept has been revolted through advances in the information and communication technologies. The benefits attributed to B2B e-commerce that have been identified include: (1) reduction or elimination of transaction costs,[20] (2) facilitation of industry coordination,[18] and (3) promotion of information flow, market transparency, and price discovery.[19] In this way, the implementation of B2B e-commerce in supply chains results is reducing the Forrester Effect by bringing better coordination of the supply chain, reducing stock levels at all stages, cutting costs.

5. Critical Success Factors of B2B Solutions

The B2B literature can be classified into three research streams based on different theoretical paradigms taking different assumptions.[24] One views the adoption of EDI as an innovation adoption, another as an information system implementation, and the third as an organizational behavior, with respect to inter-organizational relationships.

The adoption of innovations' paradigm assumes that the adopting organizations perceive B2B solutions as innovations developed by a third party (B2B is an external innovation). The attributes of the innovation (i.e. its relative advantage, its compatibility, etc) determine to a large extent its adoption or rejection. As a consequence, the diffusion of B2B within one or more industry sectors depends on the technology itself.

According to the organizational behavior' paradigm, there are certain organizational factors that play a significant role in the adopting behavior. Particularly, a business may have criteria such as cost, return on investment, contribution to competitive advantage, etc., when evaluating a certain B2B technology, but there are other factors as well that impinge upon its adoption, e.g. the top management support and availability of resources.

According to the critical mass' paradigm, B2B technologies are considered to be collective innovations, thus their adoption depends on the collaboration among potential adopters if any adopting organization is to receive any benefit. The critical mass theorists argue that the adopting organizations base their decisions on their perceptions of what the group is doing. Their decisions are influenced by how many others have already adopted the innovation, how much others have committed themselves and/or who has participated. In contrast to adoption of innovations' paradigm, the attributes of the innovation while important are insufficient to explain adopting behavior. Table 2 lists these paradigms and the associated factors in detail.

According to the above research streams, the critical success factors of B2B solutions can be classified into two levels, the corporate level and the supply level, which incorporate two critical areas: the value of B2B solutions and their limitations (Table 3).

5.1. *Strategy: Cooperate to Compete*

As a direct result of critical mass theory, enterprises are bounded by the behaviour of the group(s) they form. Companies always face the trilemma, cooperate, compete, or merge. Cooperation in supply chains seems the best alternative to those supporting the argument that competition occurs between supply chains not between companies. In this respect, strategic alliances, vertical coordination, and other forms of co-operative actions between enterprises will be pre-requisite to achieve competitive advantage.

5.2. *Commitment to Customer Service*

Although SCM aims at cost savings due to greater integration between supply partners, its success depends on all partners sharing the same standards of customer service. Cost savings give a competitive advantage to

Table 2. Summary of factors impinging upon the adoption of EDI

Theory	Factors	Description
Adoption of Innovations	Compatibility	The degree to which EDI is perceived as being consistent with existing technologies (technological compatibility) and operations (operational compatibility)
	Complexity	The degree to which EDI is perceived as relatively difficult to understand and use
	Cost	Cost includes implementation, and operational, transaction costs
	Observability	Visibility of EDI's results.
	Relative advantage	The degree to which EDI is perceived better than the system it supersedes
	Trialability	Availability to experiment with EDI on a limited basis
Organizational Behavior	Championing	The existence of a charismatic individual who supports EDI to overcome plausible resistances towards its adoption
	Competitive advantage necessity	The desire to gain an advantage over competition as a result of EDI adoption the pressure to adopt EDI as a result of competition.
	Inadequate resources	Lack of resources often restrict SMEs from adopting EDI
	Limited education	Personnel might need further training in EDI systems
	Organizational size	Size is commonly measured in terms of number of employees, revenues, and profits.
	Organizational readiness in SME	The availability of the needed organizational resources for EDI adoption
	Productivity	An increase of productivity will be the result of lowering inventories levels, reducing transaction costs, and facilitating supply chain management.
	Top management support	In large corporations top management often has to support initiatives like EDI adoption
Critical mass	Dependency	Being in a position not able to exert control over transactions.
	External pressure	EDI is adopted as a result of pressure from business environment (trading partners, suppliers, customers)
	Power	The capacity of an organization to exert influence on another organization to act against its will.

Table 3. Critical Success Factors of B2B Solutions

Functional Areas	Corporate level	Supply Level
Value	Strategy "Co-operate to Compete"	Win-Win Strategy
Limitations	Commitment to Customer Service	Common Applications

the supply partners only when they deliver those products that fulfill consumers growing demand for service, speed, and customization. Due to the high involvement of partners in this initiative and the complexities of SCM implementation, top management commitment is considered mandatory to achieve this goal. Top management should not only support common goals but has a commitment to customer service. Furthermore, supply partners need to share strategic information such as stock levels, demand and stock forecasts, inventory and production scheduling. The implementation of B2B e-commerce should be in line with those commitments in order to achieve greater operational compatibility. Particularly, by considering B2B e-commerce an external innovation to supply partners, its success depends on its compatibility with the supply chain objectives and operations. B2B solutions that forge commitment to customer service would have a higher degree of acceptance among supply partners.

5.3. *Win-Win Strategy*

No Enterprise would be willing to involve into a cooperative scheme if there were no overt gains. Innovation adoption theory states that adopting organizations (i.e. the supply partners) have to perceive a relative advantage of the innovation (i.e. B2B solutions) over previous or alternative solutions (i.e. supply chain without B2B applications). As B2B solutions are adopted by more than one supply partner, eventually, all adopting partners should rip the benefits of B2B e-commerce. Alternatively, partners are adopting a collaborative win-win strategy.

5.4. *Common Applications*

Quick Information Exchanges are mandatory to cope with industrial dynamics of the chain. B2B solutions should seamlessly be adopted by supply partners. This is consistent to innovation adoption theory which states that the innovation should be characterized by technological compatibility. However, given that most companies run their in house software applications, the technological compatibility of B2B solutions should be considered a critical barrier due to significant migration costs. There are a vast number of technical difficulties for current B2B solutions to overcome in order to integrate seamlessly the current applications of supply partners. B2B solutions need to offer more than data and documents integration to fully support process and workflow integration across the supply chain. For instance, common applications allow a supplier to search and browse his partner system to track the approval or buying process of an order or a product. Web-based applications have the permit these applications to run The ability to make this process visible is one of the most significant developments of the Internet.

6. Discussion and Recommendations

This chapter has presented a critical success factors approach which can be used to plan the development of Business-to-Business E-commerce solutions for supply chain management. There is increased consensus about supply chain management and the benefits which it can bring to today's industrial environment. Yet we still see very few examples of successful supply chain management in practice and in particular B2B solutions to support inter-enterprise collaboration. One of the reasons identified for this, is the lack of a framework of the factors that impinge upon the adoption and implementation of information and communication technologies across enterprises.

Adapting an enterprise to include B2B solutions for supply chain management is a profound problem. E-business solutions are attractive for their impetus to enhance customer service, eliminate waste, streamline inventory, and cut processing costs. However, B2B solutions are not straightforward implemented as they influence critical strategic decisions. The review of three research areas of B2B solutions (innovation adoption, organizational behavior, and critical mass) revealed that the critical success factors of B2B solutions are (i) strategy "co-operate to compete", (ii) win-win strategy, (iii) commitment to customer service, and (iv) common applications. B2B e-commerce solutions need to be consistent with the SCM objectives for efficient distribution channels, cost reductions, and enhancing customer value. The core advantage of B2B e-commerce applications is the consistency with SCM objectives. However, supply partners need to overcome substantial technical obstacles such as the lack of common standards and migration of software applications. Still in an embryonic stage, B2B e-commerce solutions to SCM need further empirical research in order to shed light in specific managerial aspects. What factors prevent partners from transforming the relative advantages of B2B solutions into competitive advantages? Which factors remain critical across industries and contexts? In what terms can B2B solutions be re-invented to meet industry and market standards?

References

1. P. Barnes-Vieyra and C. Claycomb, Business-to-Business E-Commerce: Models and Managerial Decisions. *Business horizons* (May-June 2001).
2. D. J. Bowersox and D. C. Closs. *Logistical Management: The Integrated Supply Chain Process.* McGraw-Hill Series in Marketing, NY: The McGraw-Hill Companies (1996).
3. A.C. Boynton and R.W. Zmud. An assessment of critical success factors. *Sloan Management Review*, Vol. **25** No. 4, pp. 17-27 (1984).
4. M. C. Cooper, D. M Lambert, and J. D. Pagh. Supply Chain Management: More than a new name for Logistics. *The International Journal of Logistics Management*, Vol. **8**, No. 1, pp. 1-14 (1997).
5. J. C. Crag and R. M. Grant. *Strategic Management.* West Publishing, St Paul, MN (1993).
6. C. Fine. *Clockspeed: Winning Industry Control in the Age of Temporary Advantage.* Perseus Books, Reading, MA (1998).
7. J. W. Forrester. *Industrial Dynamics.* MIT Press, Cambridge, MA (1960).
8. T. Guimaraes. Ranking critical success factors. *Proceedings of the Fifth International Conference on Information Systems*, Calgary, Alberta (1984).
9. J. Jimenez and Y. Polo. The international diffusion of EDI. *Journal of Internet Banking and Commerce*, Vol. **1**, No. 4 (1996).
10. D. Kardaras and E. Papathanassiou. The development of B2C E-commerce in Greece: current situation and future potential. *Internet Research: Electronic Networking Applications and Policy*, Vol. **10**, No. 4, pp. 284-294 (2000).
11. L. K. Kay, W. L. Thomas, and G. James. Critical success factors in captive, multi-line insurance agency sales. *Journal of Personal Selling and Sales Management*, Vol. **15** No. 1, Winter, pp. 17-33 (1995).
12. H. L. Lee, V. Padmanabhan and S. Whang. Information distortion in a supply chain: the bullwhip effect. *Management Science*, Vol. **43**, No.4, pp. 546-558 (1997).
13. J. K. Leidecker and A. V. Bruno. Identifying and using critical success factors. *Long Range Planning*, Vol. **17** No. 1, pp. 23-32 (1984).
14. J. Martin. *Information Engineering: Book II: Planning and Analysis*, Prentice-Hall, Englewood Cliffs, NJ (1990).
15. R. Mason-Jones and D. R. Towill. Using the Information Decoupling Point to Improve Supply Chain Performance. *The International Journal of Logistics Management*, Vol. **10**, No. 2, pp. 13-26 (1999).
16. J. T. Mentzer, W. DeWitt, J. S. Keebler, S. Min, N. W. Nix, C. D. Smith,

Z. G. Zacharia. Defining Supply Chain Management. *Journal of Business Logistics* (Fall, 2001).
17. K. Myoung, S. Park, K. Yang, D. Kang, H. Chung. A supply chain management process modelling for agricultural marketing information system. EFITA, 3rd conference of the European Federation for Information Technology in Agriculture, Food and the environment, Montpellier, France, June 18-20, 409-414 (2001).
18. R. Nicolaisen. How will agricultural e-markets evolve? *USDA Outlook Forum*. Washington DC, 22-23 February (2001).
19. B. Poole. How will agricultural e-markets evolve? *USDA Outlook Forum*, Washington DC, 22-23 February (2001).
20. M. Porter. Strategy and the Internet. *Harvard Business Review*. Vol. **79**, No. 2, pp. 63-78 (2001).
21. J. F. Rochart. Chief executives define their own data needs. *Harvard Business Review*. Vol. 57, No. 2, March-April, pp. 81-92 (1979).
22. R. H. Thompson, K. B. Manrodt, M. C. Holcomb, G. Allen, and R. Hoffman. The Impact of e-Commerce on Logistics: Logistics@ Internet Speed. Year 2000 Report on Trends and Issues in Logistics and Transportation, Cap Gemini Ernst & Young and The University of Tennessee (2000).
23. D. R. Towill. Time compression and supply chain management: a guided tour. *Supply Chain Management*, Vol. **1**, No. 1, pp. 15-27 (1996).
24. I. P. Vlachos. Paradigms of the Factors that Impinge upon Business-to-Business e-Commerce Evolution. *International Journal of Business and Economics*. (forthcoming) (Fall, 2002).
25. I. P. Vlachos, C. I. Costopoulou, B. D. Mahaman, and A. B. Sideridis. A Conceptual Framework For E-Commerce Development Between African Countries & European Union. *EFITA 2001, 3rd conference of the European Federation for Information Technology in Agriculture, Food and Environment*, Montpellier-France, June 18th - 21st, pp. 491-496 (2001).
26. J.G.A.J. Van Der Vorst, A. J. M. Beulens, P. De Wit, W. Van Beek. Supply Chain Management in Food Chains: Improving Performance by Reducing Uncertainty. *International Transactions in Operational Research*, Vol. 5, No. 6, pp. 487-499 (1998).

CHAPTER 11

TOWARDS THE IDENTIFICATION OF HUMAN, SOCIAL, CULTURAL AND ORGANIZATIONAL REQUIREMENTS FOR SUCCESSFUL E-COMMERCE SYSTEMS DEVELOPMENT

A. S. Andreou

Department of Computer Science,
University of Cyprus,
75 Kallipoleos Str., P.O.Box 20537, CY1678, Nicosia, Cyprus
E-mail: aandreou@ucy.ac.cy

S. M. Mavromoustakos

Department of Computer Science,
University of Cyprus,
75 Kallipoleos Str., P.O.Box 20537, CY1678, Nicosia, Cyprus
E-mail: stephano@logosnet.cy.net

C. N. Schizas

Department of Computer Science,
University of Cyprus,
75 Kallipoleos Str., P.O.Box 20537, CY1678, Nicosia, Cyprus
E-mail: schizas@ucy.ac.cy

E-commerce systems' poor and incomplete design fail to meet users expectations and businesses goals. A major factor of failure of these systems is ignoring important requirements that result from human, cultural, social and organizational factors. The present work introduces a new Web engineering methodology for performing requirements elicitation through a model called SpiderWeb. This is a cross-relational structure comprised of three main axons: Country Characteristics, User Requirements and Application Domain. The purpose of this model is to provide a simple way for analysts to identify these hidden requirements which otherwise could be missed or given little attention. Factors gathering is performed based on a certain form of ethnography analysis, which is conducted in a short-scale and time-preserving manner, taking into consideration the importance of immediacy in deploying e-commerce applications. Two e-commerce systems were developed and evaluated. The first was based on the proposed SpiderWeb methodology and the second on the WebE process. Finally, a survey of purchase preference was conducted demonstrating and validating the applicability and effectiveness of the SpiderWeb methodology.

1. Introduction

According to Nielsen Net Ratings[13] and eMarketer[6] research groups, the number of people with home Internet access worldwide is currently near five hundred millions and e-commerce transactions are estimated to reach two trillion dollars in 2002. While these numbers seem impressive, 30 percent of the enterprises with Web sites do not derive a competitive advantage from their use. Another study of commercial Web sites[9] showed that only 15 percent of e-commerce businesses were successful in selling online.

Researchers have recently demonstrated the importance of human, social, cultural, and organizational (HSCO) factors in e-commerce engineering, proving that these constitute significant factors that, if ignored, will lead to poor system design and plunge from their business goals. Examples can be found in the work by Fraser and Zarkada-Fraser,[7] who have illustrated that ethnic groups follow different decision-making in determining the Web site they prefer to buy from, and that significant differences exist between cultures. Furthermore, Olsina, *et al.*[15] examined the quality of six academic operational sites to understand the level of fulfillment of essential quality characteristics, given a set of functional and non-functional requirements from the viewpoint of students. The latter work proposed a quality requirement tree specifically for academic domains, classifying the elements that might be part of a quantitative evaluation, comparison and ranking process.

While there is a plethora of research works in e-commerce, there is lack of methods for revealing HSCO factors, which otherwise stay well hidden within the working environment analyzed. The risk of missing the requirements resulting from these factors leads us to propose a new methodology to uncover and analyze HSCO factors, as well as to translate them to system requirements. The methodology utilizes a new model called SpiderWeb, aspiring at recording critical factors that must be incorporated as functional or non-functional features in the e-commerce application under development. Taking into consideration the importance of immediacy in deploying e-commerce applications, an oriented form of ethnography analysis is introduced, which can be conducted in a non-time consuming manner to identify requirements sourcing from HSCO factors, based on a certain informational profile developed via focus questions.

The structure of this chapter is as follows: Section 2 provides a description of the SpiderWeb model and its main axons, introduces ethnography analysis as an information gathering methodology for e-commerce applications, and defines the model within the context of the system life cycle. Section 3 briefly describes two e-commerce applications, one developed based on the proposed methodology and the other on the WebE process.[16] This section also demonstrates analytically the use and effectiveness of the SpiderWeb model in practice and finally, it provides the results of user evaluation of the two systems. Section 4 sums up the findings of the chapter and provides some concluding remarks.

2. The SpiderWeb Methodology

The purpose of the SpiderWeb model is to visualize and classify valuable requirement components for the better identification of critical factors that will lead to successful development. The model categorizes system requirements into three main axons: The Country Characteristics, the User Requirements, and the Application Domain axon (Figure 1). Each axon includes certain components, which are directly connected and interrelated. The SpiderWeb axons are also interdependent, allowing the sharing of same, similar, or different characteristics among each other (table 1).

2.1. *The SpiderWeb Model*

A description of each of the axons of the SpiderWeb model is as follows.

2.1.1. *Country Characteristics*

An e-commerce application must be tailored-made for each country or region of countries. In requirements analysis phase the emphasis should be put on the range of countries on which the e-commerce application will target and give special attention to the specific characteristics of the region for successful system development. These characteristics include:

- Demographics:
  It is well known that human behavior varies according to gender and age. Therefore, these issues can significantly affect system design. The Web engineer or project manager must specify and design the e-commerce application based on the targeted population. In

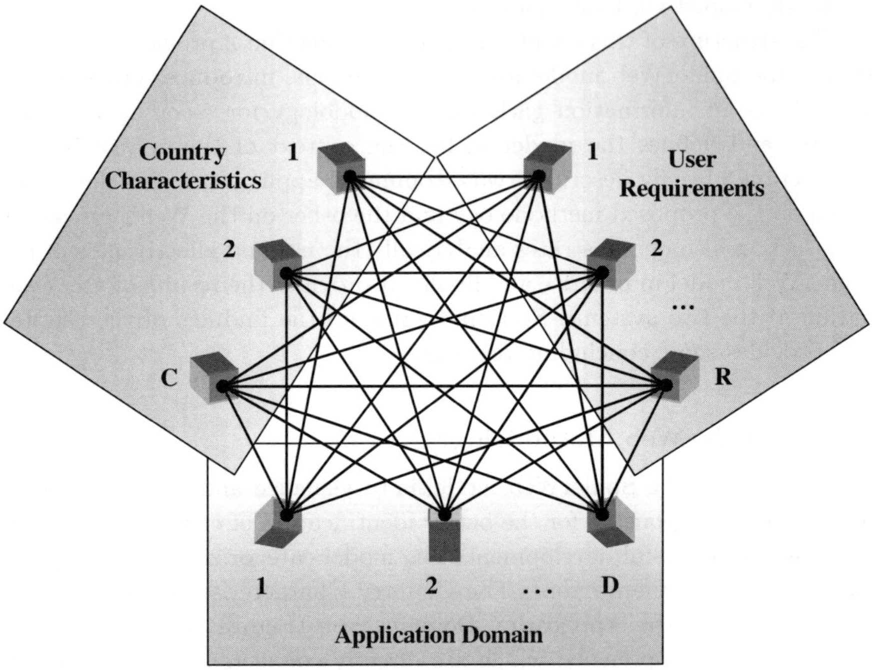

Fig. 1. The SpiderWeb Model

addition, introducing new products and services to a region it is important to have access to the various channels of distribution for achieving short-term and long-term organizational goals.

- Social characteristics:
  The analyst/developer must examine the educational system, the literacy level, as well as the languages spoken within the population, in order for the e-commerce application to be designed in such a way that will accommodate diverged features. Religion plays a significant role in politics, culture and economy in certain countries. Thus, the analyst must investigate whether religion affects the system design and to what degree.
- Legal characteristics:
  The political system and legislation among countries vary; therefore one must investigate political stability and all the relevant laws prior to the development of an e-commerce application. National and international laws must be analyzed to guide the system to-

Table 1. Axon categorization of the SpiderWeb Model

AXON	COMPONENT	DECOMPOSITION
Country Characteristics	Demographics	Gender, age
	Social characteristics	Language, literacy, religion
	Legal characteristics	International and domestic laws
	Technical characteristics	Web access, type of technology
User Requirements	Usability	Understandability, learnability, operability, playfulness
	Functionality	Suitability, accuracy, compliance, interoperability, security
	Reliability	Fault tolerance, crash frequency, recoverability, maturity
	Efficiency	Time behavior, resource behavior
	Timeliness	Business goals
	Maintainability	Analyzability, changeability, stability, testability
Application Domain	Informational	Online newspapers, electronic books, newsletters
	Interactive	Registration forms, customized presentations, online games
	E-commerce/Transactional	Electronic shopping, online banking
	Workflow	Online planning and scheduling systems, status monitoring
	Collaborative work environments	Distributed authoring systems, collaborative design tools
	Online communities marketplaces	Chat groups, online auctions
	Web portals	Online intermediaries, electronic shopping malls

wards alignment and compliance upon full operation.

- Technical characteristics:
  Identifying the technology level of each targeted country will help the Web engineer to decide on the type of technology and resources to use. Countries with advanced technologies and high Web usage are excellent candidates for an e-commerce application. On the other hand, countries new in the Internet arena will need time to

adapt to this challenging electronic environment before taking the risk of doing business on-line.

2.1.2. *User Requirements*

The User Requirements axon of the SpiderWeb model follows the general software quality standards as defined by the ISO 9126 and the Web engineering guidelines proposed by Olsina.[15] Each component is decomposed into several features that must be separately addressed to fulfill these user needs:

- *Usability*
  Issues like understandability, learnability, friendliness, operability, playfulness and ethics are vital design factors that Web engineers cannot afford to miss. The system must be implemented in such a way to allow for easy understanding of its functioning and behavior even by non-expert Internet users. Aesthetics of user-interface, consistency and ease-of-use are attributes of easy-to-learn systems with rapid learning curve. E-commerce systems, by keeping a user profile and taking into consideration human emotions, can provide related messages to the user, whether this is a welcome message or an order confirmation note, thus enhancing the friendliness of the system. Playfulness is a feature that should be examined to see whether the application requires this characteristic, and if so, to what extent. E-commerce systems must reflect useful knowledge looking at human interactions and decisions.
- *Functionality*
  The system must include all the necessary features to accomplish the required task(s). Accuracy, suitability, compliance, interoperability and security are issues that must be investigated in designing an e-commerce system to ensure that the system will perform as it is expected to. The e-commerce application must have searching and retrieving capabilities, navigation and browsing features and application domain-related features.[15]
- *System Reliability*
  Producing a reliable system involves understanding issues such as fault tolerance, crash frequency, recoverability and maturity. The system must maintain a specified level of performance in case of software faults with the minimum crashes possible. It also must have the ability to re-establish its level of performance. A system

must consistently produce the same results, and meet or even exceed users' expectations. The e-commerce application must have correct link recognition, user input validation and recovery mechanisms.

- *Efficiency*
  An e-commerce system's goal is usually to increase productivity, decrease costs, or a combination of both. Users expect the system to run in an efficient manner in order to support their goals. System's response-time performance, as well as page and graphics generation speed, must be high enough to satisfy user demands. Fast access to information must be examined also throughout the system life to ensure that users' requirements are continuously met on one hand, and that the system remains competitive and useful on another.
- *Maintainability*
  Some crucial features related to maintaining an e-commerce application is its analyzability, changeability, stability, and testability. The primary target here is to collect data that will assist designers to conceive the overall system in its best architectural and modular form, from a future maintenance point of view. With the rapid technological changes especially in the area of Web engineering, as well as the rigorous users' requirements for continuous Web site updates, easy system modifications and enhancements, both in content and in the way this content is presented, are also success factors for the development and improvement of an e-commerce system.
  Another important area the researcher must concentrate on is the timeliness of the content (i.e. the information processed within the system), the functionality (i.e. the services offered by the system) and the business targets (i.e. the business goals using the system) the e-commerce system must exhibit. Timeliness is examined through a cultural prism aiming at identifying certain human, social, and organizational needs in all three of its coordinates, as most of the applications exhibiting a high rate of change often depend highly on ethos and customs of different people in different countries (i.e. electronic commerce systems).

2.1.3. *Application Domain*

The Web engineer should investigate users satisfaction on existing e-commerce applications and their expectations on visiting an online store.

He should also identify the driving factors that stimulate users to purchase online. Emphasis should also be given on users concerns, feelings, trust and readiness of using and purchasing through an e-commerce system.

2.2. *The SpiderWeb Information Gathering Methodology*

The analysis of the axon components of the SpiderWeb model presented in the previous part aimed primarily at providing the basic key concepts that developers must utilize to collect proper system requirements. These concepts will be used as guidelines for the significant process of gathering critical information that may affect the functional and non-functional behavior of the system under development. We propose the use of an oriented form of ethnography analysis conducted in a small-scale time-wise manner for collecting and analyzing information for the three axons described before.

Ethnography originates from anthropology where it was primarily used in sociological and anthropological research as an observational analysis technique, during which anthropologists study primitive cultures.[17] Today, this form of analysis constitutes a valuable tool in the hands of software engineers by utilizing techniques, such as observations, interviews, video analyses, questionnaires and other methods, for collecting HSCO factors. In a design context, ethnography aims to provide an insight understanding of these factors to support the design of computer systems.[12,18,3,4,19] This approach offers great advantages in the system development process by investigating HSCO factors and exploring human activity and behavior that otherwise software engineers would have missed. Examples can be seen in several studies performed in a variety of settings, including underground control rooms,[10] air traffic control,[11] police,[1] banking,[5] film industry,[17] and emergency medicine.[2]

Having in mind on one hand, that ethnography analysis is time consuming by nature and the immediacy constraint in deploying e-commerce applications,[14] on the other, we propose a short-scale form of ethnography analysis, focusing on cognitive factors. Our proposition lays on examining the existing working procedures of the client organization, either manual or computerized, together with the consumers' behavior. Specifically, the working environment of the organization and its employees, as well as a group of customers currently doing business transactions with the organization are set as targeted population of the analysis, utilizing this shortened form of ethnography on the three axons of our model. The short-scale ethnography analysis may include observations, interviews, historical and empirical data, as well as questionnaires. Emphasis is given on focus questions produced in the form of questionnaires. These questions are distributed among the targeted group or are used as part of the interviewing process, and the answers are recorded, analyzed and evaluated.

Data collection mechanisms, as well as the kind of information for analyzing each primary component in the axons of the proposed model, are defined in the SpiderWeb methodology via a profile shell that Web engineers must develop before requirements analysis starts. Each component is associated with suggested focus questions provided in tables 2 through 4. It must be noted, that these are a proposed set of key questions for the analyst to use as guidelines, but he may also enhance the set with other application specific questions he may regard equally essential for the application under development.

Table 2. Focus questions for collecting HSCO factors on the Country Characteristics axon of the SpiderWeb model

Country Characteristics	**Focus Questions**
Demographics	What is the gender and age of the targeted population?
	What are the channels of distribution?
	Are the neighboring countries open for electronic trade of goods and services?
Social Characteristics	What are the main languages spoken in the region?
	What is the religion of the targeted population?
	What is the literacy percentage grouped by gender and age?
	What is the level of efficiency of the educational system with respect to the Web?
Legal	Is there political stability in the area?
	Are there any laws that prohibit the electronic sale of certain goods?
Technical	What is the percentage of the targeted population with Web access, by gender and age?
	What is the Web access rate of increase?
	What is the average transmission speed to browse the Internet?

2.3. *The SpiderWeb Methodology and the Web Engineering Process*

The SpiderWeb methodology can be incorporated in the Web engineering (WebE) process,[16] as an add-on feature to enhance the development of

Table 3. Focus questions for collecting HSCO factors on the User Requirements axon of the SpiderWeb model

User Requirements	Focus Questions
Usability	How do expert and non-expert Internet users understand the system?
	Are easy-to-learn systems too complicated for expert users?
	How do users perceive content layout and how does this affect user retention?
	How does the system handle the conflicting requests for maximum or minimum playfulness?
	How does the content layout (colors, menus, consistency) affect Web usage?
	What is the level of sensitivity in ethical issues among the targeted user-group and how does this affect the way they interact with the Web?
	What is the level of trust for ensuring privacy?
	How can on-line shopping be more entertaining than in-store shopping?
Functionality	How do users feel with registering to a Web application is a prerequisite for accessing its content?
	What is the required level of security of functions, for individuals to provide their credit card for on-line purchases?
	What is the maximum bearable time for users to wait in search for information before dropping the site?
Maintainability	How often will users need to see content updates?
	Are market conditions matured for such a system?
	How do people accept system changes?
Reliability	What is the acceptable fault tolerance that will not drive away existing users?
Efficiency	At what degree users expect to decrease their costs? Can these expectations be met?

e-commerce applications (Figure 2).

The WebE process includes six phases: a) Formulation, b) Planning, c) Analysis, d) Engineering, e) Page Generation & Testing, and f) Customer Evaluation:

- Formulation – Defines the tasks and goals of the e-commerce application and specifies the length of the first increment.
- Planning – Estimates the total project cost and the risks associated with it, and sets a timeframe for the implementation of the first increment as well as the process of the next increments.
- Analysis – Identifies all the system and user requirements together with system content.
- Engineering – It involves two parallel tasks: (i) Content design and

Table 4. Focus questions for collecting HSCO factors on the Application Domain axon (E-commerce/Transactional system) of the SpiderWeb model

Focus Questions
Are users satisfied with the current e-commerce sites? What are their recommendations for improvement?
What do users expect to find, shopping in an e-commerce application versus shopping in a traditional store?
How does a user behavior change when using long versus short registration forms?
Are users ready for e-commerce, both in b2b and b2c?
What are the users' feelings and trust on doing business on-line?
What are the users' concerns and doubts on security, product delivery, efficiency, and company legitimacy?
What types of auctions users are accustomed to?
How easily users are affected by outside factors in their shopping decisions?

production, and (ii) Architectural, navigation, and interface design.

- Page Generation & Testing – Development task using automated tools for the creation of the e-commerce application, applets, scripts, and forms.
- Customer Evaluation – Evaluates each task and proposes new modifications and expansions that need to be incorporated to the next increment.

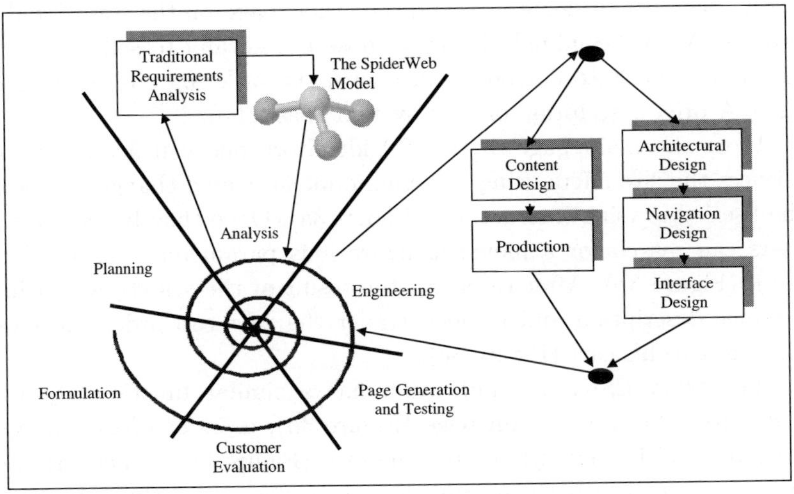

Fig. 2. The SpiderWeb Model within the WebE Process

The SpiderWeb model can be a valuable tool within the WebE process to enhance the development of e-commerce applications. During the Planning phase the SpiderWeb methodology time and cost must be estimated and added to the total of the application. During the Analysis phase, the analyst following the classical approach studies the current system and processes and defines functional and non-functional requirements. The SpiderWeb model is invoked next, performing short-scale ethnography analysis to obtain HSCO factors. These factors are then translated into functional and non-functional requirements. Requirements management follows, which deletes any duplication of the ones already found using the traditional method, or resolves conflicts resulting from contradictory requirements. After updating the system requirements, their final form is used in the Engineering phase to support the e-commerce application development. The Web engineer designs the e-commerce application's structure, the navigation mechanisms, the interface and the content, based on the results obtained from the previous phase.

3. Validation of the SpiderWeb Methodology

Two e-commerce applications were developed, one based on the SpiderWeb methodology (E-VideoStore), and the other based on the traditional WebE process (MOVIESonline). Their purpose is basically to sell or rent videotapes and DVDs to customers in Cyprus by ordering on-line, and in real time. A brief description of each system is as follows:

From the main page of the E-VideoStore one can access any of the services the site offers using a dynamic menu, which changes according to the user's browsing instructions (Figure 3a). Once they become members, users can search for a movie using various parameters (i.e. title, actors, etc.), (Figure 3b). After receiving the result of the search, which includes a movie description and a video trailer the user can order the movie by placing it to his cart (Figure 3c).

The MOVIES*online* application has a similar functionality with E-Videostore. From the main page (Figure 3d), users can log-in and search for movies either by title or by category (Figure 3e). When users decide on the movie they wish to order they can place it on their shopping cart (Figure 3f).

We will first demonstrate the steps of the SpiderWeb methodology followed during the development of the E-VideoStore system and next we will present a comparison between the two systems in terms of user-preference.

Due to chapter size limitations we will present only part of the E-VideoStore analysis, omitting the preceding part including the traditional analysis activities and the subsequent part involving the requirements management process.

3.1. *Analysis of the E-VideoStore Project Using the SpiderWeb Methodology*

A short-scale ethnography analysis was performed which included historical and empirical data, observation activities, as well as questionnaires, to identify HSCO factors of the E-VideoStore application. Emphasis was given on focus questions that were distributed to the customers and employees of a traditional video store. Our objective was twofold: (a) to understand the existing working procedures and (b) to identify HSCO factors from both employees and customers. Our research ended after three days and resulted the collection and incorporation of the following HSCO factors/requirements:

Country Characteristics

- The number of Cypriot Internet users is rapidly increasing even though they are relatively newcomers. Most of them are young in age; therefore older ones are not Web experienced and need to be trained. The system incorporates help features to support the learnability process of inexperienced users.
- The targeted population is men and women of ages 18 to 50 that are already familiar with the Internet, or they are possible candidates for using it in the near future. The E-VideoStore application is characterized by user-friendliness that aids the understandability and learnability process of the application since the customers are of different age, education, and Web experience.
- The democratic political system and the general legal system in Cyprus, are supported by strong ethical behavior and conservative culture, therefore the law prohibits the sale and rental of pornographic material. Since the law applies also to online business, the E-VideoStore application excludes movies of any sexual content.
- Since this is a new application for Cyprus' standards, there is no existing competition available yet. Thus, on one hand, its quick development and deployment is a primary target for market entrance and on the other hand, the management must be prepared to face the fact that people will be somehow skeptical and cautious to use it.

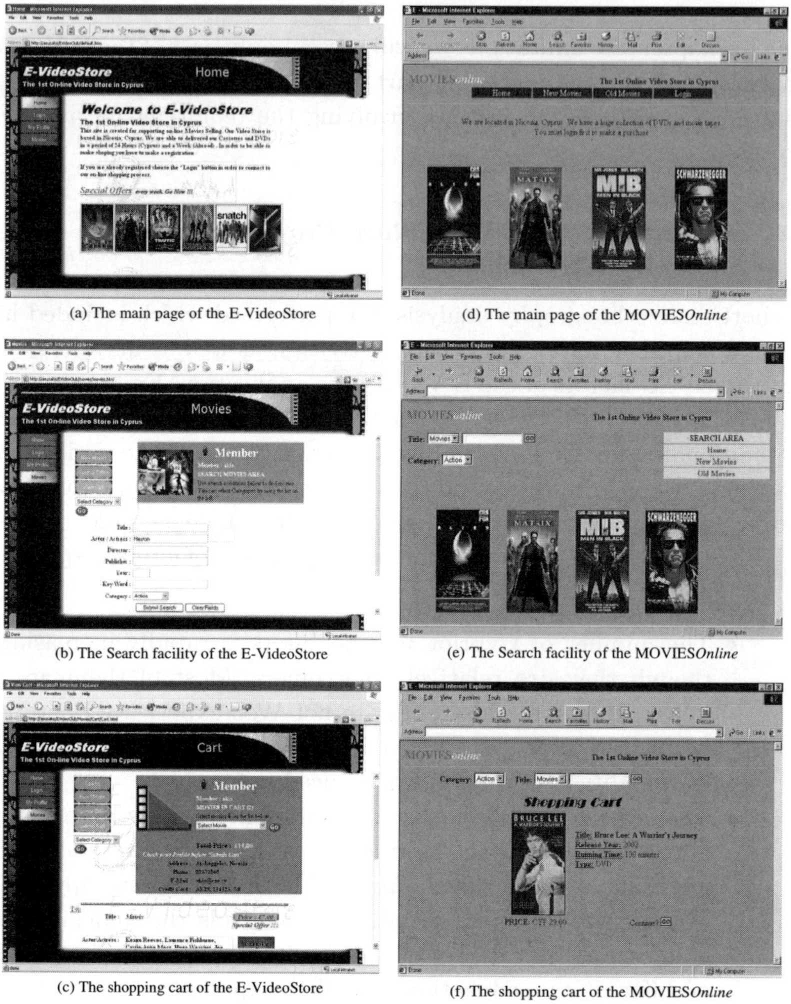

(a) The main page of the E-VideoStore

(d) The main page of the MOVIES*Online*

(b) The Search facility of the E-VideoStore

(e) The Search facility of the MOVIES*Online*

(c) The shopping cart of the E-VideoStore

(f) The shopping cart of the MOVIES*Online*

Fig. 3. Two e-commerce systems for videotapes and DVD rental/purchase: The E-VideoStore (a, b and c) and the MOVIESonline (d, e and f).

- While Greek is the national language, Cypriots are accustomed to the English language and the English way of doing business, that is, in a fast, accurate, legal and well-organized manner. The E-VideoStore application is developed in an efficient way supporting both Greek and English languages, with well-organized and ergonomic way of presenting and ordering the available movie stock.

- Cypriots, although traditional and conservative by nature, are technically oriented in practice. Emphasis is given on the development of an easy-to-use e-commerce application that utilizes technologically advanced features like real-time multimedia presentations.
- Cyprus has fast telecommunication lines; therefore the E-VideoStore application can afford high data traffic. While these improved the system's efficiency, they also allowed for adopting playfulness features through the use of movie trailers.
- Cyprus is a small country with a relatively small market. However, the application targets at setting a new standard in the selling and renting of videotapes and DVDs with on-line ordering, and aspires at expanding in Greece and the Middle-East (for sales only). Awareness of this target aided the process of future maintenance of the E-VideoStore application to expand in foreign markets in terms of design, both in content and in payment procedures.

User Requirements

- Cypriots are not patient and are easily frustrated. This is handled in the application offering advanced searching capabilities, such as searching by title, producer, year and category. Searching affords easy, fast, and accurate location of relevant information. In addition, the E-VideoStore application runs on fast Web servers with access speed of less than 12 seconds for each of its page.
- The Cypriot customers are hesitant on surfing the Internet, as they believe it would be difficult using it. The E-VideoStore application is developed to provide easy-to-use navigation to enhance the understandability, learnability and friendliness. The friendliness of the system is also enriched through customized messages to the user, such as welcoming, thanking, and confirming messages, by storing his profile on a running session.
- Cypriot customers are also skeptical when using an e-commerce system, as they believe the site will not provide them all the necessary functions and capabilities as their in-store shopping. The E-VideoStore application includes all purchasing procedures usually followed in a traditional video store, enhanced by advanced functions (i.e. movie description and trailer). In addition, the application is designed in a simple, yet dynamic, reliable and efficient way, to enhance its functionality and performance.
- Cypriots are impressed by high quality graphics and content layout.

The right choice of colors and fonts together with proper graphics and a movie trailer for the members to watch, the E-VideoStore design emphasized on attractiveness, providing users a friendly and playful way of selecting the movie of their preference.

- The Cypriot customers wish to have early access to new movie releases. The E-VideoStore application is developed to update its content frequently and dynamically, based on processes allowing fast and easy downloading.
- Watching a movie is an entertainment need; therefore customers expect the e-commerce application to be an entertaining one too. Thus, the E-VideoStore is designed with attractive graphics, including the movie trailers to improve its interactivity and playfulness in general.
- The customers like to have special offers and discounts. The E-VideoStore application includes a "special offers" message on the main page that can be frequently updated (e.g. once or twice a week) offering customers discounts.
- Cypriots are also skeptical and reluctant to use their credit cards on-line. While the transactions are operated on a secured server, the sense of security and privacy is also visible throughout the system to enhance consumer trust, with proper messages and disclaimers.
- While there is no on-line competition yet, the site has to compete with businesses serving the people in the traditional way, thus users expect to receive competitive prices.
- Cypriots are willing to adopt high technology and the Internet for their daily transactions and movie reservations.

Application Domain

- E-VideoStore is an e-commerce application whose purpose is to sell/rent videotapes and DVDs online.
- The system provides all functions a traditional store offers, plus movie insights through a movie description and a movie trailer.
- While customers are somehow afraid of online fraud they are willing to gradually enter the e-business agora. The e-commerce application runs on secure Web servers and the site provides messages of security to enhance users' trust on purchasing online.
- Transactional Web systems must be efficient enough to handle customers' requests; therefore the E-VideoStore application is developed to run on fast Web servers with minimum resources.

3.2. *Results*

The two systems were demonstrated to a hundred people in groups of twenty. The target population included both male and female, with average age of 20-30 years old, being regular movie buyers or renters, and having little or no Web (Internet) experience. A questionnaire including thirty questions was provided to identify the website these users would prefer to buy from (table 5). The results of the questionnaire showed that 72% of the population answered they would prefer to buy or rent movies from the E-VideoStore, while the rest 28% from the MOVIES*online.*

Table 5. A representative subset of questions of the questionnaire provided to evaluate the two e-commerce systems

Sample Questions
Which site made you feel more secure and why?
From which site did you like the graphics, colors and content layout best?
How did you find the navigation and searching capabilities of the sites (poor, fair, good, very good)?
Do you think playfulness is an important factor for selecting a site? If so, which site do you prefer and why?
What did you like most and what did you like least in each site?
Did you feel frustrated at some point in any of the sites (pain in the eyes, backaches, fatigue, slow speed)?
With which site did you feel more comfortable and why?
What feature did you find most helpful or significant in each system? Why?
Which site do you think resembles traditional video purchase/rental?
If you are currently an online purchaser which site would you prefer to buy or rent from?
If you are not an online purchaser, but you feel it's probable to be one in the near future, what site would you choose? What is the reason of your choice?

As one can see from table 5, the participants were asked to justify their answers to several questions. Some of the most popular responses are listed below:

- The E-VideoStore was easier to use than the MOVIES*online*
- Using the E-VideoStore I could find a movie easily by several parameters, while with MOVIES*online* I could search only by title and category
- I liked the movie trailers in the E-VideoStore
- While I liked the use of many colors in the MOVIES*online*, I believe the background color should have less brightness
- The MOVIES*online* had only a limited help facility, while the E-

VideoStore provided a detailed help screen
- The MOVIES*online* gave me the impression that it offered less movies than the E-VideoStore
- I prefer the E-VideoStore because it is like shopping the way I am used to but in a modern way
- I got the impression that E-VideoStore was more secure than MOVIES*online* because in the former case the system's authentication and monitoring of sensitive data was stricter.

The general conclusion drawn from this survey was that users preferred the E-VideoStore over the MOVIES*online* system, appealing mostly the features incorporated in the former based on the HSCO factors that the SpiderWeb methodology identified.

4. Conclusion

A new methodology for identifying significant human, social, cultural, and organizational requirements was proposed to contribute to the enhancement of developing e-commerce applications. The methodology introduces the SpiderWeb model, which includes three main axons: Country Characteristics, User Requirements and Application Domain. Each axon interacts with the other two to provide a user profile. The strength of this model is in developing an understanding of how these critical HSCO factors can affect the system development process. These factors are identified and collected through a special form of ethnography analysis conducted in small-scale to meet the hard time constraints posed by the feature of immediacy that characterizes e-commerce applications development. The analysis uses specially prepared focus questions, aspiring at incorporating related functional and non-functional characteristics when designing and implementing an e-commerce system.

We have developed two e-commerce systems one based on the proposed SpiderWeb methodology (E-VideoStore application) and the second on the WebE process (MOVIES*Online*). A survey of purchase preference was conducted showing that 72% of potential consumers preferred the E-VideoStore, while 28% the MOVIES*Online* application. We have successfully demonstrated how the SpiderWeb methodology revealed critical characteristics that once incorporated into an e-commerce system enhanced its quality characteristics and strengthened its future market value.

The proposed methodology does not, by any means, aspire to substitute existing practices of requirements analysis reported in the international literature of Web Engineering, but rather to complete and enhance them.

References

1. S. Ackroyd, R. Harper, J.A. Hughes, and D. Shapiro. *Information Technology and Practical Police Work.* Milton Keynes: Open University (1992).
2. S.A. Andreou. Tackling the identification of human-computer issues through ethnography analysis: The Application Oriented Ethnography Profile. *Proceedings of the PC-HCI 2001: Panhellenic Conference with International Participation on Human-Computer Interaction*, Greece (forthcoming) (2001).
3. P. Beynon-Davies. Ethnography and information systems development: ethnography of, for and within IS development. *Information and Software Technology,* **39**, pp. 531-540 (1997).
4. L. J. Ball and T. C. Ormerod. Applying ethnography in the analysis and support of expertise in engineering design. *Design Studies,* **21**, 403-421 (2000).
5. S. Blythin, M. Rouncefield, and J.A. Hughes. Never mind the ethno stuff—what does all this mean and what do we do now? Ethnography in the commercial world. *Interactions,* **4**, 38-47 (1997).
6. EMarketer (2002). eStat Database. http://www.emarketer.com
7. C. Fraser and A. Zarkada-Fraser. Cultural differences in HCI and telepresence – A comparison of e-commerce buying behavior in Greek and Anglo Australian women. In N. Avouris & N. Fakotakis, Eds. *Proceedings of the Advances in Human – Computer Interaction I – PC HCI 2001.* Patras, Greece: Typorama Publ. (2001).
8. Gardner Group, (2000). http://www.gardner.com
9. R. P. Heath. Design a killer web site. *American Demographics*, 50-55 (1997).
10. C. Heath and P. Luff. Collaboration and control: crisis management and multimedia technology in London Underground control rooms. *Computer Supported Cooperative Work,* **1**, 69-94 (1992).
11. J.A. Hughes, D. Randall and D. Shapiro. Faltering from ethnography to design. In J. TURNER & R. KRAUT, Eds. *Proceedings of the ACM Conference on Computer Supported Cooperative Work—CSCW'92,* pp.115-122. Toronto, Canada: ACM Press (1992).
12. E. Hutchins. *Cognition in the wild.* Cambridge, MA: MIT Press (1995).
13. Nielsen Net Ratings (2002). January 2002 global Internet index average usage. http://www.nielsen-netratings.com
14. K. Norton. Applying Cross Functional Evolutionary Methodologies to Web Development, *Proceedings of the First ICSE Workshop on Web Engineering*, ACM (1999).
15. L. Olsina, D. Godoy, G. Lafuente, G. Rossi. Specifying Quality Characteristics and Attributes for Websites, *ICSE99 Web Engineering Workshop*, Los Angeles, USA (1999).
16. R. S. Pressman. *Software engineering: A practitioner's approach.* London: McGraw-Hill (2000).

17. J. Simonsen and F. Kensing. Using ethnography in contextual design. *Communications of the ACM,* **40**, 7, pp. 82-88 (1997).
18. I. Sommerville. *Software Engineering, 5th Ed.* pp. 94-96, Addison-Wesley (1996).
19. S. Viller and I. Sommerville. Ethnographically informed analysis for software engineers. *International Journal of Human-Computer Studies,* **53**, pp. 169-196 (2000).

CHAPTER 12

TOWARDS INTEGRATED WEB-BASED ENVIRONMENT FOR B2B INTERNATIONAL TRADE: MALL2000 PROJECT CASE

R. Nikolov

Sofia University "St. Kliment Ohridski",
Faculty of Mathematics and Informatics
E-mail: roumen@fmi.uni-sofia.bg

B. Lomev

Sofia University "St. Kliment Ohridski",
Faculty of Economics and Business Administration
E-mail: lomev@feb.uni-sofia.bg

S. Varbanov

Institute of Mathematics and Informatics - BAS
E-mail: varbanov@fmi.uni-sofia.bg

1. Introduction

The chapter describes an approach for development of an integrated B2B e-commerce environment that supports the "One-Stop Trade" international trade model. Most of the existing web-based platforms that facililiate Small and Medium Enterprises (SMEs) in inernational trade are offering mainly services like consultancy and training (*Agora - One-Stop-Shops for SMEs*,[a] *The Academic Library as an "One Stop Shop" Information Provider for the SMEs.*[b]). Brokering, one of the most important services in a global trade environment that provides the backbone of the sale, has not been developed at an appropriate level. There are some web platforms offering a

[a]http://www.ecotec.com/sharedtetriss/projects/files/agora.html#AGORA%20Project%20Products
[b]http://educate.lib.chalmers.se/IATUL/proceedcontents/fullpaper/kbhpap.html

restricted brokering system that provides *some additional services*, such as finding new business partners, business information support. However such systems are not based on the concept of uniform standardization,[c] or do not include opportunities to specify a concrete good/commodity to buy or sell.[d] In all cases, there is no existing e-business web platform that can lead a SME on the way from brokering, negotiating and contracting, through payment and logistics, to customs services and final accomplishment of a trade deal.

The modern information technologies allow building an environment that facilitates international trade by supporting electronically the major trade process stages: brokering, negotiating, contracting, payment, logistics, customs services, etc. We recognize the most important features of such environment to be:

- Multilingual web-based support for international trade.
- Integrated services that support the major stages of international trade transaction.
- Implementation of World-wide accepted standards unifying international trade process.
- Usage of modern XML-based technologies for trade-related information representation.

The most distinctive advantage of this approach is that it provides "a closed business cycle" starting from finding the most appropriate supplier, through contracting and shipping the goods, to the sales process of publishing a sales offer, delivering it only to interested parties, providing ready techniques for contracting it, and mediation towards shipping it. The process of concluding a trade deal will be substantially shortened and become less expensive and easier to perform.

Short overview on the existing e-commerce standards, Mall2000 E-commerce System and its possible extension towards One-Stop Trade Environment are presented below.

[c] http://www.bizeurope.com
[d] http://www.e-trade-center.com

2. B2B E-commerce – Existing Standards for Product and Document Description

The development and dissemination of standards for international trade facilitation and electronic data exchange is within the scope of several non-governmental and business organizations. The leading role in this process play some specialized United Nations bodies, such as UN/CEFACT *(United Nations Centre for Facilitation of Administration, Commerce and Transport)* and UNECE *(United Nations Ecomonic Commission for Europe).* The major steps towards building an unified international trade standard are briefly described below.

2.1. *EDI*

UN/EDIFACT defines EDI as *"the computer-to-computer transmission of (business) data in a standard format."* [UN/EDIFACT, 1995]. This definition reveals the basic principles of EDI:

- *computer-to-computer*: no human intervention should be required
- *business data in a standard format*: namely electronic business documents (which generally closely resemble their conventionally printed counterpart) conforming the specifications.

EDI is based on the concept of *transactions*, that comprise *messages* (business documents) in predefined formats. A message consists of *data segments*, which themselves are a collection of *data elements* –basic units of information. The main purpose of EDI standard is to specify message structures (in terms of their constituent parts) for the cooperation of different types of business processes between two companies. Typical examples are invoices and purchase orders. Although EDI plays substantial role for facilitating business data exchange, it has some significant shortcomings, e.g: it defines not only message formats, but also communication protocols and hardware requirements; it is not well suited for Internet environment, but rather for expensive Value Added Networks.

EDI uses two major standards X12 in USA and EDIFACT in Europe:

- X.12 was originally developed by ANSI, the American National Standards Institute, but is currently maintained by the not-for-profit organisation DISA, the *Data Interchange Standards Association.*
- UN/EDIFACT means *United Nations Electronic Data Interchange for Administration, Commerce and Transport* and is managed by

CEFACT *(Centre for Facilitation of Administration, Commerce and Transport)* and UNECE *(United Nations Ecomonic Commission for Europe)*.

2.2. *SOAP*

SOAP *(Simple Object Access Protocol)* is an initiative of Microsoft, DevelopMentor, Userland Software, IBM and Lotus. It is a lightweight mechanism for exchanging XML-marked up data over the Internet, a very heterogeneous, distributed environment. On the one hand, this information can consist of a request or response, with appropriate parameters, for some application logic on the receiving side. Therefore, SOAP's Request/Response model is often said to be an RPC *(Remote Procedure Call)* protocol 64. On the other hand, this standard is also applicable for more general, "EDI-style" document-exchange. The full specification can be found on http://www.w3.org/TR/SOAP.

2.3. *UDDI*

UDDI *(Universal Description, Discovery and Integration)* is an initiative that has evolved from a collaboration between Microsoft, IBM and Ariba on several standards initiatives such as XML, SOAP, cXML and BizTalk. It claims to accelerate the growth of B2B eCommerce by enabling businesses to discover each other, and define how they interact over the internet and share information using web services.

UDDI uses a distributed registry as a common mechanism to publish web service descriptions. In order to be easily accepted, UDDI makes use of established standards (HTTP, XML, SOAP), to which companies offering and using web services will usually already be acquainted.

UDDI is mainly but not necessarily dealing with RPC-style messaging, for access to application functionality that is exposed over the Internet.

2.4. *ebXML*

ebXML[e] means "electronic business XML", and its ambition is to become a *global standard for electronic business.* UN/CEFACT and OASIS are two major non-profit, international organizations, that are developing ebXML. These two organizations can also be considered as a group of vertical and

[e]http://www.ebxml.org

horizontal industry consortia and standardization bodies (IETF, OMG, DISA, CommerceNet), governmental agencies, companies and individuals from all over the world. They all support the standards their respective organization sets and thus forming quite large user base.

ebXML can be considered as a "unified global EDI standard", for companies of *all* sizes, both large international companies and SMEs (Small and Medium Enterprises), in *every* industrial sector. The major ebXML's design goal is to lower the entry barrier to electronic business for SMEs and consequently to provide smooth interoperability, in the form of compact applications that smaller companies can plug in to their information systems. This implies that the ebXML business processes must be detailed and specific enough for immediate application. ebXML sees itself as *complementary* (not competitive) to other B2B initiatives such as the SOAP and UDDI standards and the RosettaNet and OAGIS frameworks. As compared to other frameworks, ebXML is neither a *vertical* (industry-specific) standard, nor merely *horizontal* (cross-industry) one. ebXML is in a process of development and dissemination.

2.5. *UNSPSC*

One alternative approach to B2B standardisation is UNSPSC (Universal Standard Products and Service Codes) developed by the Electronic Commerce Code Management Association (ECCMA).[f] UNSPSC is a schema that classifies and identifies commodities. It is used both in sell and buy side catalogs and as a standardized account code in analyzing expenditure as well. The UNSPSC strategy is to align with the vertical industries although this initiative was started by actors who are not focused on a vertical need and general standardization bodies.

3. Mall2000 - B2B E-Commerce System

Mall2000 (Mall for Online Business beyond the Year 2000) is a project funded by the European Commission under INCO COPERNICUS Programme (Project No. 977041). It implements an early prototype of the "One-Stop Trade" international trade model. The distinctive feature of the project is the XML-based implementation of part of the UN/CEFACT standards.[1]

[f] http://www.eccma.org

Partners of the project consortium are: University of Sofia - Dept. of IT (Bulgaria), Darmstadt University of Technology - Dept. of CS (Germany), Virtech Ltd. (Bulgaria), Stylo srl. - Bologna (Italy), Object Technology Deutschland (Germany), Compaq Computer Corporation - Galway (Ireland), Institute of Informatics - FAST Technical University of Brno (Czech Republic) and Directnet Consult Ltd. - Brno (Czech Republic).

Mall2000 web site provides a set of business-to-business services to small and medium enterprises (SME) in Bulgaria, the Czech Republic and other CEE countries for business contacts with partners in the European Union. The web site is an Internet-based client-server application. Its services are accessible by a standard web browser.

Mall2000 utilizes two widely accepted standards:

- The "Harmonized Commodity Description and Coding System" (HS) adopted by Customs administrations world- wide as the method for the classification of goods and recommended by *UN/CEFACT's* Recommendations 30;
- The Statistical Classification of the Economic Activities (NACE Rev.1) in the European Union, which provides accepted descriptions of economic activities (industries) and groups of economic activities that are most commonly presented in an economy.

3.1. *Mall2000 Users and Services*

The users of Mall2000 are divided into three categories, depending upon the type of services they can use: suppliers, consumers (both referred to as *subscribers*), and visitors.

Suppliers are users who *act on behalf of a company they represent* and wish to enter into trade relations or otherwise establish business contacts through the Mall2000 site. They are the users with extended access rights. Suppliers can:

- Register the company they represent and both comprehensively present it by opening a web- site (Front Desk) and shortly present it by creating an electronic Business Card, hosted on the Mall2000 server;
- Publish offers to sell goods/ commodities by a given date; (HS used);
- Receive notification by email if another subscriber has published a request, intending to buy a good/commodity specified by the

supplier in their offer; (HS used);
- Publish requests to buy goods / commodities by a given date; (HS used);
- Receive notification by email if another supplier has published an offer, intending to sell a good/commodity specified by the supplier in their request; (HS used);
- Use a Search Wizard for instantly locating offers currently available in the database (HS used);
- Search for the registered companies in Mall2000 (and browse their Front Desks and Business Cards) (NACE Rev.1 used);
- Use the Currency calculator.

Consumers are also users who *act on behalf of a company they represent* and wish to enter into trade relations or otherwise establish business contacts through the Mall2000 site. In contrast to suppliers, however, they have fewer access rights. Consumers can:

- Register the company they represent and only shortly present it by creating an electronic Business Card hosted on the Mall2000 server;
- Publish requests to buy goods / commodities by a given date; (HS used);
- Receive notification by email if a supplier has published an offer, intending to sell a good/commodity specified by the consumer in their request; (HS used);
- Use a Search Wizard for instantly locating offers currently available in DB (HS used);
- Search for the registered companies in Mall2000 (and browse their Front Desks or Business Cards) (NACE Rev.1 used);
- Use the Currency calculator.

Visitors *do not act on behalf of a company.* They are users who wish only to browse the site and receive personalized information about the services offered by Mall2000. Visitors can:

- Search for registered companies in Mall2000 (and browse their Front Desks or Business Cards);
- Use the Currency calculator.

Mall2000 currently contains information about over 500 Bulgarian companies, operating in 30 business activities. These companies are not subscribers of Mall2000, but the information about them is accessible to all prospective subscribers.

3.2. *Basic Web Technologies Used in Mall2000*

More information about the way the technologies summarized below are exploited in Mall2000 can be found in Refs. 2, 3.

Our approach for encoding the particular Service/Product offers and requests – the instances of ontology concepts – is to use XML as a data format (see Ref. 4 for an excellent and in depth introduction to XML). XML is used mainly as a format for transferring offers and requests between the client and the server. The actual XML documents are stored in a series of related database tables and columns. Before storing a document it will be parsed and broken into fragments, which are then stored into the tables. Later, when a particular document is requested, it is reassembled from the fragments stored in the database and delivered to the client.

In order to implement the storage and retrieval of XML documents in the way outlined above, we currently recognize the need of **Oracle XML SQL Utility (XSU).**

3.2.1. *XSU*

While XML provides an enabling framework for a wide array of applications, it is only an enabling technology—it is not an application in itself. Until there is an agreed-upon schema or DTD, applications cannot use XML to reliably exchange or render data.

XML has been tightly linked to the Internet for a number of significant reasons. Because the content of an XML document is simply text, exchanging documents is easy over existing Internet protocols, across operating systems, and through firewalls. This capability gives rise to two major application areas—delivering content to a wide range of Internet-enabled devices and interchanging e-business data.

Many applications benefit from having their data reside in databases and querying these databases when data is required. An XML-enabled database benefits from being able to have these queries return data already marked up in XML in accordance with the database schema. The XML SQL Utility is a set of Java classes that accept these application queries, passing them through JDBC to the database and returning the resulting data in an XML format corresponding to the database schema of the query. As a complementary process, the XML SQL Utility can also accept an XML document conformant to the database schema and save the data untagged in the database across this schema.

In addition to reading and writing XML data into JDBC-enabled databases, the XML SQL Utility can create the DTD that represents the queried database schema. This DTD can then be used in application development with Oracle's Class Generators.

Saving XML-Formatted Data

Once a schema is created in the database, the XML SQL Utility can begin saving data as long as the XML-formatted data conforms to the DTD generated from the schema. The XML SQL Utility provides the ability to map the XML documents to table rows. The storage uses a simple mapping of element tag names to columns with XML strings converted to the appropriate data types through default mappings. If the XML element has child elements, it is mapped to a SQL object type.

To save the XML-formatted data, the XML SQL Utility initiates an insert statement binding all the values of the elements in the VALUES clause of the insert statement. The contents of each row element are mapped to a separate set of values.

Extending the XML SQL Utility

While the XML SQL Utility currently supports both DOM and String outputs, it can be extended to support other forms, including SAX.

The core functions are wrapped with an abstract layer, **OracleXMLDocGen**, to generate the XML document. In the current implementation, this abstract class is extended by **OracleXMLDocGenDOM** to generate a DOM output, and by **OracleXMLDocGenString** to generate a String output of the XML document. Additional classes can extend the OracleXMLDocGen class to support other representations.

3.2.2. *Java Server Pages*

Java Server Pages or JSP for short is Sun's solution for developing dynamic web sites. JSP provide excellent server side scripting support for creating database driven web applications. JSP enable the developers to directly insert Java code into jsp file. This makes the development process very simple and its maintenance easy. JSP technology is efficient, it loads the necessary Java classes into the web server memory on receiving the request very first time and the subsequent calls are served within a very short period of time.

Typical functional components are the database, as a convenient way to store the data, and JDBC to provide excellent database connectivity in heterogeneous database environment.

These technologies permit easy integration of the new and extended functionality planned to be developed, namely multilingual support, interface to external services (banking, logistics, consultations, etc.), secure data interchange, WAP access.

4. Towards One-Stop Trade Environment

The experience of developing and testing Mall2000 system lead us to the following conclusions:

- The utilization of international trade standards and XML is a fruitful approach that leads to an easy to implement and use functionality.
- To achieve high effectiveness and attractiveness of the system, integrated services that support all major stages of international trade transaction have to be provided.
- The European context demands multi-lingual interface and service content.
- Security and data protection have to be reliably assured.
- Ensuring interface to mobile devices is also attractive feature for the European customers.

This outlines the directions of expanding Mall2000 functionality towards a trade environment that can lead the customer all the way from brokering, negotiating and contracting, through payment and logistics, to customs services, to accomplish a trade deal – a goal quite achievable by applying the following existing solutions and technologies.

4.1. *Multilanguage User–Interface Support*

A commonly used method for providing multilanguage support on the web is Site Replication. The information content in the default language of the web-site resides in the root folder of the site, according to this approach. To provide an interface and content in another language, replication of the entire site into another directory is necessary. This method has serious disadvantages: any bug that is cleared on the main site needs to be cleared in all the other language sites; in case that 3 languages are supported, the work involved in any maintenance/bug-fixing/content-changing task increases 3-fold.

In contrast, Mall2000 environment will offer multilanguage user-interface support using XML and XSL methodologies by applying the Selective Dynamic Content Generation approach. In this method the core user-interface text (used in standard Menus, Forms' labels, System Messages, etc.) is stored in a database. Every page carries a variable (a session variable or a query string) to identify which language the site is to be displayed in. Based on that, the content is pulled out from the respective tables for the language chosen, and displayed. What still remains to be translated, is the user-entered text (such as the Free Description of goods/commodities offered or requested), which will be achieved by means of the above Site Replication approach. (This is why the Dynamic Content Generation is selective). The advantage of the Dynamic Content Generation is its flexibility, which allows introducing a new language version of the site easily, without replicating code. The way several well-known multilanguage sites, such as: Google.com, Sourceforge.net, etc., are organized, follows closely the Dynamic Content Generation approach.

4.2. *Data Exchange between Different Systems*

The rapid rise of internet-based business processes, such as e-commerce, has generated a huge increase in the number of business solutions requiring interoperability - reliable, repeatable, two-way exchange of data between different systems. The need for standards for the exchange of data between different systems and applications is a familiar topic for many professionals - and one to which the XML, with its XSU utility may, to a large extend, provide the answer. The XSU technology will be used in our system for building connection to other information systems that may have different data structures (use different protocols) - those used in Banks, Shipping Agencies, etc. so that correct data exchange between all these systems is guaranteed.

4.3. *Security and Data Protection*

Network payment is a key task for realization of e-commerce, and safety electronic transaction is the base of participating in e-commerce. Currently the key technology to ensure the safety of an electronic transaction comprises Security Socket Layer (SSL) and Safety Electronic Transaction (SET). SSL is the protocol that encodes the whole session among computers and provides the safe communication service on Internet. Secure servers using SSL to encrypt card details as they are sent back and forth provide a level of security at least equivalent to traditional means of card use. SET protocol aims to offer a solution for business by way of credit card payment among the customer, the supplier and the bank. SET was developed by international organizations of Visa and MasterCard and now it has won support from many large internal companies like IBM, HP, Microsoft, Netscape, etc.

For ensuring security and data protection, the above standard solutions will be used in our system.

4.4. *Mobile Internet*

The Wireless Application Protocol (WAP) is a composite of protocols used by three of the world's biggest mobile phone manufacturers - Nokia, Ericsson, Motorola, and a browser company called Unwired Planet (now Phone.com). WAP allows a mobile phone to retrieve information from the Internet via a server installed in the mobile phone network. It was created in order to have a standard that would reach the most end-users, and would be most agreeable to service providers. WAP will be adopted to provide the possibility of mobile access to the Mall2000 site functionality.

References

1. R. Nikolov. Mall2000 a Business-to-Business E-Commerce System. *Proceedings of the Second SEE Conference in E-Commerce*, Sofia (2000).
2. S.D. Kounev and K. Nikolov. The Analysis Phase in the Development of E-Commerce. *TOOLS EE '99 Conference*, Blagoevgrad, 1-4 June, 1999.
3. S.A. Angelov, K. Nikolov. Object-Oriented System Development and Electronic Commerce: Development of Brokering Service. *TOOLS EE '99 Conference*, Blagoevgrad, 1-4 June, 1999.
4. A.M. Rambhia. *XML Distributed Systems Design*, 1st edition, published by Sams, March 4, 2002, ISBN: 0672323281.

CHAPTER 13

PORTFOLIO OPTIMIZATION WITH DRAWDOWN CONSTRAINTS

A. Chekhlov

TrendLogic Associates, Inc.;
One Fawcett Place, Greenwich, Ct 06830;
E-mail: achekhlov@trendlogic.com

S. Uryasev

University of Florida, ISE,
P.O. Box 116595, 303 Weil Hall Gainesville, FL 32611-6595;
E-mail: uryasev@ufl.edu

M. Zabarankin

University of Florida, ISE,
P.O. Box 116595, 303 Weil Hall Gainesville, FL 32611-6595;
E-mail: zabarank@ufl.edu

We propose a new one-parameter family of risk measures, which is called Conditional Draw-down-at-Risk (CDaR). These measures of risk are functionals of the portfolio drawdown (underwater) curve considered in an active portfolio management. For some value of the tolerance parameter β, the CDaR is defined as the mean of the worst $(1-\beta)*100\%$ drawdowns. The CDaR risk measure includes the Maximal Drawdown and Average Drawdown as its limiting cases. For a particular example, we find the optimal portfolios for a case of Maximal Drawdown, a case of Average Drawdown, and several intermediate cases between these two. The CDaR family of risk measures is similar to Conditional Value-at-Risk (CVaR), which is also called Mean Shortfall, Mean Access loss, or Tail Value-at-Risk. Some recommendations on how to select the optimal risk measure for getting practically stable portfolios are provided. We solved a real life portfolio allocation problem using the proposed measures.

1. Introduction

Optimal portfolio allocation is a longstanding issue in both practical portfolio management and academic research on portfolio theory. Various methods have been proposed and studied (for a recent review, see, for example, Ref. 6). All of them, as a starting point, assume some measure of portfolio risk.

From a standpoint of a fund manager, who trades clients' or bank's proprietary capital, and for whom the clients' accounts are the only source of income coming in the form of management and incentive fees, losing these accounts is equivalent to the death of his business. This is true with no regard to whether the employed strategy is long-term valid and has very attractive expected return characteristics. Such fund manager's primary concern is to keep the existing accounts and to attract the new ones in order to increase his revenues. A particular client who was persuaded into opening an account with the manager through reading the disclosure document, listening to the manager's attractive story, knowing his previous returns, etc., will decide on firing the manager based, most likely, on his account's drawdown sizes and duration. In particular, it is highly uncommon, for a Commodity Trading Advisor (CTA) to still hold a client whose account was in a drawdown, even of small size, for longer than 2 years. By the same token, it is unlikely that a particular client will tolerate a 50% drawdown in an account with an average- or small-risk CTA. Similarly, in an investment bank setup, a proprietary system trader will be expected to make money in 1 year at the longest, i.e., he cannot be in a drawdown for longer than a year. Also, he/she may be shut down if a certain maximal drawdown condition will be breached, which, normally, is around 20% of his backing equity. Additionally, he will be given a warning drawdown level at which he will be reviewed for letting him keep running the system (around 15%). Obviously, these issues make managed accounts practitioners very concerned about both the size and duration of their clients' accounts drawdowns.

First, we want to mention Ref. 7, where an assumption of log-normality of equity statistics and use of dynamic programming theory led to an exact analytical solution of a maximal drawdown problem for a one-dimensional case. A subsequent generalization of this work for multiple dimensions was done in Ref. 3. In difference to these works, which were looking to find a time-dependent fraction of "capital at risk", we will be looking to find a constant set of weights, which will satisfy a certain risk condition over a period of time. We make no assumption about the underlying probability distribution, which allows considering variety of practical applications. We primarily concentrate on the portfolio equity curves over a particular past history path, which, effectively, makes the risk measures not stochastic but historical. Being perfectly aware of this insufficiency, we leave the issue of predictive power of a constant set of weights for future research, trying to introduce and test the new approach in this simplified version. To some extend we consider a setup similar to the index tracking problem [4] where an index historical performance is replicated by a portfolio with constant weights.

In this chapter, we have introduced and studied a one-parameter family of risk measures called Conditional Drawdown-at-Risk (CDaR). This measure of risk quantifies in aggregated format the number and magnitude of the portfolio drawdowns over some period of time. By definition, a drawdown is the drop in the portfolio value comparing to the maximum achieved in the past. We can define drawdown in absolute or relative (percentage) terms. For example, if at the present time the portfolio value equals \$9M and the maximal portfolio value in the past was \$10M, we can say that the portfolio drawdown in absolute terms equals \$1M and in relative terms equals 10%. For some value of the tolerance parameter β, the β-CDaR is defined as the mean of the worst $(1-\beta) * 100\%$ drawdowns experienced over some period of time. For instance, 0.95-CDaR (or 95%- CDaR) is the average of the worst 5% drawdowns over the considered time interval. The CDaR risk measure includes the average drawdown and maximal drawdown as its limiting cases. The CDaR takes into account both the size and duration of the drawdowns, whereas the maximal drawdown measure concentrates on a single event – maximal account's loss from its previous peak.

CDaR is related to Value-at-Risk (VaR) risk measure and to Conditional Value-at-Risk (CVaR) risk measure studied in Ref. 13. By definition, with respect to a specified probability level β, the β-VaR of a portfolio is the lowest amount α such that, with probability β, the loss will not exceed α in a specified time τ (see, for instance, Ref. 5), whereas the β-CVaR is the conditional expectation of losses above that amount α. The CDaR risk measure is similar to CVaR and can be viewed as a modification of the CVaR to the case when the loss-function is defined as a drawdown. CDaR and CVaR are conceptually closely related percentile-based risk performance measures. Optimization approaches developed for CVaR can be directly extended to CDaR. Ref. 11 considers several equivalent approaches for generating return-CVaR efficient frontiers; in particular, it considers an approach, which maximizes return with CVaR constraints. A nice feature of this approach is that the threshold, which is exceeded $(1-\beta)*100\%$, is calculated automatically using an additional variable (see details in Refs. 11, 13) and the resulting problem is linear. CVaR is known also as Mean Excess Loss, Mean Shortfall,[4,10] or Tail Value-at-Risk.[2] A case study on the hedging of a portfolio of options using the CVaR minimization technique is included in [11]. Also, the CVaR minimization approach was applied to credit risk management of a portfolio of bonds.[1] A case study on optimization of a portfolio of stocks with CVaR constraints is considered in Ref. 11.

Similar to the Markowitz mean-variance approach,[9] we formulate and solve the optimization problem with the return performance function and CDaR constraints. The return-CDaR optimization problem is a piece-wise linear convex optimization problem (see definition of convexity in Ref. 12), which can be reduced to a linear programming problem using auxiliary variables. Explanation of the procedure for reducing the piece-wise linear convex optimization problems to linear programming problems is beyond the scope of this chapter. In formulating the optimization problems with CDaR constraints and reducing it to a linear programming problem, we follow ideas presented in Ref. 11. Linear programming allows solving large optimization problems with hundreds of thousands of instruments. The algorithm is fast, numerically stable, and provides a solution during one run (without adjusting parameters like in genetic algorithms or neural networks). Linear programming approaches are routinely used in portfolio optimization with various criteria, such as mean absolute deviation,[8] maximum deviation,[14] and mean regret.[4] The reader interested in other applications of optimization techniques in the finance area can find relevant papers in Ref. 15.

2. General Setup

Denote by function $w(\mathbf{x},t)$ the uncompounded portfolio value at time t, where portfolio vector $\mathbf{x}=(x_1,x_2,\ldots,x_m)$ consists of weights of m instruments in the portfolio. The drawdown function at time t is defined as the difference between the maximum of the function $w(\mathbf{x},t)$ over the history preceding the point t and the value of this function at time t

$$f(\mathbf{x},t)=\max_{0\le\tau\le t}\{w(\mathbf{x},\tau)\}-w(\mathbf{x},t). \tag{1}$$

We consider three risk measures: (i) Maximum Drawdown (MaxDD),

(ii) Average Drawdown (AvDD), and (iii) Conditional Drawdown-at-Risk (CDaR). The last risk measure, Conditional Drawdown-at-Risk, is actually a family of performance measures depending upon a parameter β. It is defined similar to Conditional Value-at-Risk studied in Ref. 2 and, as special cases, includes the Maximum Drawdown and the Average Drawdown risk measures.

Maximum drawdown on an the interval $[0, T]$, is calculated by maximizing the drawdown function $f(\mathbf{x}, t)$, i.e.,

$$\mathbf{M}(\mathbf{x}) = \max_{0 \le t \le T} \{f(\mathbf{x}, t)\}. \tag{2}$$

The average drawdown is equal to

$$\mathbf{A}(\mathbf{x}) = \frac{1}{T} \int_0^T f(\mathbf{x}, t)\, dt. \tag{3}$$

For some value of the parameter $\beta \in [0, 1]$, the CDaR, is defined as the mean of the worst $(1 - \beta) * 100\%$ drawdowns. For instance, if $\beta = 0$, then CDaR is the average drawdown, and if $\beta = 0.95$, then CDaR is the average of the worst 5% drawdowns. Let us denote by $\alpha(\mathbf{x}, \beta)$ a threshold such that $(1 - \beta) * 100\%$ of drawdowns exceed this threshold. Then, CDaR with tolerance level β can be expressed as follows

$$\Delta_\beta(\mathbf{x}) = \frac{1}{(1-\beta)T} \int_\Omega f(\mathbf{x}, t)\, dt, \quad \Omega = \{t \in [0, T] : f(\mathbf{x}, t) \ge \alpha(\mathbf{x}, \beta)\}. \tag{4}$$

Here, when β tends to 1, CDaR tends to the maximum drawdown, i.e. $\Delta_1(\mathbf{x}) = \mathbf{M}(\mathbf{x})$.

To limit possible risks, depending upon our risk preference, we can impose constraints on the maximum drawdown given by (2)

$$\mathbf{M}(\mathbf{x}) \le \nu_1 C,$$

on average drawdown given by (3)

$$\mathbf{A}(\mathbf{x}) \le \nu_2 C,$$

on CDaR given by (4)

$$\Delta_\beta(\mathbf{x}) \le \nu_3 C,$$

or combine several constraints together

$$\mathbf{M}(\mathbf{x}) \le \nu_1 C, \quad \mathbf{A}(\mathbf{x}) \le \nu_2 C, \quad \Delta_\beta(\mathbf{x}) \le \nu_3 C, \tag{5}$$

where the constant C represents the available capital and the coefficients ν_1, ν_2 and ν_3 define the proportion of this capital which is "allowed to be lost". Usually,

$$0 \le \nu_1 \le 1, \quad 0 \le \nu_2 \le 1, \quad 0 \le \nu_3 \le 1. \tag{6}$$

Suppose that the historical returns for m portfolio instruments on interval $[0, T]$ are available. Let vector $\mathbf{y}(t) = (y_1(t), y_2(t), \dots, y_m(t))$ be a set of uncompounded cumulative net profits for m portfolio instruments at a time moment t. The cumulative portfolio value then equals $w(\mathbf{x}, t) = \sum_{k=1}^{m} y_k(t)\, x_k = \mathbf{y}(t) \cdot \mathbf{x}$.

The average annualized return $R(\mathbf{x})$ over a period $[0, T]$, which is a linear function of $\mathbf{x}$, is defined as follows

$$R(\mathbf{x}) = \frac{1}{Cd}\, w(\mathbf{x}, t) = \frac{1}{Cd}\, \mathbf{y}(t) \cdot \mathbf{x}, \tag{7}$$

where d is the number of years in the time interval $[0, T]$.

For the case considered, the so-called technological constraints on the vector $\mathbf{x}$ need to be imposed. Here, we assume that they are given by the set of box constraints:

$$X = \left\{ \mathbf{x} : \, x_{\min} \le x_k \le x_{\max}, \ \forall\, k = \overline{1, m} \right\}. \tag{8}$$

for some constant values of $x_{\min}$ and $x_{\max}$.

Our objective is to maximize the return $R(\mathbf{x})$ subject to constraints on various risk performance measures and technological constraints (8) on the portfolio positions.

3. Problem Statement

Maximization of the average return with constraints on maximum drawdown can be formulated as the following mathematical programming problem

$$\max_{\mathbf{x}\in X} R(\mathbf{x}) \\ s.\,t. \quad \mathbf{M}(\mathbf{x}) \leq \nu_1 C. \tag{9}$$

Maximization of the average return with constraints on the average drawdown can be formulated as follows

$$\max_{\mathbf{x}\in X} R(\mathbf{x}) \\ s.\,t. \quad \mathbf{A}(\mathbf{x}) \leq \nu_2 C. \tag{10}$$

Analogously, maximization of the average return with constraints on CDaR can be formulated as follows

$$\max_{\mathbf{x}\in X} R(\mathbf{x}) \\ s.\,t. \quad \Delta_\beta(\mathbf{x}) \leq \nu_3 C. \tag{11}$$

Similar to [2], the problems (9), (10), (11) can be reduced to linear programming problems using some auxiliary variables.

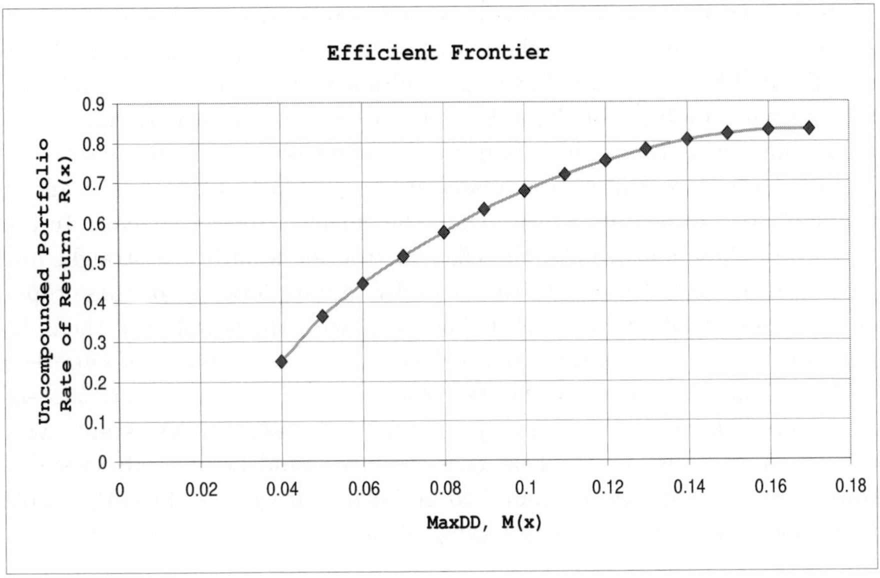

Fig. 1. Efficient frontier for the MaxDD problem (rate of return versus MaxDD).

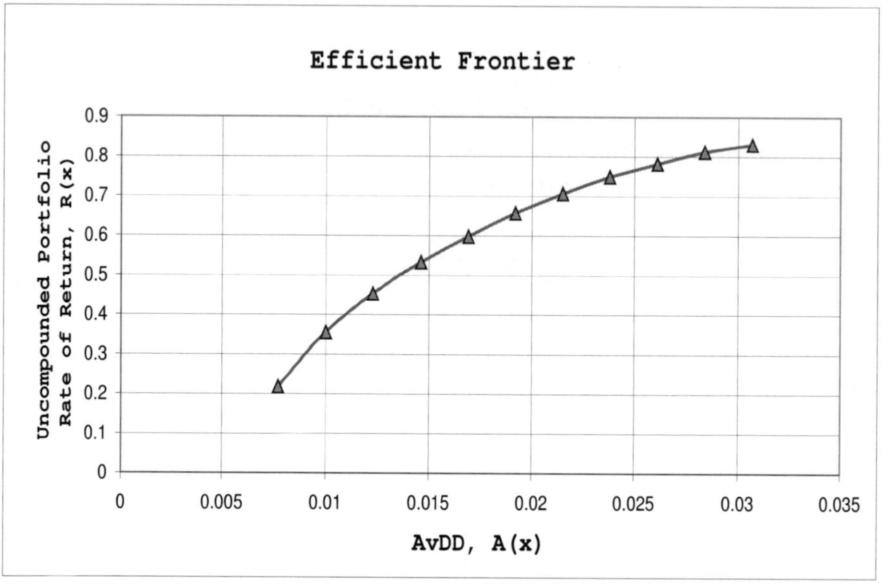

Fig. 2. Efficient frontier for the AvDD problem (rate of return versus AvDD).

It is necessary to mention several issues related to technological constraints (8). In our case, we chose $x_{\min} = 0.2$ and $x_{\max} = 0.8$. This choice was dictated by the need to have the resultant margin-to-equity ratio in the account within admissible bounds, which are specific for a particular portfolio. These constraints, in this futures trading setup is analogous to the "fully-invested" condition from classical Sharpe-Markowitz theory,[1] and it is namely this condition, which makes the efficient frontier concave. In the absence of these constraints, the efficient frontier would be a straight line passing through (0,0), due to the virtually infinite leverage of these types of strategies. Another subtle issue has to do with the stability of the optimal portfolios if the constraints are "too lax". It is a matter of empirical evidence that the more lax the constraints are – the better portfolio equity curve you can get through optimal mixing – and the less stable with respect to walk-forward analysis these results would be. The above set of constraints was empirically found to be both leading to sufficiently stable portfolios and allowing enough mixing of the individual equity curves.

4. Discrete Model

By dividing interval $[0, T]$ into N equal intervals (for instance, trading days)

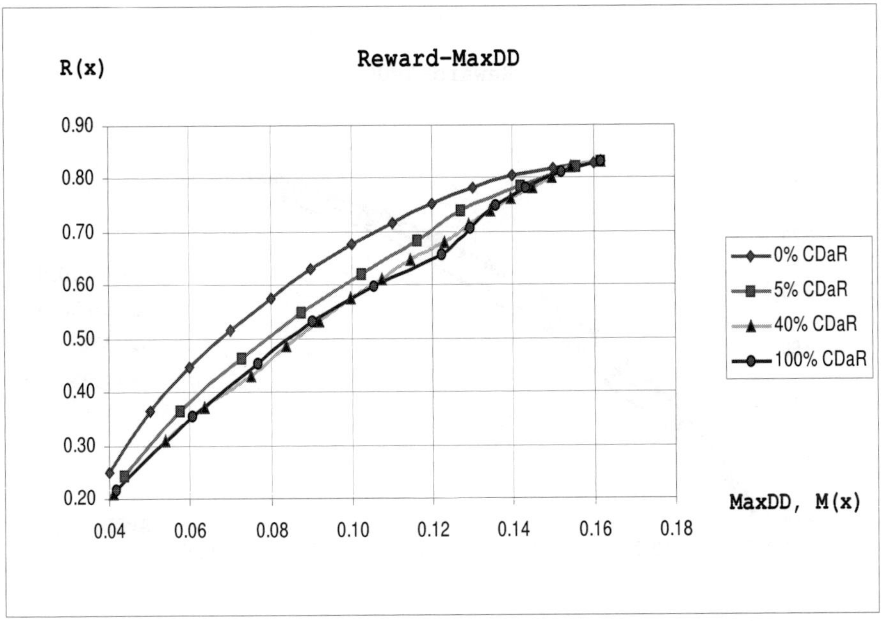

Fig. 3. Reward-MaxDD graphs for optimal portfolios with $(1-\beta) = 0$, 0.05, 0.4 and 1 CDaR constraints (rate of return versus MaxDD). The frontier is efficient only for the case with $(1-\beta) = 0$ CDaR constraints, which corresponds to the MaxDD risk measure.

$$t_i = i \cdot \frac{T}{N}, \quad i = \overline{1, N}, \tag{12}$$

we create the discrete approximations of the vector function $\mathbf{y}(t)$

$$\mathbf{y}(t_i) = \mathbf{y}_i, \tag{13}$$

the drawdown function

$$f_i(\mathbf{x}) = \max_{1 \le j \le i} \{\mathbf{y}_j \cdot \mathbf{x}\} - \mathbf{y}_i \cdot \mathbf{x}, \tag{14}$$

and the average annualized return function

$$R(\mathbf{x}) = \frac{1}{Cd}\, \mathbf{y}_N \cdot \mathbf{x}. \tag{15}$$

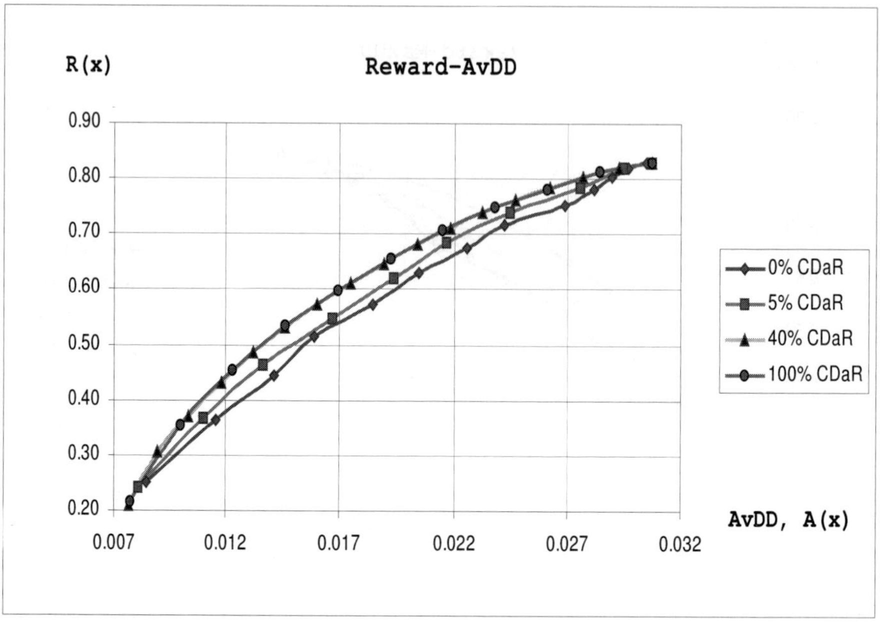

Fig. 4. Reward-AvDD graphs for optimal portfolios with $(1-\beta) = 0$, 0.05, 0.4 and 1 CDaR constraints (rate of return versus AvDD). The frontier is efficient only for the case with $(1-\beta) = 1$ CDaR constraints, which corresponds to the AvDD risk measure.

For the discrete time case, problems (9), (10) and (11) can be accordingly reformulated. The optimization problem with constraint on maximum drawdown is given below

$$\begin{aligned} &\max_{\mathbf{x}} \tfrac{1}{Cd}\, \mathbf{y}_N \cdot \mathbf{x} \\ &s.\,t. \quad \max_{1\le i\le N} \{ \max_{1\le j\le i} \{\mathbf{y}_j \cdot \mathbf{x}\} - \mathbf{y}_i \cdot \mathbf{x}\} \le \nu_1\, C, \\ &\qquad\quad x_k \in [x_{\min}, x_{\max}], \quad \forall\, k = \overline{1,m}. \end{aligned} \tag{16}$$

The optimization problem with constraint on average drawdown can be written as follows

$$\begin{aligned} &\max_{\mathbf{x}} \tfrac{1}{Cd}\, \mathbf{y}_N \cdot \mathbf{x} \\ &s.\,t. \quad \tfrac{1}{N} \sum_{i=1}^{N} \{ \max_{1\le j\le i} \{\mathbf{y}_j \cdot \mathbf{x}\} - \mathbf{y}_i \cdot \mathbf{x}\} \le \nu_2\, C, \\ &\qquad\quad x_k \in [x_{\min}, x_{\max}], \quad \forall\, k = \overline{1,m}. \end{aligned} \tag{17}$$

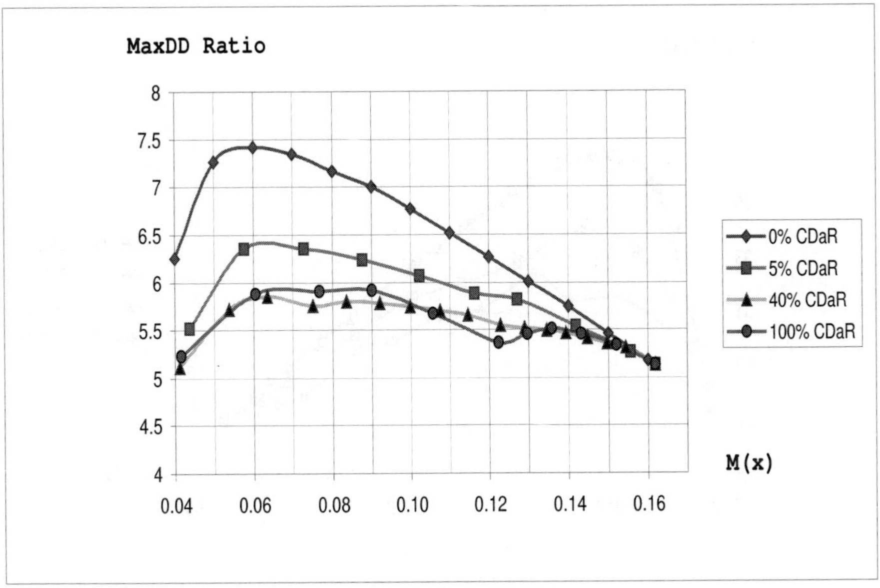

Fig. 5. MaxDDRatio graphs for optimal portfolios with $(1-\beta) = 0$, 0.05, 0.4 and 1 CDaR constraints (MaxDDRatio versus MaxDD). The maximum MaxDDRatio is achieved in the case with $(1-\beta) = 0$ CDaR constraints, which corresponds to the MaxDD risk measure.

Following the approach for Conditional Value-at-Risk (CVaR) [2], it can be proved that the discrete version of the optimization problem with constraint on CDaR may be stated as follows

$$\begin{aligned}
&\max_{\mathbf{x}} \tfrac{1}{Cd}\, \mathbf{y}_N \cdot \mathbf{x} \\
&s.\,t. \quad \alpha + \tfrac{1}{(1-\beta)N} \sum_{i=1}^{N} (\{\max_{1\le j\le i} \{\mathbf{y}_j \cdot \mathbf{x}\} - \mathbf{y}_i \cdot \mathbf{x}\} - \alpha)^+ \le \nu_3\, C, \\
&\qquad\quad x_k \in [x_{\min}, x_{\max}], \quad \forall\, k = \overline{1,m},
\end{aligned} \tag{18}$$

where we use the notation $(g)^+ = \max\{0, g\}$. An important feature of this formulation is that it does not involve the threshold function $\alpha(\mathbf{x}, \beta)$. An optimal solution to the problem (18) with respect to $\mathbf{x}$ and α gives the optimal portfolio and the corresponding value of the threshold function.

The problems (16), (17), and (18) have been reduced to linear programming problems using auxiliary variables and have been solved by the CPLEX solver (inputs are prepared with C++ programming language). An alternative verification of the solutions was obtained via solving simi-

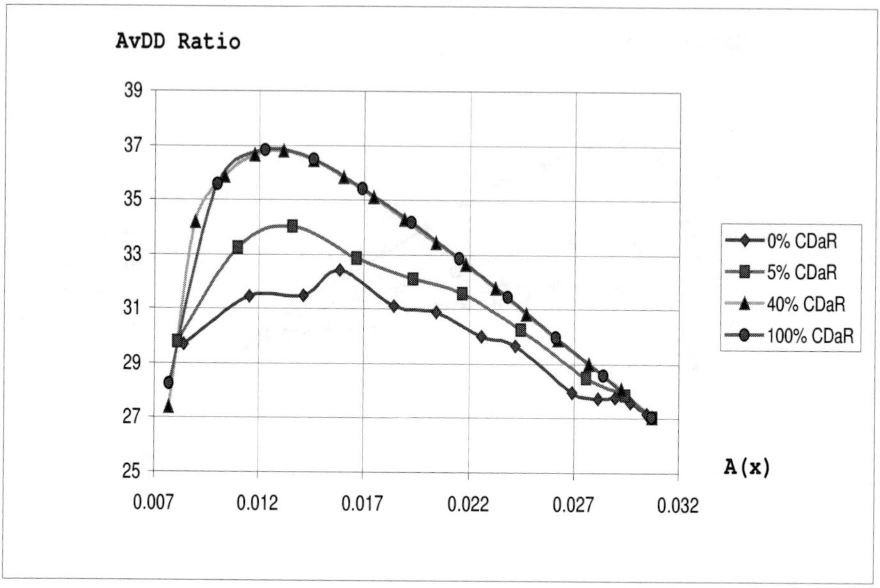

Fig. 6. AvDDRatio graphs for optimal portfolios with $(1 - \beta) = 0$, 0.05, 0.4 and 1 CDaR constraints (AvDDRatio versus AvDD). The maximum AvDDRatio is achieved in the case with $(1 - \beta) = 1$ CDaR constraints, which corresponds to the AvDD risk measure.

lar optimization problems using a more general Genetic Algorithm method implemented in VB6, discussion of which is beyond the present scope.

5. Results

As the starting equity curves, we have used the equity curves generated by a characteristic futures technical trading system in $m = 32$ different markets, covering a wide range of major liquid markets (currencies, currency crosses, U.S. treasuries both short- and long-term, foreign long-term treasuries, international equity indices, and metals). The list of market ticker symbols, provided in the results below, is mnemonic and corresponds to the widely used data provider, FutureSource.

The individual equity curves, when the market existed at the time, covered a time span of 1/1/1988 through 9/1/1999. The equity curves were based on \$20M backing equity in a margin account and were uncompounded, i.e. it was assumed that the amount of risk being taken, was always based of the original \$20M, not taking the money being made or lost into account.

The problem, then, is to find a set of weights $\mathbf{x} = (x_1, x_2, \ldots, x_m)$, such that it solves the minimization problems (16), (17), or (18). Let us denote the problem (16) as the MaxDD problem, the problem (17) as the AvDD problem, and the problem (18) as the β-CDaR problem. We have solved the above optimization problems for cases of $(1-\beta) = 0$, 0.05, 0.1, 0.2, 0.4, 0.6, 0.8 and 1. As we have noted before, cases of $(1-\beta) = 0$ and $(1-\beta) = 1$ correspond to MaxDD and AvDD problems, respectively.

Table 1. Solution results for the MaxDD problem. The solution achieving maximal Reward/Risk ratio is boldfaced.

Risk, %	4.0	5.0	**6.0**	7.0	8.0	9.0	10.0	11.0	12.0	13.0	14.0	15.0	16.0	17.0
Reward, %	25.0	36.3	**44.5**	51.4	57.3	63.0	67.7	71.7	75.2	78.0	80.4	81.9	82.9	83.0
Reward/Risk	6.26	7.27	**7.42**	7.34	7.16	7.00	6.77	6.52	6.27	6.00	5.74	5.46	5.18	4.88

Tables 1–2 and 3–4 provide the list of markets and corresponding sets of optimal weights for MaxDD and AvDD problems. Tables 5–6 provide the weights for the case with $(1-\beta) = 0.05$ CDaR. In these tables, the solution achieving maximal Reward/Risk ratio is boldfaced. Note that the smallest value of risk is chosen in such a way that the solutions to the optimization problem still exist. This means that each problem does not have a solution beyond the upper and lower bounds of the risk range covered (the whole efficient frontier is shown). Notions of risk and rate of return are expressed in percent with respect to the original account size, i.e. \$20M.

Efficient frontiers for problems reward-MaxDD and reward-AvDD, are shown in Figures 1 and 2, respectively. We do not show efficient frontiers for CDaR measure on separate graphs (except for MaxDD and AvDD). However, we show on Figure 3 the reward-MaxDD graphs for portfolios optimal with $(1-\beta) = 0$, 0.05, 0.4 and 1 CDaR constraints. As it is expected, the case with $(1-\beta) = 0$ CDaR corresponding to MaxDD has a concave efficient frontier majorating other graphs. The reward is not maximal for each level of MaxDD when we solved the optimization problems with $(1-\beta) = 0.05$, 0.4 and 1 CDaR constraints. Viewed from the reference point of MaxDD problem, $(1-\beta) < 1$ solutions are uniformly "worse". However, none of these solutions are truly better or worse than others from a mathematical standpoint. Each of them provides the optimal solution in its own sense. Some thoughts on which might be a better solution from a practical standpoint are provided below. Similar to Figure 3, Figure 4 depicts the reward-AvDD graphs for portfolios optimal with $(1-\beta) = 0$, 0.05, 0.4 and 1 CDaR constraints. The case with $(1-\beta) = 1$ CDaR corresponding to AvDD has a concave efficient frontier majorating other graphs.

As in classical portfolio theory, we are interested in a portfolio with a maximal Reward/Risk ratio, i.e., the portfolio where the straight line coming through (0,0) becomes tangent to the efficient frontier. We will call the Reward/Risk ratios for Risk defined in terms of problems (16), (17), and

Table 2. Optimal portfolio configuration corresponding to Table 1.

AAO	0.20	0.25	**0.25**	0.28	0.21	0.39	0.68	0.80	0.69	0.80	0.80	0.80	0.80	0.80
AD	0.20	0.40	**0.74**	0.80	0.80	0.80	0.80	0.80	0.80	0.80	0.80	0.80	0.80	0.80
AXB	0.20	0.37	**0.32**	0.47	0.63	0.80	0.55	0.64	0.80	0.80	0.80	0.80	0.80	0.80
BD	0.20	0.20	**0.20**	0.20	0.62	0.41	0.53	0.56	0.80	0.80	0.80	0.80	0.80	0.80
BP	0.20	0.20	**0.20**	0.20	0.20	0.20	0.20	0.20	0.22	0.51	0.77	0.80	0.80	0.80
CD	0.25	0.59	**0.80**	0.80	0.80	0.80	0.80	0.80	0.80	0.80	0.80	0.80	0.80	0.80
CP	0.62	0.80	**0.77**	0.80	0.80	0.80	0.80	0.80	0.80	0.80	0.80	0.80	0.80	0.80
DGB	0.20	0.80	**0.80**	0.80	0.80	0.80	0.80	0.80	0.80	0.80	0.80	0.80	0.80	0.80
DX	0.20	0.20	**0.20**	0.20	0.20	0.20	0.20	0.20	0.63	0.80	0.80	0.80	0.80	0.80
ED	0.20	0.20	**0.20**	0.20	0.20	0.20	0.20	0.20	0.20	0.35	0.74	0.80	0.80	0.80
EU	0.20	0.20	**0.20**	0.80	0.80	0.80	0.80	0.80	0.80	0.80	0.80	0.80	0.80	0.80
FV	0.20	0.20	**0.39**	0.58	0.52	0.50	0.54	0.80	0.80	0.80	0.80	0.80	0.80	0.80
FXADJY	0.27	0.58	**0.77**	0.80	0.80	0.80	0.80	0.80	0.80	0.80	0.80	0.80	0.80	0.80
FXBPJY	0.20	0.20	**0.20**	0.20	0.20	0.20	0.53	0.80	0.80	0.80	0.80	0.80	0.80	0.80
FXEUBP	0.20	0.28	**0.29**	0.32	0.34	0.65	0.72	0.80	0.80	0.80	0.80	0.80	0.80	0.80
FXEUJY	0.20	0.20	**0.41**	0.80	0.80	0.80	0.80	0.80	0.80	0.80	0.80	0.80	0.80	0.80
FXEUSF	0.33	0.20	**0.25**	0.30	0.73	0.80	0.80	0.80	0.80	0.80	0.80	0.80	0.80	0.80
FXNZUS	0.20	0.20	**0.20**	0.20	0.20	0.20	0.20	0.20	0.20	0.20	0.20	0.27	0.80	0.80
FXUSSG	0.20	0.20	**0.20**	0.20	0.20	0.20	0.20	0.28	0.21	0.43	0.72	0.80	0.80	0.80
FXUSSK	0.20	0.80	**0.80**	0.65	0.73	0.70	0.60	0.35	0.20	0.20	0.20	0.80	0.80	0.80
GC	0.20	0.20	**0.20**	0.20	0.20	0.20	0.20	0.20	0.20	0.20	0.20	0.57	0.80	0.80
JY	0.20	0.23	**0.34**	0.25	0.37	0.80	0.80	0.80	0.80	0.80	0.80	0.80	0.80	0.80
LBT	0.20	0.35	**0.62**	0.80	0.80	0.80	0.80	0.80	0.80	0.80	0.80	0.80	0.80	0.80
LFT	0.20	0.20	**0.20**	0.20	0.39	0.63	0.80	0.80	0.80	0.80	0.80	0.80	0.80	0.80
LGL	0.20	0.20	**0.20**	0.20	0.20	0.20	0.20	0.37	0.80	0.80	0.80	0.80	0.80	0.80
LML	0.20	0.27	**0.36**	0.46	0.51	0.60	0.78	0.80	0.80	0.80	0.80	0.80	0.80	0.80
MNN	0.20	0.30	**0.42**	0.45	0.44	0.80	0.80	0.80	0.77	0.80	0.80	0.80	0.80	0.80
SF	0.20	0.20	**0.37**	0.39	0.52	0.52	0.63	0.75	0.80	0.80	0.80	0.80	0.80	0.80
SI	0.20	0.20	**0.20**	0.20	0.20	0.20	0.20	0.20	0.20	0.20	0.20	0.20	0.40	0.80
SJB	0.49	0.74	**0.80**	0.80	0.80	0.80	0.80	0.80	0.80	0.80	0.80	0.80	0.80	0.80
SNI	0.20	0.56	**0.67**	0.69	0.78	0.80	0.80	0.80	0.80	0.80	0.80	0.80	0.80	0.80
TY	0.20	0.20	**0.23**	0.32	0.60	0.69	0.80	0.80	0.80	0.80	0.80	0.80	0.80	0.80

Table 3. Solution results for the AvDD problem. The solution achieving maximal Reward/Risk ratio is boldfaced.

Risk, %	0.77	1.00	**1.23**	1.46	1.50	1.69	1.92	2.15	2.38	2.61	2.84	3.07
Reward, %	21.7	35.6	**45.3**	53.3	54.5	59.9	65.7	70.6	74.8	78.2	81.2	83.0
Reward/Risk	28.2	35.6	**36.8**	36.5	36.3	35.4	34.2	32.9	31.4	30.0	28.6	27.0

(18) as MaxDDRatio, AvDDRatio, and CDaRRatio which, by definition, are

Table 4. Optimal portfolio configuration corresponding to Table 3.

AAO	0.20	0.46	**0.61**	0.77	0.80	0.80	0.80	0.80	0.80	0.80	0.80	0.80
AD	0.21	0.57	**0.80**	0.80	0.80	0.80	0.80	0.80	0.80	0.80	0.80	0.80
AXB	0.20	0.20	**0.23**	0.55	0.62	0.80	0.80	0.80	0.80	0.80	0.80	0.80
BD	0.20	0.20	**0.20**	0.20	0.20	0.20	0.20	0.20	0.20	0.52	0.80	0.80
BP	0.20	0.20	**0.20**	0.20	0.20	0.20	0.20	0.20	0.43	0.80	0.80	0.80
CD	0.20	0.37	**0.54**	0.80	0.80	0.80	0.80	0.80	0.80	0.80	0.80	0.80
CP	0.24	0.60	**0.80**	0.80	0.80	0.80	0.80	0.80	0.80	0.80	0.80	0.80
DGB	0.33	0.80	**0.80**	0.80	0.80	0.80	0.80	0.80	0.80	0.80	0.80	0.80
DX	0.20	0.20	**0.20**	0.20	0.20	0.20	0.20	0.20	0.20	0.30	0.71	0.80
ED	0.20	0.30	**0.35**	0.33	0.32	0.21	0.31	0.44	0.70	0.75	0.80	0.80
EU	0.20	0.20	**0.20**	0.20	0.20	0.20	0.46	0.80	0.80	0.80	0.80	0.80
FV	0.20	0.20	**0.37**	0.50	0.53	0.76	0.80	0.80	0.80	0.80	0.80	0.80
FXADJY	0.20	0.20	**0.20**	0.31	0.33	0.42	0.57	0.73	0.80	0.80	0.80	0.80
FXBPJY	0.20	0.20	**0.32**	0.49	0.50	0.69	0.80	0.80	0.80	0.80	0.80	0.80
FXEUBP	0.20	0.20	**0.29**	0.53	0.58	0.77	0.80	0.80	0.80	0.80	0.80	0.80
FXEUJY	0.20	0.59	**0.80**	0.80	0.80	0.80	0.80	0.80	0.80	0.80	0.80	0.80
FXEUSF	0.29	0.62	**0.80**	0.80	0.80	0.80	0.80	0.80	0.80	0.80	0.80	0.80
FXNZUS	0.20	0.20	**0.20**	0.20	0.20	0.27	0.80	0.80	0.80	0.80	0.80	0.80
FXUSSG	0.20	0.20	**0.20**	0.40	0.48	0.71	0.80	0.80	0.80	0.80	0.80	0.80
FXUSSK	0.20	0.74	**0.80**	0.80	0.80	0.80	0.80	0.80	0.80	0.80	0.80	0.80
GC	0.20	0.20	**0.20**	0.20	0.20	0.20	0.20	0.20	0.20	0.20	0.20	0.79
JY	0.20	0.38	**0.62**	0.80	0.80	0.80	0.80	0.80	0.80	0.80	0.80	0.80
LBT	0.20	0.52	**0.80**	0.80	0.80	0.80	0.80	0.80	0.80	0.80	0.80	0.80
LFT	0.20	0.20	**0.20**	0.20	0.20	0.20	0.20	0.20	0.20	0.36	0.46	0.80
LGL	0.20	0.20	**0.20**	0.20	0.20	0.20	0.29	0.48	0.65	0.80	0.80	0.80
LML	0.20	0.20	**0.21**	0.34	0.34	0.49	0.64	0.80	0.80	0.80	0.80	0.80
MNN	0.20	0.20	**0.20**	0.20	0.20	0.20	0.42	0.80	0.80	0.80	0.80	0.80
SF	0.20	0.20	**0.38**	0.50	0.54	0.67	0.80	0.80	0.80	0.80	0.80	0.80
SI	0.20	0.20	**0.20**	0.20	0.20	0.20	0.20	0.20	0.20	0.20	0.20	0.80
SJB	0.23	0.67	**0.80**	0.80	0.80	0.80	0.80	0.80	0.80	0.80	0.80	0.80
SNI	0.20	0.33	**0.47**	0.62	0.66	0.72	0.80	0.80	0.80	0.80	0.80	0.80
TY	0.20	0.20	**0.20**	0.20	0.20	0.20	0.20	0.32	0.69	0.77	0.80	0.80

Table 5. Solution resilts for the CDaR problem with $(1-\beta) = 0.05$. The solution achieving maximal Reward/Risk ratio is boldfaced.

Risk, %	3.0	3.2	3.7	3.8	3.9	4.0	**5.0**	6.0	7.0	8.0	9.0	10.0	11.0	12.0
Reward, %	24.2	27.2	33.3	34.4	35.5	36.6	**46.3**	54.7	62.1	68.4	73.9	78.6	82.0	83.0
α, %	2.55	2.64	3.10	3.18	3.27	3.36	**4.26**	5.13	6.02	6.81	7.66	8.61	9.57	9.98
Reward/Risk	8.06	8.50	8.99	9.04	9.09	9.14	**9.26**	9.12	8.86	8.55	8.21	7.86	7.45	6.92

Table 6. Optimal portfolio configuration corresponding to Table 5.

AAO	0.20	0.21	0.30	0.32	0.33	0.34	**0.49**	0.54	0.69	0.80	0.80	0.80	0.80	0.80
AD	0.24	0.36	0.60	0.64	0.68	0.69	**0.80**	0.80	0.80	0.80	0.80	0.80	0.80	0.80
AXB	0.20	0.20	0.20	0.20	0.20	0.20	**0.20**	0.20	0.33	0.46	0.80	0.80	0.80	0.80
BD	0.20	0.20	0.20	0.20	0.20	0.20	**0.20**	0.60	0.69	0.67	0.80	0.80	0.80	0.80
BP	0.20	0.20	0.20	0.20	0.20	0.20	**0.20**	0.20	0.20	0.20	0.20	0.20	0.80	0.80
CD	0.20	0.20	0.29	0.31	0.32	0.33	**0.49**	0.64	0.80	0.80	0.80	0.80	0.80	0.80
CP	0.23	0.34	0.41	0.44	0.46	0.51	**0.80**	0.80	0.80	0.80	0.80	0.80	0.80	0.80
DGB	0.50	0.71	0.80	0.80	0.80	0.80	**0.80**	0.80	0.80	0.80	0.80	0.80	0.80	0.80
DX	0.20	0.20	0.20	0.20	0.20	0.20	**0.20**	0.20	0.20	0.20	0.31	0.80	0.80	0.80
ED	0.20	0.20	0.20	0.20	0.20	0.20	**0.26**	0.27	0.31	0.28	0.28	0.48	0.64	0.80
EU	0.20	0.20	0.23	0.26	0.30	0.31	**0.80**	0.80	0.80	0.80	0.80	0.80	0.80	0.80
FV	0.20	0.20	0.20	0.23	0.25	0.30	**0.47**	0.47	0.56	0.73	0.80	0.80	0.80	0.80
FXADJY	0.20	0.22	0.33	0.34	0.35	0.36	**0.49**	0.69	0.80	0.80	0.80	0.80	0.80	0.80
FXBPJY	0.20	0.20	0.20	0.20	0.20	0.20	**0.20**	0.32	0.50	0.73	0.80	0.80	0.80	0.80
FXEUBP	0.20	0.20	0.29	0.31	0.34	0.34	**0.43**	0.39	0.46	0.76	0.80	0.80	0.80	0.80
FXEUJY	0.20	0.35	0.68	0.72	0.74	0.77	**0.80**	0.80	0.80	0.80	0.80	0.80	0.80	0.80
FXEUSF	0.20	0.20	0.28	0.30	0.31	0.29	**0.38**	0.59	0.80	0.80	0.80	0.80	0.80	0.80
FXNZUS	0.20	0.20	0.20	0.20	0.20	0.20	**0.20**	0.20	0.20	0.20	0.20	0.20	0.77	0.80
FXUSSG	0.20	0.20	0.20	0.20	0.20	0.20	**0.20**	0.37	0.59	0.75	0.80	0.80	0.80	0.80
FXUSSK	0.20	0.20	0.22	0.22	0.24	0.25	**0.61**	0.80	0.80	0.80	0.79	0.80	0.80	0.80
GC	0.20	0.20	0.20	0.20	0.20	0.20	**0.20**	0.20	0.20	0.20	0.20	0.20	0.20	0.80
JY	0.31	0.35	0.42	0.43	0.45	0.47	**0.75**	0.80	0.80	0.80	0.80	0.80	0.80	0.80
LBT	0.20	0.20	0.20	0.20	0.20	0.20	**0.47**	0.80	0.80	0.80	0.80	0.80	0.80	0.80
LFT	0.20	0.20	0.20	0.20	0.20	0.20	**0.25**	0.28	0.43	0.58	0.66	0.76	0.80	0.80
LGL	0.20	0.20	0.20	0.20	0.20	0.20	**0.20**	0.20	0.20	0.20	0.27	0.66	0.80	0.80
LML	0.20	0.20	0.20	0.20	0.20	0.20	**0.20**	0.20	0.31	0.52	0.69	0.74	0.80	0.80
MNN	0.20	0.20	0.20	0.20	0.20	0.20	**0.20**	0.34	0.74	0.80	0.80	0.80	0.80	0.80
SF	0.20	0.20	0.20	0.20	0.20	0.20	**0.20**	0.54	0.80	0.80	0.80	0.80	0.80	0.80
SI	0.20	0.20	0.20	0.20	0.20	0.20	**0.20**	0.20	0.20	0.20	0.20	0.20	0.58	0.80
SJB	0.47	0.57	0.71	0.74	0.77	0.80	**0.80**	0.80	0.80	0.80	0.80	0.80	0.80	0.80
SNI	0.21	0.22	0.29	0.29	0.30	0.33	**0.58**	0.80	0.80	0.80	0.80	0.80	0.80	0.80
TY	0.20	0.20	0.20	0.20	0.20	0.20	**0.20**	0.20	0.20	0.39	0.70	0.80	0.80	0.80

$$\text{MaxDDRatio} = \frac{R(\mathbf{x})}{M(\mathbf{x})}, \quad \text{AvDDRatio} = \frac{R(\mathbf{x})}{A(\mathbf{x})}, \quad \text{CDaRRatio} = \frac{R(\mathbf{x})}{\Delta_\beta(\mathbf{x})}.$$

The charts of MaxDDRatio and AvDDRatio quantities are shown in Figures 5 and 6 for the same cases of $(1-\beta)$ as in Figures 3 and 4.

We have solved optimization problem (18) for cases of $(1-\beta) = 0$, 0.05, 0.1, 0.2, 0.4, 0.6, 0.8 and 1. Let us note that already the case of $(1-\beta) = 0.05$ (see Table 3), which considers minimization of the worst 5% part of the underwater curve, is producing a set of weights significantly different from the $(1-\beta) = 0$ case (MaxDD problem), and $(1-\beta) = 0.05$ CDaR case includes several tens of events over which the averaging

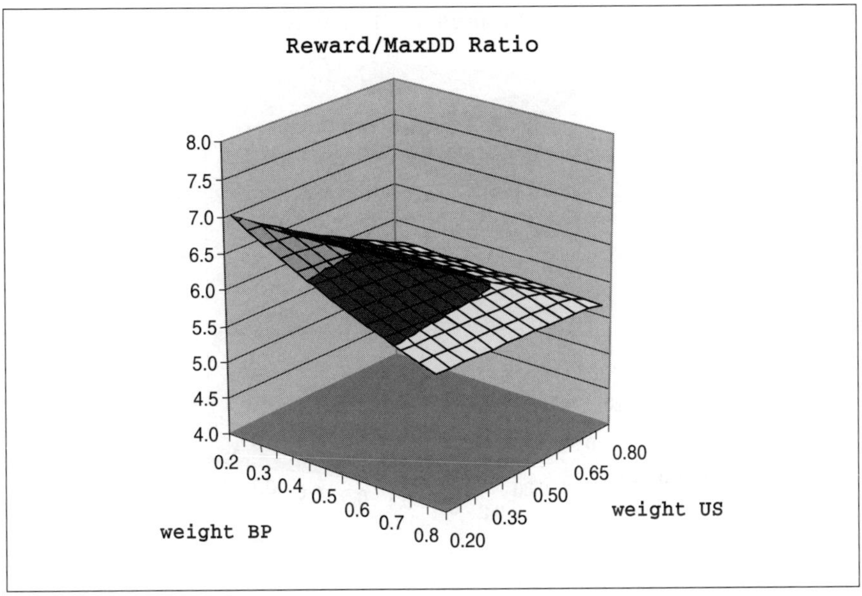

Fig. 7. Example of Reward to Risk ratio of two instruments. The risk is defined by the value of portfolio MaxDD.

was performed. We consider that optimization with $(1 - \beta) = 0.05$ or 0.1 constraints produces a more robust portfolio than the optimization with MaxDD or AvDD constraints. CDaR solution takes into account many significant drawdowns, comparing to the case with MaxDD constraints, which considers only the largest drawdown. Also, CDaR solution is not dominated by many small drawdowns like the case with AvDD constraints.

We have also made an alternative check of our results via solving the related nonlinear optimization problems corresponding to problems (16)-(18). These problems have optimized the corresponding drawdown ratios defined above within the same set of constraints. Verification was done using Genetic Algorithm-based search software. We were satisfied to find that this procedure has produced the same sets of weights for the optimal solutions.

6. Conclusions

We have introduced a new CDaR risk measure, which, we believe, is useful for the practical portfolio management. This measure is similar to CVaR

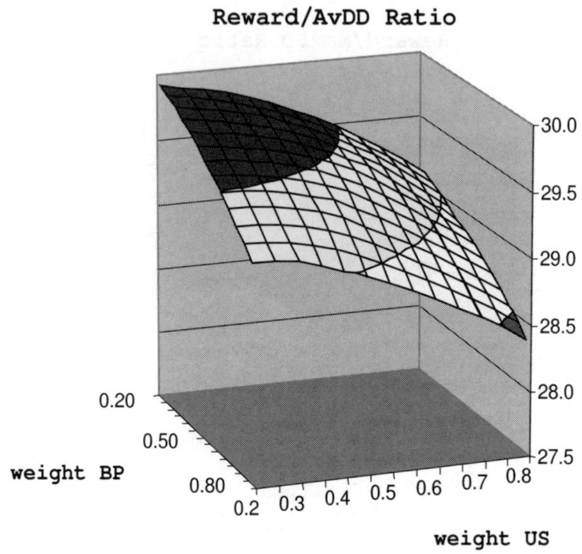

Fig. 8. Example of Reward to Risk ratio of two instruments. The risk is defined by the value of portfolio AvDD. Using MaxDD leads to nonsmooth picture while, using AvDD, which is an integrated characteristic, determines the smooth ratio. Solutions based on using CDaR or AvDD seem to be more robust than those obtained by using MaxDD.

risk measure and has the MaxDD and AvDD risk measures as its limiting cases. We have studied Reward/Risk ratios implied by these measures of risk, namely MaxDDRatio, AvDDRatio, and CDaRRatio. We have shown that the portfolio allocation problem with CDaR, MaxDD and AvDD risk measures can be efficiently solved. We have posed and for a real-life example, solved a portfolio allocation problem. These developments, if implemented in a managed accounts' environment will allow a trading or risk manager to allocate risk according to his personal assessment of extreme drawdowns and their duration on his portfolio equity.

We believe that however attractive the MaxDD approach is, the solutions produced by this optimization may have a significant statistical error because the decision is based on a single observation of maximal loss. Having a CDaR family of risk measures allows a risk manager to have control over the worst $(1-\beta) * 100\%$ of drawdowns, and due to statistical averaging within that range, to get a better predictive power of this risk measure in the future, and therefore a more stable portfolio. Our studies indicate that when considering CDaR with an appropriate level (e.g., $\beta = 0.95$, i.e., optimizing over the 5% of the worst drawdowns), one can get a more stable weights allocation than that produced by the MaxDD problem. A detailed study of this issue calls for a separate publication.

Acknowledgments

Authors are grateful to Anjelina Belakovskaia, Peter Carr, Stephan Demoura, Nedia Miller, and Mikhail Smirnov for valuable comments which helped to improve the chapter.

References

1. F. Andersson and S. Uryasev. Credit Risk Optimization With Conditional Value-At-Risk Criterion. *Research Report 99-9. ISE Dept., University of Florida*, August. (1999) (Revised version submitted to the journal of *Mathematical Programming* can be downloaded: www.ise.ufl.edu/uryasev/and_mp.pdf)
2. P. Artzner, F. Delbaen, J.M. Eber, and D. Heath. Coherent Measures of Risk. *Mathematical Finance*, 9, 203-228 (1999).
3. J. Cvitanic and I. Karatzas. On Portfolio Optimization Under "Drawdown" Constraints. *IMA Lecture Notes in Mathematics & Applications* **65**, 77-88 (1995).
4. R.S. Dembo and A.J. King. Tracking Models and the Optimal Regret Distribution in Asset Allocation. *Applied Stochastic Models and Data Analysis.* Vol. 8, 151-157 (1992).
5. Ph. Jorion. *Value at Risk : A New Benchmark for Measuring Derivatives Risk.* Irwin Professional Pub. (1996).
6. R.C. Grinold and R.N. Kahn. *Active Portfolio Management*, McGraw-Hill, New York (1999).
7. S. J. Grossman and Z. Zhou. Optimal Investment Strategies for Controlling Drawdowns, *Mathematical Finance,* **3**, 241-276 (1993).
8. H. Konno and H. Yamazaki. Mean Absolute Deviation Portfolio Optimization Model and Its Application to Tokyo Stock Market. *Management Science*, **37**, 519-531 (1991).
9. H.M. Markowitz. Portfolio Selection. *Journal of Finance.* **7**(1), 77-91 (1952).
10. H. Mausser and D. Rosen. Beyond VaR. From Measuring Risk to Managing Risk. *ALGO Research Quarterly.* **1**(2), 5-20 (1999).
11. J. Palmquist, S. Uryasev, and P. Krokhmal. Portfolio Optimization with Conditional Value-At-Risk Objective and Constraints. *Research Report 99-14*, ISE Dept., University of Florida (1999) (can be downloaded: www.ise.ufl.edu/uryasev/pal.pdf).
12. R.T. Rockafellar. *Convex Analysis.* Princeton Mathematics, Vol. 28, Princeton Univ. Press (1970).

13. R.T. Rockafellar and S. Uryasev. Optimization of Conditional Value-at-Risk. *The Journal of Risk,* accepted for publication (2000) (can be downloaded: www.ise.ufl.edu/uryasev/cvar.pdf).
14. M.R. Young. A Minimax Portfolio Selection Rule with Linear Programming Solution. *Management Science.* **44**(5), 673-683 (1998).
15. W.T. Ziemba and J.M. Mulvey, eds. *Worldwide Asset and Liability Modeling*, Cambridge Univ. Pr. (1998).

CHAPTER 14

PORTFOLIO OPTIMIZATION USING MARKOWITZ MODEL: AN APPLICATION TO THE BUCHAREST STOCK EXCHANGE

C. Viju

Mediterranean Agronomic Institute of Chania,
Dept. of Economic Sciences, Management, Marketing and Finance,
E-mail: crina@maich.gr

G. Baourakis

Mediterranean Agronomic Institute of Chania,
Dept. of Economic Sciences, Management, Marketing and Finance,
E-mail: crina@maich.gr

A. Migdalas

Technical University of Crete,
Dept. of Production Engineering & Management,
E-mail: migdalas@ergasya.tuc.gr

M. Doumpos

Technical University of Crete,
Dept. of Production Engineering & Management,
E-mail: dmichael@ergasya.tuc.gr

P. M. Pardalos

University of Florida,
Dept. of Industrial & Systems Engineering,
E-mail: pardalos@ufl.edu

The Bucharest Stock Exchange, with all its economical, social and political problems and sudden ups and downs, is a good reflection of the transition period that emerging economy is currently undergoing. This study focuses on the use of an appropriate methodology for constructing efficient stock portfolios in an extremely unstable market that makes the trade-off between risk and return even more difficult to achieve. The objective is set in order to assess the market behavior: employing the Markowitz model, to construct a set of optimum portfolios under a number of varying constraints and to compare them with the market portfolio. The results obtained are presented in the chapter along with a discussion of the main problems encountered due to the particular features of a stock market in a state of transition.

Keywords: Portfolio construction, portfolio optimization, Markowitz model, expected return, risk, efficient frontier.

1. Introduction

Portfolio selection and management has been one of the major fields of interest in the area of finance for almost the last 50 years. Generally stated, portfolio selection and management involves the construction of a portfolio of securities (stocks, bonds, treasury bills, mutual funds, financial derivatives, etc.) that maximizes the investor's utility. The term "construction" of a portfolio refers to the allocation of a known amount of capital to the securities under consideration. Generally, portfolio construction can be realized as a two-stage process:

(1) In the first stage of the process, the investor needs to evaluate the available securities that constitute possible investment opportunities on the basis of their future perspectives. This evaluation leads to the selection of a reduced set consisting of the best securities.
(2) Once this compact set of the best securities is specified in the first stage, the investor needs to decide on the allocation of the available capital to these securities. The allocation should be performed so that the resulting portfolio best meets the investor's policy, goals and objectives.

The existing research on the portfolio selection and management problem can be organized into three major categories:

(1) The studies focusing on the securities' risk/return characteristics. These studies are primarily conducted by financial researchers in order to specify the determinants of risk and return in investment decisions. The most well known examples of studies within this category include Sharpe's study on the capital asset pricing model (CAPM),[38] Ross' study on the arbitrage pricing theory (APT)[37] and the Black-Scholes study on option valuation.[1]

(2) The studies focusing on the development of methodologies for evaluating the performance of securities according to different performance measures. These studies can be further categorized into two groups:

- The first group includes studies on the modelling and representation of the investor's policy, goals and objectives in a mathematical model, usually of a functional form. This model aggregates all the pertinent factors describing the performance of the securities to produce an overall evaluation of the securities that complies with the policy of the investor. The securities with the highest overall evaluation according to the developed model are selected for portfolio construction purposes in a latter stage of analysis. The developed model usually has the form of a utility function following the general framework of portfolio theory. According to this model, the investor is interested in constructing a portfolio that maximizes his/her utility.
- The second group involves studies regarding the forecasting of securities' prices. The objective of this forecasting-based approach is to develop models that are able to provide accurate predictions on the future prices of the securities from historical time-series data.

(3) The studies on the development of methodologies for portfolio construction. These methodologies follow an optimization perspective, usually in a multiobjective context. This complies with the nature of the portfolio construction problem. Indeed, portfolio construction is a multiobjective optimization problem, even if it is considered in the traditional mean-variance framework. Within this framework, the investor is interested in constructing a portfolio that maximizes the expected return and minimizes the risk of the investment. This is a two-objective optimization problem. Furthermore, considering that actually both return and risk are multidimensional, it is possible to extend the traditional mean-variance framework so that all pertinent risk and return factors are considered.

Following this line of research, the construction of portfolios within this extended optimization framework, can be performed through multiobjective mathematical and goal programming techniques.

Most recently, researchers have implemented dynamic investment models to study long-term problems and to improve performance. Four approaches are available for solving the stochastic models: 1) solve a sequence of single period optimization problems;[14] 2) employ a multi-stage stochastic program;[3,19] 3) solve the problem via stochastic control methods cite4,5,6 and 4) set up a stochastic simulation by means of selected decision rules and optimize the Monte-Carlo simulation.[30,35] They can lead to non-convex optimization models, requiring extensive searching to find a global optimal solution, which generate inferior results. Financial planning models grow in complexity as a direct function of time periods. Stochastic programs are among the most difficult in numerical computations. First, the model's size can be enormous, depending upon the number of periods and decision variables (for stochastic programs), or the size of the state space (for stochastic control models). Second, the computational costs are so high as to be impractical for many users.[31]

The following research proposes the use of Markowitz mean-variance model to construct a set of efficient portfolios. The performance of this approach was explored using daily data for a period of three years, 1999 – 2001, from Bucharest Stock Exchange. In order to perform the analysis we used the basic Markowitz model in two situations. In its basic form, this model requires to determine the composition of a portfolio of assets, which minimizes risk while achieving a predetermined level of expected return. In the first case, we considered that the aim of the investor is to minimize the level of risk without putting any constraints for the level of the expected return, and, in the second one, we minimize the level of risk considering a minimum guaranty level of expected return. No short sales are allowed in the basic Markowitz model and, also, we do not consider transaction costs. The next step in the analysis was to compare the constructed portfolios with the market indices and to test their future performance.

The rest of the chapter is organized as follows. The next section is devoted to the main features of the Markowitz Mean-Variance Theory, the limitations and the alternatives for this model. Section 3 focuses on presenting the methodology used in our research. The main characteristics of the Bucharest Stock Exchange are described in Section 4. Section 5 illustrates the obtained results of the empirical study and, finally, Section 6 summarizes the main conclusions reached and discusses issues for future research.

2. Markowitz Mean-Variance Theory

Until the 1950s, the majority of investment analysts and portfolio managers believed that the best strategy in forming portfolios was the one that incorporates investments or assets that exhibit the highest expected return into a portfolio. The main disadvantage of this strategy was the ignorance of the risk dimension during the selection process. Therefore, the investors' primary objective was the maximization of the expected return.

Harry Markowitz initiated the field, called modern portfolio theory, by proposing a simple quadratic program for selecting a diversified portfolio of securities.[25,26]

Markowitz identified the trade-off facing the investor: risk versus expected return. Mean-variance theory is an important model of investments based on decision theory. The investment decision is not merely which securities to own, but how to allocate the investor's wealth amongst securities. This is the problem of Portfolio Selection.

The fundamental assumption underlying the Markowitz approach to portfolio analysis is that investors are basically risk-averse. This means simply that investors must be compensated with higher return in order to accept higher risk. Consequently, given a choice, for example, between two securities with equal rates of return, an investor will select the security with the lower level of risk, thereby rejecting the higher-risk security. In more technical terms, this assumption means that investors maximize expected utility rather than merely trying to maximize expected return. Utility, a measure of satisfaction, considers both risk and return.

Presuming risk-aversion, Markowitz then developed a model of portfolio analysis that can be summarized as follows:[12]

first, the two relevant characteristics of a portfolio are its expected return and some measure of the dispersion of possible returns around the expected return; the variance is analytically the most tractable;

second, rational investors will choose to hold efficient portfolios - those that maximize expected returns for a given degree of risk or, alternatively and equivalently, minimize risk for a given expected return;

third, it is theoretically possible to identify efficient portfolios through proper data analysis of information for each security on expected return, variance of return and the interrelationship between the return for each security and that for every other security as measured by the covariance.

2.1. *Asset Allocation versus Equity Portfolio Optimization*

Asset allocation and equity portfolio optimization are the two most popular applications of mean – variance optimization.[28] In both cases, the optimization finds optimal allocations of capital to maximize expected return and minimize risk, subject to various constraints.

In asset allocation studies, the number of risky assets rarely exceeds 50 and is typically in the range of 3-20 and the number of constraints is relatively small. Sample means, variances and correlations, based on monthly, quarterly or annual historic data, may be used as a starting point for optimization input estimates.

An equity portfolio optimization generally includes many securities. Domestic equity optimizations typically include 100-500 stocks, and international equity optimizations include 4.000-5.000 stocks. Also, equity portfolio optimization includes many constraints on portfolio characteristics, industry or sector membership, and trading cost restrictions. Modern financial theory provides a rich framework for defining expected and residual return for equities.

2.2. *Required Model Inputs*

In order to use the Markowitz full covariance model for portfolio construction, the investor must obtain estimates of the returns and the variances and the covariances of returns for the securities in the universe of interest. Considering the case of N stocks, there is a need not only for N return estimates and N variance estimates, but also $\frac{N(N-1)}{2}$ for covariance estimates, for a total of $[2N + \frac{N(N-1)}{2}]$ estimates.

While the Markowitz model is the most comprehensive one, it has proven to be of relatively little use in solving practical problems of analysing universes with large numbers of securities, mainly because of the overwhelming burden of developing input estimates for the model.[12]

2.3. *Limitations of the Markowitz Model*

Although the mean-variance theory is solid and its use has greatly enhanced the portfolio management process, it is difficult to use it properly. Uncritical acceptance of mean-variance theory output can result in portfolios that are unstable, counterintuitive and, ultimately, unacceptable. The limitations of mean-variance theory are:

The effects of estimation error - If the inputs are free of estimation error, mean-variance optimization is guaranteed to find the optimal or efficient portfolio weights. However, because the inputs are statistical estimates (typically created by analysing historical data), they cannot be devoid of error. This inaccuracy will lead to overinvestment in some asset classes and underinvestment in others. Estimation error can also cause an efficient portfolio to appear inefficient. One approach to limit the impact of estimation error is to use *constrained* optimization. In a constrained optimization the user sets the maximum or minimum allocation for a single asset or group of assets.

Unstable solutions - A related problem with mean-variance optimization is that its results can be *unstable*; that is, small changes in inputs can result in large changes in portfolio contents. In order to minimize dramatic changes in recommended portfolio composition *sensitivity analysis* can be used. The goal is to identify a set of asset class weights that will be close to efficient under several different sets of plausible inputs.

Reallocation costs - Depending on the asset classes within two portfolios and the magnitude of the quantities involved, it may be quite costly to implement a reallocation of one portfolio in order to reach the same expected return as the other one. The correct policy may be to retain the current allocation despite its lack of optimality.

Scepticism of the uninitiated - Many investors use mean-variance optimization, but they invest on the basis of trading, and they do not understand allocation systems because this theory is complex and it includes statistics, optimization, and modern portfolio theory.

Political fallout - The use of mean-variance optimization for asset allocation may be against the interests of some employees within a money management firm.

2.4. *Alternatives to the Markowitz Model*

Even though many authors have raised serious objections regarding the efficiency of the mean-variance (MV) method,[28] the analysis shows that the alternatives often have their own serious limitations and that mean-variance efficiency is far more robust than is appreciated.

Non-variance risk measures - One non-variance measure of risk is the *semi-variance or semi-standard deviation of return.* In this risk measure, only returns below the mean are included in the estimate of variability because the variance of returns above the mean is not considered risk by the investors. Many other *non-variance measures* of variability are also available. Some of the more important include the mean absolute deviation and range measures.

Utility function optimization - For many financial economists, maximizing expected utility of terminal wealth is the basis for all rational decision making under uncertainty. Markowitz mean-variance efficiency is strictly consistent with expected utility maximization only if either of the two conditions hold: normally distributed asset returns, or quadratic utility function. The normal distribution assumption is unacceptable to most analysts and investors. Because a quadratic function does not increase monotonously as a function of wealth, from some point on, expected quadratic utility declines as a function of increasing wealth. Consequently, MV efficiency is not strictly consistent with expected utility maximization.

Multi-period objectives - Markowitz mean-variance efficiency is formally a single-period model for investment behaviour. Many institutional investors have long-term investment horizons on the order of 5, 10 or 20 years. Some ways to address long-term objectives is to base mean-variance efficiency analysis on long-term units of time or to consider the multi-period distribution of the geometric mean of return.

Monte Carlo financial planning - In a Monte Carlo financial planning study, a computer model simulates the random functioning of a fund and changes in its liabilities over time. It is argued that plan funding status and cash flow objectives are more meaningful than the mean-variance efficiency of a feasible portfolio.

Linear programming optimization - Linear programming portfolio optimization is a special case of quadratic programming. The most significant difference is that linear programming does not include portfolio variance. There are two possibilities to exclude portfolio variance from the model. First, the objective is to maximize expected equity portfolio return subject to a variety of linear equality and inequality constraints on portfolio structure. The second possibility is to assume that the risk function is given by the absolute deviation of the rate of return, instead of the standard deviation (or variance) employed by Markowitz. The mean-absolute deviation was first proposed by Ref. 21 as an alternative to the classical mean-variance model. It has been demonstrated in Ref. 21 that this model can generate an optimal portfolio much faster than the mean-variance model since it can be reduced to a linear programming problem. Also, it is shown in Ref. 21 that mean-variance and mean-absolute deviation model usually generate similar portfolios. Further, it is proved in Ref. 32 that those portfolios on the mean-absolute deviation efficient frontier correspond to efficient portfolios in the sense of the second degree stochastic dominance, regardless of the distribution of the rate of return, which is not valid for the portfolios on the mean-variance efficient frontier.[22]

3. Methodology

Mathematically, the basic Markowitz model can be formulated as follows:

$$\min \sum_{i=1}^{n} \sum_{j=1}^{n} C_{ij} x_i x_j$$

subject to:

$$\sum_{i=1}^{n} r_i x_i \geq \bar{R}_i$$

$$\sum_{i=1}^{n} x_i = 1$$

$$x_i \geq 0,$$

where r_i defines the expected rate of return of asset i, $\bar{R}_i$ is the minimum level of return for the portfolio, C_{ij} is the covariance between asset i and asset j and x_i is the fraction of the portfolio value invested in asset i. The objective function minimizes the variance/covariance term, which in turn minimizes the risk of the portfolio.

The first constraint that we have in this model specifies the minimum level of return expected from the portfolio and the second constraint, called the budget constraint, requires 100% of the budget to be invested in the portfolio. The nonnegativity constraints express that no short sales are allowed. By specifying a level of expected return on the portfolio, the above quadratic model computes the corresponding minimum risk of the portfolio.

The methodology used in order to find the optimal portfolio is the following: we filter the historical data, selecting the best sample, which was implemented in the mean-variance model. Taking into consideration the level of risk aversion, the optimization process was performed and in the end we obtained the Markowitz efficient frontier.

4. Characteristics of Bucharest Stock Exchange

Before we discuss the results of our analysis, we should point out some characteristics of the Bucharest Stock Exchange.

Figure 1 presents the evolution of the listed companies at the Bucharest Stock Exchange. Currently, there are 118 member firms and 65 listed companies of which 19 are listed in the first tier.

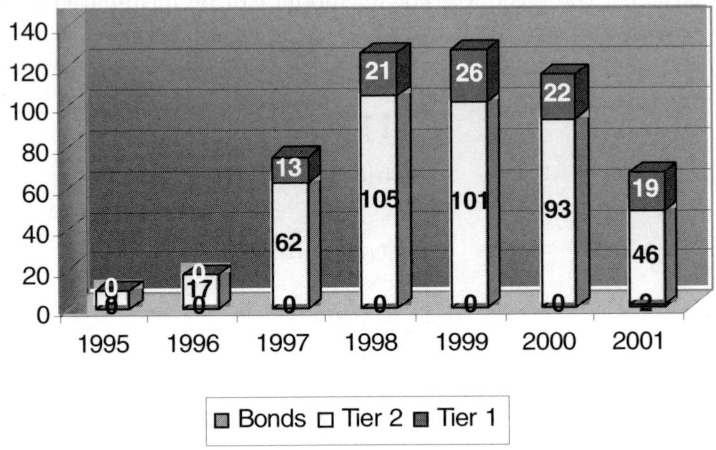

Fig. 1. The evolution of listed issuers at Bucharest Stock Exchange

The Romanian economy continued, in 2001, the improving trend of the previous year and this has been reflected in the performances of the macro-economic indices. Regarding inflation, the official estimate was at 30% (year-on-year) at the end of 2001, as compared to 42% at the end of 2000. However, the foreign direct investments decreased by 5% compared to 2000.

As we can see in Fig. 2 and 3, in 2001, the Bucharest Stock Exchange recorded both a quantitative and a qualitative progress. The total value traded in 2001 increased by 105.8% in nominal terms and by 51% in real terms.

The Bucharest Stock Exchange capitalization rose last year by 23.5% representing 3.56% of the GDP for the year 2001.

Compared to the previous years, in 2001, the Bucharest Stock Exchange had a positive evolution and it is worth mentioning that, for the first time, the market capitalization increased in real terms over the level recorded in 1997 (the top performing year of the Bucharest Stock Exchange).

In the Fig. 4 and 5, we can see the poor performance of the BSE compared to the other three emerging European stock exchanges.

The poor performance of the Bucharest Stock Exchange is mainly because of the very bad economical situation, which in turn was partly due

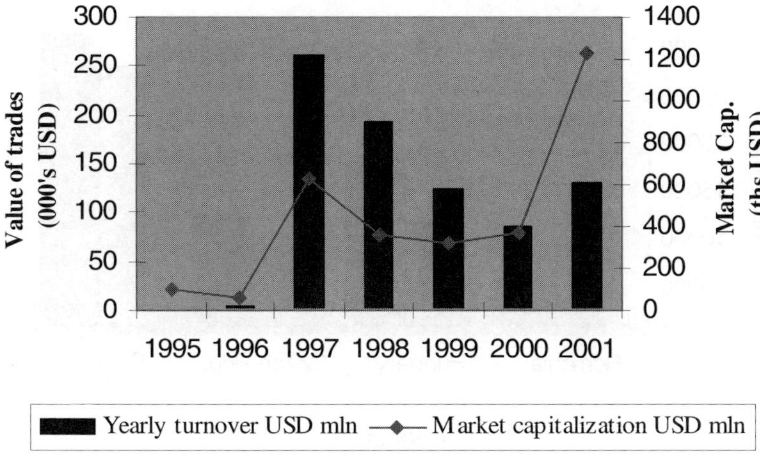

Fig. 2. Evolution of market capitalization and trading value

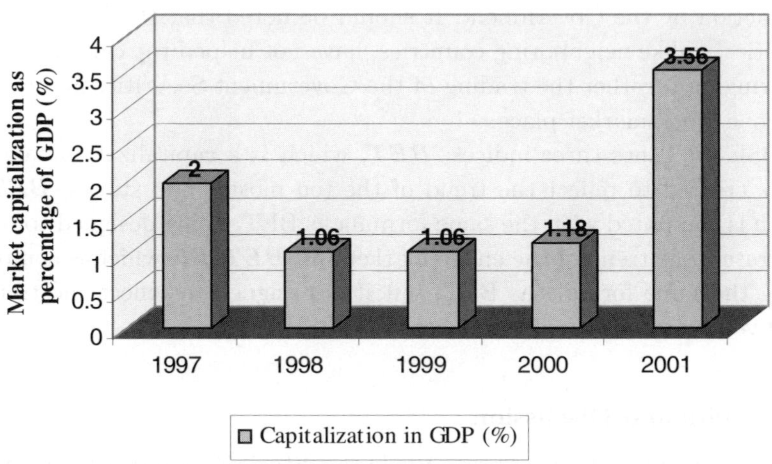

Fig. 3. Market capitalization as percentage of GDP

to uncontrollable or external events such as the blocking of the Danube during NATO action in Kosovo and the drought of 2000. The major cause,

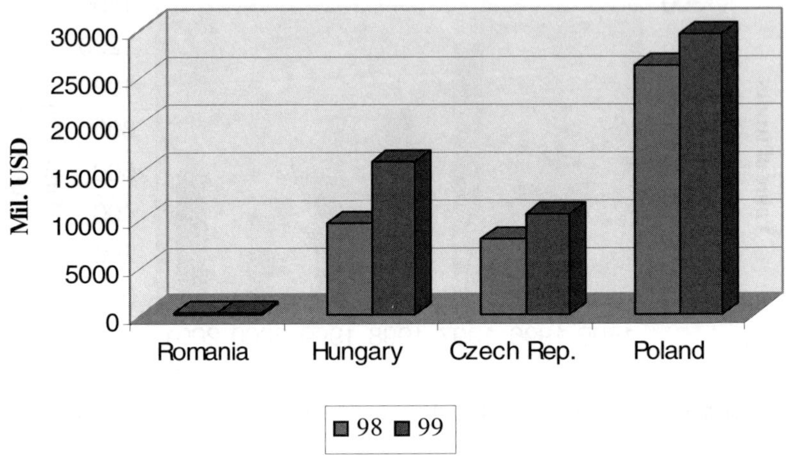

Fig. 4. Market Capitalization in Eastern European Countries

though, of the poor economic environment was the lack of clear direction and action by the Government. It should be noted that the Romanian authorities, unlike neighboring countries, have not helped the capital markets by bringing together the trading of the Government Securities and equities into one stock market place.

BSE publishes three indices: ***BET,*** which is a capitalization weighted index created to reflect the trend of the ten most liquid stocks, ***BET-C,*** which is computed with the same formula as BET, being designed to reflect accurately the trend of the entire market and ***BET-FI,*** which is computed using the same formula as BET, but it is designed to reflect the trend of the five Investment Companies listed on BSE.

5. Results and Discussion

As was previously discussed, the Bucharest Stock Exchange is an emerging market. We should point out that the main focus of the analysis is on finding an appropriate optimization methodology for constructing optimum stock portfolios and not on showing how one can make profits by investing in an emerging stock market.

The purpose of our study is:

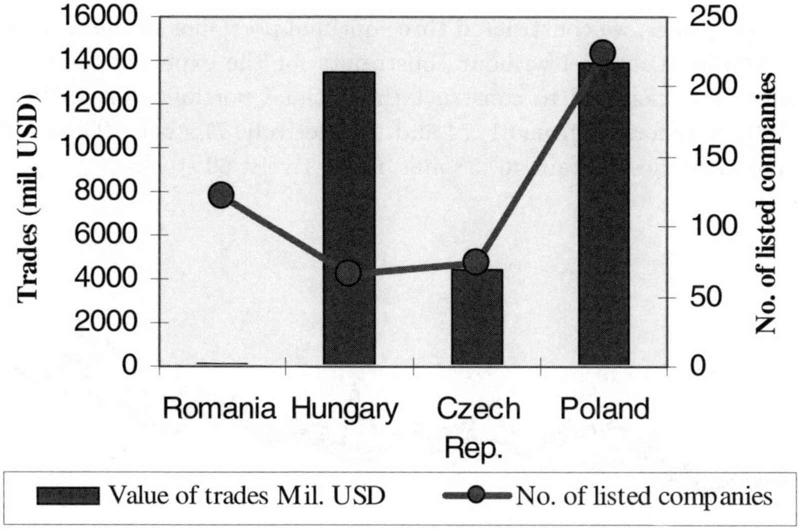

Fig. 5. Trade values and listed companies in Eastern European Countries in 1999

- to construct three optimal portfolios :
  - one for 1999 in a sample of 61 stocks;
  - one for 2000 in a sample of 74 stocks;
  - one for 2001 in a sample of 77 stocks;
- using the basic Markowitz model without constraints;
- to compare the constructed optimal portfolios with the market indices;
- to construct a set of efficient portfolios for each year using the same Markowitz model, with additional constraints for the expected return;
- to test the future performance of the constructed portfolios and to compare them with the market indices;

The stocks used in our analysis were frequently traded on the Bucharest Stock Exchange over the period 1999-2001.The methodology adopted is similar to the one developed by Markowitz (1952) as described earlier. Daily return intervals were selected between the years 1999-2001 for conducting our analysis. The data was initially filtered and pre-processed in order to remove spurious records from the time series.

5.1. *Minimum Risk Portfolios*

As stated earlier, we constructed three optimal portfolios in the case of the basic Markowitz model without constraints for the expected return. The samples of stocks used to construct the optimal portfolios for 1999, 2000 and 2001 were formed from 61, 74 and, respectively, 77 stocks. The resulted optimal portfolios contain 46, 58 and, respectively, 69 stocks.

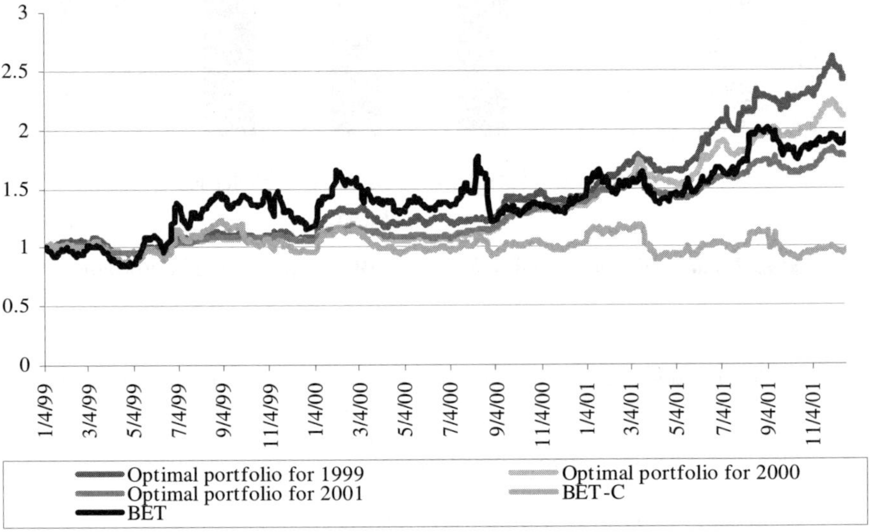

Fig. 6. Comparison between the evolution of the constructed optimal portfolios for 1999, 2000 and 2001 until the end of 2001 and the two market indices

Table 1. The daily expected return and standard deviation for the optimal portfolios constructed in 1999, 2000 and 2001 and for the two market indices at the end of 2001

2001	**Optimal portfolio for 2001**	**Optimal portfolio for 1999**	**Optimal portfolio for 2000**	**BET_C**	**BET**
Expected Daily Return	0.07716%	0.12319%	0.10116%	0.00619%	0.10521%
Standard Deviation	0.52246%	0.96480%	0.62804%	1.36372%	1.84349%

Note: expected returns and standard deviations estimated using daily time-series data for the period 1999- 2001.

Fig. 6 and Table 1 show the behavior of the 2001 constructed optimal portfolio and, also, the expected evolution of the 1999 and 2000 optimal portfolios until the end of 2001. We, also, illustrate the comparison between these constructed portfolios and the two market indices. The optimal portfolio constructed in 2001 has the lowest level of risk compared to the market indices. The index BET_C has a high level of risk and a very small expected return compared to the three Markowitz portfolios. The results indicate that the BET index increased by **97.046%** and the BET_C index decreased by **2.517%** until the end of 2001. The optimal portfolio constructed in 2001 increased until the end of the year by **76.444%.** So, we can reach the conclusion that the efficient portfolio outperforms the BET_C index, but it performs less than the BET index. The Markowitz optimal portfolio for 2001 track very closely the BET_C index until the end of 2000 and the BET index in 2001.

The 1999 and 2000 constructed portfolios have a positive evolution until the end of 2001. As we can see in the table, the investor's choice depends on his level of risk aversion. If he invests in the 1999 or 2000 portfolio he will obtain a higher level of risk and a higher level of return than he would by investing in the 2001 optimal portfolio. The results indicate that the optimal portfolio for 1999 increased by **144.862%** and the optimal portfolio for 2000 increased by **110.703%** until the end of 2001. It is also interesting to note the return that an investor would get if he has chosen to invest in the 1999 optimal portfolio until the end of 2000 and in the 2000 optimal portfolio from the beginning of 2001 until the end of 2001. The results show that the investor would have achieved a return of **140.284%**, if he had decided to follow this investment strategy until the end of 2001, involving portfolio restructuring on an annual basis.

5.2. *Portfolios with Minimum Expected Return Constraints*

In the previous part, we analyzed the results of the Markowitz mean-variance optimization method without considering any constraint on the minimum expected return. In this section, some additional results are presented involving the construction of optimal risk portfolios when constraints are imposed on their expected return.

We performed five simulations for each year considering different values for the minimum expected return. Each simulation produces an efficient portfolio determined by the different values for the minimum expected return. We chose these values taking into consideration the average daily return of the samples of stocks used in the analysis.

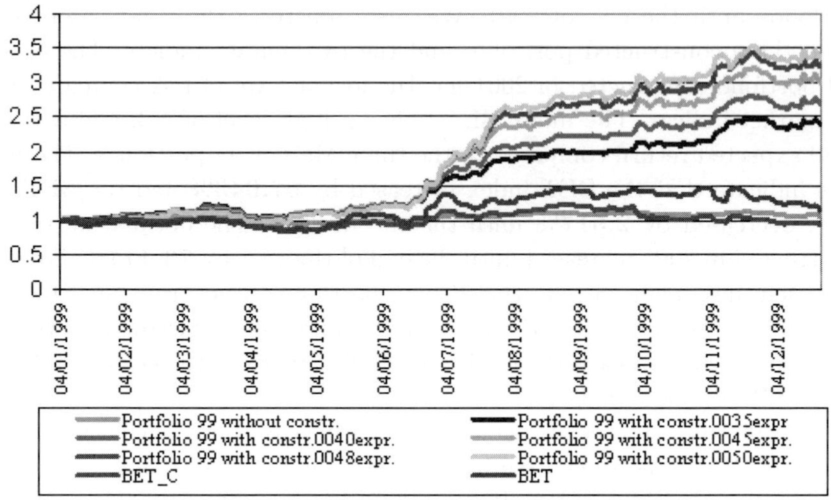

Fig. 7. The efficient portfolios for 1999

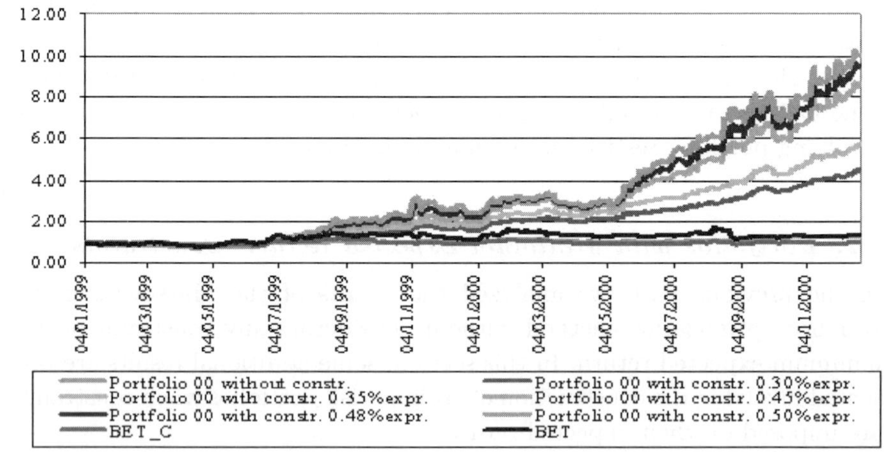

Fig. 8. The efficient portfolios for 2000

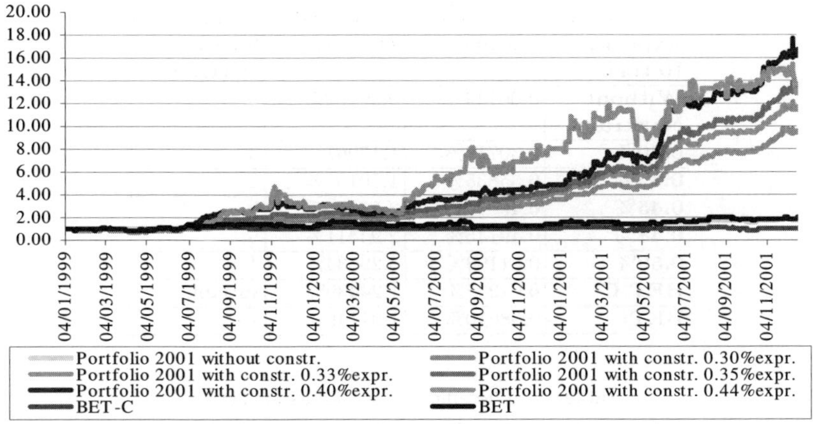

Fig. 9. The efficient portfolios for 2001

Table 2. The efficient portfolios for 1999

Minimum Expected Return	Expected Return	Standard Deviation	Number of Stocks in Portfolio
Without constraint	0.03530%	0.64155%	46
0.35%	0.35417%	1.32685%	22
0.40%	0.40476%	1.54706%	15
0.45%	0.45536%	1.82730%	12
0.48%	0.48571%	2.02788%	9
0.50%	0.50595%	2.18586%	8
BET_C	-0.01206%	1.28407%	variable
BET	0.07238%	1.81009%	10

We can observe in Tables 2, 3 and 4 that the constructed portfolios without any constraint have the lowest level of risk, but also the lowest level of expected return compared to the other portfolios. As the level of expected return increases, the level of risk increases and the number of stocks that form the portfolios decreases. An investor's choice depends on the degree of risk aversion. If he/she is very risk prone and chooses to invest in the most risky portfolio, than he/she gets the maximum return. If he/she is risk averse and invests in the less risky portfolio, than he/she gets the lowest return.

We should compare the performances of the constructed efficient port-

Table 3. The efficient portfolios for 2000

Minimum Expected Return	Expected Return	Standard Deviation	Number of Stocks in Portfolio
Without constraint	0.06144%	0.50482%	58
0.30%	0.30716%	1.17961%	30
0.35%	0.35835%	1.50107%	20
0.45%	0.46074%	2.52154%	8
0.48%	0.49145%	2.94411%	5
0.50%	0.51193%	3.29634%	4
BET_C	0.01295%	1.24439%	variable
BET	0.08610%	1.97146%	10

Table 4. The efficient portfolios for 2001

Minimum Expected Return	Expected Return	Standard Deviation	Number of Stocks in Portfolio
Without constraint	0.07716%	0.52246%	69
0.30%	0.30880%	1.19158%	24
0.33%	0.33968%	1.48324%	18
0.35%	0.36027%	1.74330%	15
0.40%	0.41173%	2.68718%	5
0.44%	0.45291%	4.60097%	3
BET_C	0.00619%	1.36372%	variable
BET	0.10521%	1.84349%	10

folios with the market portfolio. As mentioned in the previous part, the performance of market portfolio is measured using two indices: BET and BET_C.

Tables 2, 3 and 4 illustrate that for the indices BET_C and BET, for almost the same level of risk as the efficient portfolios, a very small, even negative, expected return could be obtained compared to the expected returns of Markowitz efficient portfolios. The results indicate that the efficient portfolios constructed for 1999 with 0.35%, 0.40%, 0.45%, 0.48% and 0.50% minimum expected return increased until the end of 1999 by **137.360 %**, **167.029%**, **199.594%**, **220.326%** and **234.355%** respectively. The efficient portfolios constructed in 2000 with 0.30%, 0.35%, 0.45%, 0.48% and 0.50% minimum expected return increased until the end of 2000 by **348.549%**, **468.334%**, **773.142%**, **869.060%** and **920.962%** respectively and the efficient portfolios for 2001 with 0.30%, 0.33%, 0.35%, 0.40% and 0.44% minimum expected return increased until the end of 2001 by **847.902%**, **1059.627%**, **1212.954%**, **1575.565%** and **1267.933%** respectively. So, comparing the performance of Markowitz portfolios with the two indices, we can conclude that the efficient Markowitz portfolios outperform the two indices BET_C and BET.

As in the other case, the behavior of our 1999 and 2000 efficient portfolios at the end of 2001 should also be noted. As we can see in figures 9 and 10, the investor's selection depends on his/her level of risk aversion: for a higher level of risk, a higher level of return.

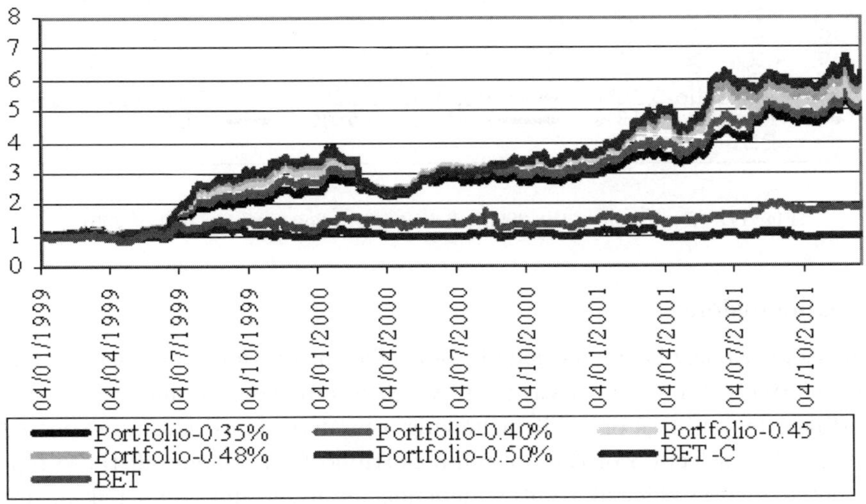

Fig. 10. Evolution of the 1999 efficient portfolios until the end of 2001

The 1999 and 2000 efficient portfolios have a positive evolution until the end of 2001. Also we can observe that the portfolios fluctuated substantially, experiencing sudden ups and downs, especially at the beginning of 2001 when all the constructed efficient portfolios suffered a sudden increase. The results show that the constructed efficient portfolios for 1999 with 0.35%, 0.40%, 0.45%, 0.48% and 0.50% minimum expected return increased until the end of 2001 by **410.716%, 431.052%, 455.868%, 490.103%** and **515.411%** respectively and the constructed efficient portfolios for 2000 with 0.30%, 0.35%, 0.45%, 0.48% and 0.50% minimum expected return increased until the end of 2001 by **581.757%, 732.241%, 1003.124%, 1114.751%** and **1193.911%** respectively.

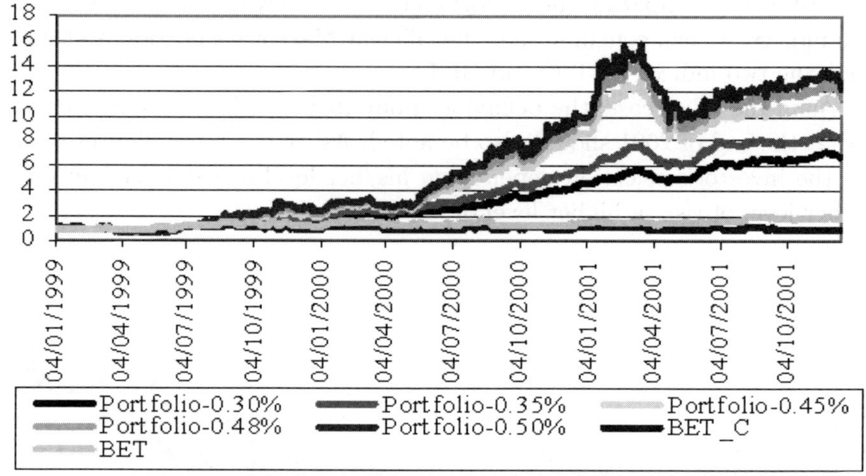

Fig. 11. Evolution of the 2000 efficient portfolios until the end of 2001

6. Conclusions

As has previously been mentioned, some efficient portfolios of assets for the period 1999-2001 were constructed, which minimizes risk that can be undertaken by an investor, while achieving a predetermined level of expected return. The selection of the optimum portfolio from among all those represented by the efficient frontier depends upon the investor's risk-return preference.

The explanations of the results obtained can be interpreted from two points of view.

First, the estimators obtained from the Markowitz equations are biased and subject to variation, particularly when less than ten years of monthly data are used in the estimation process. In our analysis we used three years of daily data and it is impossible to increase the sample size to the requirements suggested here because of data inexistence and changing market conditions over such extended periods.

Second, the Bucharest Stock Exchange is an emerging and very unstable market. One of the characteristics of the emerging markets is that high return comes with high risk and many factors can trigger troubles. Currency risk represents one risk factor for emerging market investors. If the value of the dollar declines against the currency of the emerging market country, the return will be lower. Emerging market investments entail high political and liquidity risks and as such may be more volatile. Therefore, they are considered appropriate only for long-term investors with an investment time frame of 10 or more years. Along with high potential returns, the emerging markets offer diversification benefits.

If we consider other European stock markets such as Budapest, Prague and Warsaw, Government Securities are listed and traded on the Stock Exchange. This allows investors easy access to both debt and equity investment and encourages effective portfolio management. Also, significant holdings of the major state utilities have been privatized partly through listing and sales on the Exchanges. This has not occurred to any significant extent in Romania.

In spite of its theoretical interest, the basic mean-variance model is often too simplistic to represent the complexity of real-world portfolio selection problems in an adequate fashion, as: trading limitations, size of the portfolio etc. In order to enrich the model, we need to introduce more realistic constraints, such as: allow short sales, consider transaction costs, perform sensitivity analysis and portfolio hedging or to use other optimization models.

References

1. F. Black and M. Scholes. The Pricing of Options and Corporate Liabilities. *Journal of Political Economy*; Vol. **81**; No. 3 (1973).
2. P.L. Bernstein and A. Damodaran *Investment Management.* John Wiley & Sons, Inc (1998).
3. J. R. Birge and F. Louveaux. *Introduction to Stochastic Programming.* Springer-Verlag, New York (1997).
4. G. C. Chow. *Dynamic Economics: Optimization by the Lagrange Method.* Oxford University Press, Oxford (1997).
5. A. K. Dixit. *Optimization in economic Theory* 2nd edition; Oxford University Press, New York (1990).
6. A. K. Dixit and R. Pindyck. *Investment under Uncertainty* Princeton University Press, Princeton, NJ (1994).
7. R. Dobbins, S. F. Witt, and J. Fielding. *Portfolio Theory and Investment Management – An introduction to modern portfolio theory.* Blackwell Business; 1-5, 27-30, 110-117 (1994;).
8. J.E. Dennis and R.,B. Schnabel. *Numerical methods for unconstrained optimization and nonlinear equations.* Prentice Hall (1983).
9. E.J. Elton and M.J. Gruber. *Portfolio theory, 25 years after.* North-Holland, Amsterdam (1979).
10. E.J. Elton and M.J. Gruber. *Modern portfolio theory and investment analysis.* John Wiley & Sons (1981).

11. E.J. Elton and M.J. Gruber. *Modern portfolio theory and investment analysis.* John Wiley & Sons (1987).
12. J. L. Farrel, Jr. and W. J. Reinhart, *Portfolio Management – Theory and Application.* Irwin/McGraw-Hill 10-15, 18-31, 35-38, 70-77, 91-100, 100-115 (1997).
13. S. N. Levine, ed. *Financial Analyst's Handbook Methods, Theory and Portfolio Management* Dow Jones-Irwin, Inc. (1975).
14. R. R. Grauer and N. H. Hakansson. Higher return, lower risk: historical returns on long-run, actively managed portfolios of stocks, bonds and bills, 1936-78. *Financial Analysts Journal* **38** (1982).
15. J. D. Jobson. Estimation for Markowitz efficient portfolios. *Journal of the American Statistical Asssociation* **75**, no. 371, 544-54 (1981).
16. J. D. Jobson. Potential performance and tests of portfolio efficiency. *Journal of Financial Economics* **10**, 433-66 (1982).
17. J. D. Jobson, B. Korkie, and V. Ratti. Improved Estimation for Markowitz Portfolio Using James-Stein Type Estimators. *Proceedings of the American Statistical Association, Business and Economic Statistics Section* (1979).
18. J. D. Jobson and B. Korkie. Estimation for Markowitz Efficient Portfolios. *Journal of the American Statistical Association,* **75**; 544-554 (1980); .
19. P. Kall and S. Wallace. *Stochastic programming*, John Wiley & Sons, Chichester, UK (1994).
20. P. D. Kaplan. Asset Allocation Models using the Markowitz Approach. *Ibbotson Associates*, 1-2 (1998).
21. H. Konno and H. Yamazaki. Mean absolute deviation portfolio optimization model and its application to Tokyo stock market. textitManage. Sci, **37**,519-531, (1991).
22. H. Konno and A. Wijayanayake. Portfolio optimization problem under concave transaction costs and minimal transaction unit constraints. *Mathematical Programming*, **89**, 233-250 (2001).
23. H. Levy and H. Markowitz. Approximating expected utility by a function of the mean and variance. *American Economic Review* **69** (3), 308-17 (1979).
24. S. L. Lummer, M. W. Riepe, and L. B. Siegel. Taming Your Optimizer: A Guide Through the Pitfalls of Mean-Variance Optimization. Ibbotson Associates, *Global Asset Allocation: Techniques for Optimizing Portfolio Management*; J. Lederman and R.A. Klein, John Wiley & Sons, 3-8 (1994).
25. H. Markowitz. Portfolio selection. *Journal of finance*; Vol. **7**; No. 1 ,77-91 (1952).
26. Markowitz, Harry, M.; *Portfolio Selection: Efficient Diversification of Investments;* New York: John Wiley and Sons Inc.; 1959.
27. H. Markowitz. *Mean-variance analysis in portfolio choice and capital markets*, Blackwell, Cambridge, MA (1987).
28. R. O. Michaud. *Efficient Asset Management – A practical Guide to Stock Portfolio Optimization and Asset Allocation*, Harvard Business School Press, Boston, MA 11-13, 23-32 (1998).
29. R. O. Michaud. The Markowitz optimization enigma: Is optimized optimal?. *Financial Analysts Journal* **45**; no. 1; 31-42 (1989).

30. J. M. Mulvey, D. P. Rosenbaum, and B. Shetty. Strategic financial risk management and operations research. *European Journal of Operation Research* **97**, 1-16 (1997).
31. J. M. Mulvey. Introduction to financial optimization: Mathematical Programming Special Issue. *Mathematical Programming* **89**, 205-216 (2001).
32. W. Ogryczak and A. Ruszczynski. From stochastic dominance to mean-risk model. *European Journal of Operation Research* 116 (1999).
33. A. Ohuchi and I. Kaji. Algorithms for optimal allocation problems having quadratic objective function. *Journal of the Operations Research Society of Japan* **23** 64-79 (1980).
34. J. S. Pang. A new and efficient algorithm for a class of portfolio selection problems. *Operations Research* **28**, 754-67 (1980).
35. A. F. Perold and W. F. Sharpe. Dynamic strategies for asset allocation. *Financial Analysts Journal* **44**, 16-27 (1998).
36. G. C. Philippatos. *Financial Management – Theory and Techniques.* Holden – Day, Inc. (1973).
37. S. Ross. *Return, Risk and Arbitrage* In : *Risk and Return in Finance*, I. Friend and J. Bicksler (eds.), Ballinger, Cambridge, (1976).
38. W.F. Sharpe. Capital asset Prices: A theory of market Equilibrium under conditions of Risk. *Journal of Finance*, Vol. **19**, 425-442 (1964).
39. http://www.valicenti.com/invest.htm
40. http://www.optirisk-systems.com/docs/whitepaper/whitepaper.pdf
41. http://www.dlbabson.com/dlbindex/0,5624,10-25-70,00.html
42. http://europa.eu.int/comm/economy_finance/publications/phare_ace/ace_quarterly/issue9/research_results/94-0658-R_en.html
43. http://home.gwu.edu/~scsmith/koreaipo.pdf
44. http://www.bvb.ro

CHAPTER 15

A NEW ALGORITHM FOR THE TRIANGULATION OF INPUT–OUTPUT TABLES IN ECONOMICS

B. H. Chiarini

Center for Applied Optimization
Dept. of Industrial and Systems Engineering
University of Florida
303 Weil Hall, P.O. Box 116595
Gainesville, FL 32611, USA
E-mail: chiarini@ufl.edu

W. Chaovalitwongse

Center for Applied Optimization
Dept. of Industrial and Systems Engineering
University of Florida
303 Weil Hall, P.O. Box 116595
Gainesville, FL 32611, USA
E-mail: arty@ufl.edu

P. M. Pardalos

Center for Applied Optimization
Dept. of Industrial and Systems Engineering
University of Florida
303 Weil Hall, P.O. Box 116595
Gainesville, FL 32611, USA
E-mail: pardalos@ufl.edu

Developed by Leontief in the 1930s, input-output models have become an indispensable tool for economists and policy-makers. They provide a framework upon which researchers can systematically analyze the interrelations among the sectors of an economy. In an input-output model, a table is constructed where each entry represents the flow of goods between each pair of sectors. Special features of the structure of this matrix are revealed by a technique called *triangulation*. It is shown to be equivalent to the linear ordering problem (LOP), which is an $\mathcal{NP}$-hard combinatorial optimization problem. Due to its complexity, it is essential in practice to search for quick approximate (heuristic) algorithms for the linear ordering problem. In addition to the triangulation of input-output tables, the LOP has a wide range of applications in practice. In this chapter, we develop a new heuristic procedure to find high quality solutions for the LOP. The proposed algorithm is based on a Greedy Randomized Adaptive Search Procedure (GRASP), which is one of the most effective heuristics for solving combinatorial and global optimization problems to date. We propose an improved solution technique by using a new local search scheme and integrating a path-relinking procedure for intensification. We tested our implementation on the set of 49 real-world instances of input-output tables in LOLIB.[22] In addition, we tested a set of 30 large randomly-generated instances due to Mitchell.[18] Most of the LOLIB instances were solved to optimality within 0.87 seconds on average. The average gap for the Mitchell instances was 0.0173% with an average running time of 21.98 seconds. The results prove the efficiency and high-quality of the algorithm.

Keywords: Triangulation of input-output matrices, linear ordering problem, greedy randomized adaptive search procedure (GRASP), Path-Relinking.

1. Introduction

The impact of changes of an economic variable can only be analyzed by understanding the complex series of transactions taking place among the sectors of an economy. First introduced by Leontief in the early 1930s, input-output models have become an indispensable tool for economists and policy-makers in their analysis, providing a systematic description of such interrelations among the sectors.[16]

An input-output model begins by dividing the economy of a country (or region) into a specified number of sectors. Then a table is constructed, where the entries are the total transactions between every pair of sectors. The total output(input) of a sector can be obtained by summing the entries on the corresponding row(column). The resulting table thus summarizes the interdependence among the economic sectors.

Structural properties of the input-output tables may not be apparent. A particular choice in the order of the sectors used in constructing the table might conceal an otherwise evident structure. These features are revealed by a process called *triangulation*, whose objective is to find a *hierarchy* of the sectors such that those that are predominantly producers will appear first, while those that are mainly consumers will appear last.

The economic significance is that it shows how the effects of changes in final demand propagates through the sectors. Note, however, that in the use of a hierarchic ordering there is an underlying assumption that no flow exists from lower to upper sectors. In fact, every economy exhibits a certain *circularity* in the flow of goods—e.g., the metallurgy industry supplies the vehicle industry with raw metal products, while the metallurgy sector needs vehicles as part of the cost of doing business. Obviously, the flow between any two sectors is hardly symmetric.

The degree to which an economic structure "agrees" with a hierarchy of the sectors is called *linearity*. In a perfectly linear economy, the flow of goods "cascades" from the upper sectors to the lower sectors of the hierarchic ordering. If we arrange the rows and columns of the input-output matrix according to the hierarchy, such situation would be reflected by a matrix that has an upper triangular structure, that is, all entries below the diagonal would be zero. On the other hand, if there is flow of goods back to the upper sectors, then there would be positive values on the entries below the diagonal. This leads to the definition of a quantitative measure of linearity. Let n denote the number of sectors and $E = \{e_{ij}\}$ be the n-square matrix representing the input-output table. Assume that the rows and columns have been arranged according to the hierarchy. Then, the linearity of an economy is given by

$$\lambda = \frac{\sum_{i=1}^{n-1} \sum_{j=i+1}^{n} e_{ij}}{\sum_{i=1}^{n} \sum_{j=1, i \neq j}^{n} e_{ij}}. \tag{1}$$

That is, linearity is the ratio of the sum of the elements above the diagonal to the sum of all elements (except the diagonal). It follows that $\lambda = 1$ for a perfectly linear economy. Researchers have observed that large and highly developed economies tend to have a low degree of linearity—i.e., there is a high circulation in the flow of goods among sectors—whereas underdeveloped economies tend to exhibit a clearer hierarchy. Typical linearity values are 70% for a highly developed economy, and 90% for an underdeveloped economy.[16]

The introduction of input-output tables and other quantitative economic models originated a profusion of research in many areas. For instance, Dantzig's early work in the Air Force before his development of the simplex algorithm for linear programming, consisted of investigating methods to efficiently solve large systems of linear equations, motivated by the applications to input-output tables.[7,16] However, the triangulation problem described next has not been given much attention.

Triangulation. We have assumed so far the knowledge of a hierarchy of sectors. The *triangulation of an input-output table* is the process of finding such hierarchy among all possible orderings. It is clear from the discussion above that such ordering is the one that most closely resembles an upper triangular matrix, and thus has the maximum value of λ. Note that every ordering is a permutation of the sectors and it is applied to both the rows and columns of the input-output matrix. Additionally, the denominator of Eq. (1) is constant for all permutations. Therefore, we can state the *triangulation problem* as that of finding a permutation of the rows and columns such that the sum of the elements above the diagonal is maximum. Clearly, this is equivalent to a combinatorial optimization problem known as the linear ordering problem (LOP).

Finding such permutation is not an easy task. In fact, the linear ordering problem is an $\mathcal{NP}$-hard problem and as such we can only aspire to obtain approximate solutions. Furthermore, the extent to which input-output methods are useful depends on the efficiency of the computations. Limited by computational power, practitioners often have to recur to aggregation, with the consequent loss of information and accuracy, to find optimal solutions within an acceptable time. Therefore, it is essential in practice to search for quick approximate (heuristic) algorithms.

In this chapter we propose a new algorithm based on a greedy randomized adaptive search procedure (GRASP) to efficiently approximate the optimal solution of LOP. The algorithm is integrated with a path-relinking procedure and a new local search scheme.

The remainder of this chapter is organized as follows. In Section 2 we introduce the LOP, give other applications, and discuss some previous work. In Section 3 we give a detailed implementation of our algorithm, preceded by an introduction describing the GRASP and path-relinking framework. The computational experimentation is shown in Section 4. The chapter concludes with Section 5, where some practical issues are discussed.

2. The Linear Ordering Problem

The LOP is an $\mathcal{NP}$-hard combinatorial optimization problem with a wide range of applications in economics, archaeology, and scheduling. It has, however, drawn little attention compared to other closely related problems such as the quadratic assignment problem and the travelling salesman problem.

The LOP can be stated as follows. Consider a set N of n objects and a permutation $\pi : N \to N$. Each permutation $\pi = (\pi(1), \pi(2), \ldots, \pi(n))$

corresponds one-to-one to a linear ordering of the objects. Let e_{ij}, $i, j = 1, 2, \ldots, n$, be the cost of having i before j in the ordering, and E be the n-square matrix of costs. Then the *linear ordering problem* is to find a permutation π that maximizes the total cost

$$Z(\pi) = \sum_{i=1}^{n-1} \sum_{j=i+1}^{n} e_{\pi(i)\pi(j)}. \tag{2}$$

Clearly, Eq. (2) is the sum of the elements above the diagonal of a matrix A whose elements a_{ij} are those resulting from a permutation π of the rows and columns of the matrix E—i.e., $A = XEX^T$, where X is the permutation matrix associated with the permutation π.[21] In the context of its application in economics, we can restate Eq. (1) as

$$\lambda = \frac{1}{K} \max_{\pi \in \Pi} \{Z(\pi)\} \tag{3}$$

where Π is the set of all permutations and K is a positive constant representing the sum of all the entries in the matrix.

The LOP can also be interpreted as a problem in graphs. Let $G(N, A)$ be a complete directed graph with node set N and arc set $A = \{(i, j) : i, j \in N \wedge i \neq j\}$. Let e_{ij} be the weight of arc (i, j). A spanning acyclic tournament in G induces a unique linear ordering of the node set N.[13] A tournament is defined as a directed graph in which each pair of nodes is connected by exactly one arc, which is clearly necessary since either i is before j or j is before i.

The complexity of the maximum LOP can be easily proven to be $\mathcal{NP}$-hard by noticing that it is equivalent to the minimum weighted feedback arc set problem on G, which is known to be $\mathcal{NP}$-hard.[10]

The LOP has an interesting symmetry property. If a permutation $\pi = (\pi(1), \pi(2), \ldots, \pi(n))$ is an optimal solution to the maximization version, then the reverse permutation $\overline{\pi} = (\pi(n), \pi(n-1), \ldots, \pi(1))$ is an optimal solution to the minimization version. In fact, the LOP accepts a trivial $\frac{1}{2}$-approximation algorithm.[13] Let π be an arbitrary permutation and $\overline{\pi}$ its reverse. It is easy to see that $Z(\pi) + Z(\overline{\pi})$ is a constant. Choose $\hat{\pi}$ such that $Z(\hat{\pi}) = \max\{Z(\pi), Z(\overline{\pi})\}$, then we get

$$\frac{Z(\pi^*) - Z(\hat{\pi})}{Z(\pi^*)} \leq \frac{1}{2}$$

where π^* is an optimal permutation and $Z(\pi^*) > 0$. No other approximation algorithm exists.[13] It follows that any permutation is optimal in the unweighted version of the LOP.

2.1. *Applications*

Following we discuss a few applications of the LOP besides that in economics, which are of particular relevance to the present volume (see Reinelt[21] for an extensive survey).

Consider the problem of having a group of people rank n objects. Each individual in the group is asked to express their preference with respect to every possible pair. If we let e_{ij} be the number of people who preferred i to j, the solution to the corresponding LOP is the ranking that most likely reflects the preferences of the group.

A similar application can be found in the context of sports. For example, consider a tournament of n teams in which every team plays against every other team. Let e_{ij} be the score of the match between i and j if i wins, and 0 otherwise. The ranking obtained by the LOP is considered to be the one that most closely reflects the "true" performance of the teams. Still, it has not gained support for its implementation, probably because the outcome of a particular match is not closely related to the result in the ranking.

In archaeology, the LOP is used to determine the "most probable" chronological ordering of a set of artifacts recovered from different sites. Samples belonging to various time periods are given a value based on their distance to the surface. The objective is to aggregate the data and determine an ordering of the artifacts.

Finally, it is easy to see that the LOP can be used to determine the optimal sequence of jobs in a single server, where the cost of each job depends upon its position with respect to the entire schedule.

2.2. *Problem Formulations*

As with most combinatorial optimization problems, the linear ordering problem has many alternative formulations. The LOP can be expressed as an integer programming problem as follows. Let $G(N, A)$ be the complete directed graph associated with the LOP as shown in the previous section. Define

$$x_{ij} = \begin{cases} 1 & \text{if } (i,j) \in A' \\ 0 & \text{otherwise} \end{cases} \tag{4}$$

where $A' \subset A$ is the arc set of the spanning acyclic tournament on G. Then the problem (2) becomes

$$\max \sum_{(i,j)\in A} e_{ij}x_{ij} \tag{5}$$

$$\text{s.t.} \quad x_{ij} + x_{ji} = 1 \qquad \forall i,j \in N, i < j \tag{6}$$

$$x_{ij} + x_{jk} + x_{ki} \leq 2 \qquad \forall i,j,k \in N, i \neq j, i \neq k, j \neq k \tag{7}$$

$$x_{ij} \in \{0,1\}.$$

The constraints (6) define the tournament polytope. It can be proven that the 3-dicycle inequalities (7) are sufficient to prevent any cycles.[21] Together they define the linear ordering polytope. There are $2\binom{n}{2}$ variables and $\binom{n}{2} + 2\binom{n}{3}$ constraints in this formulation.

The tournament constraints (6) motivate the use of a single variable to represent the two possible ways in which every pair of nodes can be connected. Let us substitute $x_{ji} = 1 - x_{ij}$, for every $i, j \in N, j > i$, then an equivalent integer programming formulation is

$$\max \sum_{\{(i,j)\in A: i<j\}} e'_{ij}x_{ij} \tag{8}$$

$$\text{s.t.} \quad x_{ij} + x_{jk} - x_{ik} \leq 1 \qquad \forall i,j,k \in N, i < j < k \tag{9}$$

$$x_{ij} + x_{jk} - x_{ik} \geq 0 \qquad \forall i,j,k \in N, i < j < k \tag{10}$$

$$x_{ij} \in \{0,1\}$$

where $e'_{ij} = e_{ij} - e_{ji}$. This formulation has $\binom{n}{2}$ variables and $2\binom{n}{3}$ constraints.

Finally, the linear ordering problem can be formulated as a quadratic assignment problem (QAP). The distance matrix of the QAP is the matrix of weights E and the flow matrix $F = \{f_{ij}\}$ is contructed as follows, $f_{ij} = -1$ if $i < j$ and $f_{ij} = 0$, otherwise.[2]

2.3. *Previous Work*

2.3.1. *Exact Methods*

A common approach in the LOP literature is the use of cutting plane algorithms.[12,21] The goal is to obtain an approximate description of the convex hull of the solution set by introducing valid inequalities that are violated by current fractional solutions, which are added to the set of inequalities of the current linear programming problem. Reinelt[21] introduced facets induced by subgraphs. Bolotashvili *et al.*[1] extended Reinelt results introducing a generalized method to generate new facets. A complete characterization has only been obtained for very small problems ($n = 7$) (see Christof and Reinelt[6]). In fact, we know that unless $\mathcal{P} = \mathcal{NP}$ there exist an exponential number of such facets. However, research in this area has resulted in valid inequalities that improve the performance of branch-and-cut algorithms.[21]

The first authors to consider an interior point algorithm for the LOP were Mitchell and Borchers.[18,20] The solution given by a interior point algorithm is used as a starting point for a simplex cutting plane algorithm.[20]

2.3.2. *Heuristic Methods*

Most hard problems in combinatorial optimization require the use of heuristics to obtain approximate solutions due to their inherent intractability. In this case, we are interested in finding solutions that are close "enough" to the optimal value at a low computational cost.

Heuristics, as opposed to approximation algorithms, do not give a guaranteed quality of the obtained solutions. Nevertheless, the flexibility we have in developing heuristics allows us to exploit the special structure of the problem, tailoring the existing methods, and resulting in very well performing algorithms. The quality of a heuristic, however, must be validated by extensive testing.

Heuristic methods such as *GRASP, tabu search, simulated annealing, genetic search, and evolution strategies* have shown to be able to efficiently find high quality solutions to many combinatorial and global optimization problems by thoroughly exploring the solutions space. A recent survey on multi-start heuristic algorithms for global optimization is given by Martí.[17]

One of the earliest heuristics for the LOP was proposed by Chenery and Watanabe[5] in the context of the triangulation of input-output tables. Given a sector i, the ratio of total input to the total output

$$u_i = \frac{\sum_{k=1}^{n} c_{ik}}{\sum_{k=1}^{n} c_{ki}} \tag{11}$$

is used to arrange the sectors in the order of decreasing u_i. The use of Eq. (11) gives a fairly good ordering considering its simplicity.

Based on the symmetry property mentioned in Section 2, Chanas and Kobylański[4] developed a heuristic that performs a sequence of optimal insertions and reversals.

Laguna et al.[15] developed an algorithm based on tabu search. They analyzed several intensification and diversification techniques, and compared

```
Procedure GRASP(RCLSize, StoppingCondition)
    BestSolutionFound = ∅;
    while StoppingCondition not satisfied do
        x = ConstructGreedyRandomizedSolution(RCLSize);
        LocalSearch(x);
        UpdateSolution(BestSolutionFound,x);
    end;
    return BestSolutionFound;
end;
```

Fig. 1. A Generic GRASP pseudo-code

their algorithm with that of Chanas and Kobylański.

Campos et al.[3] used a scatter search approach. A correction term based on the frequency by which an object i appears in a particular position in the ordering is added to Eq. (11) to reflect previous solutions.

GRASP, which is an iterative restart approach, has proven to be one of the most effective heuristics to date. In this chapter, we developed a GRASP-based algorithm for the LOP, offering a significant improvement on the computational time and quality of solution compared to previous heuristics. In the next section, we discuss the basic principles for implementing a new local search scheme and Path-Relinking in GRASP framework.

3. A GRASP with Path-Relinking Algorithm

3.1. *Introduction to GRASP and Path-Relinking*

Since its inception by Feo and Resende in the late 1980s, GRASP has been successfully used in many applications. In 1995, the authors formally introduced GRASP as a framework for the development of new heuristics.[8] For a recent extensive annotated bibliography of GRASP applications, see Festa et al.[9]

Each iteration of GRASP consists of two phases: a *construction phase* in which we seek to obtain a feasible solution, and a *local search phase* that attempts to improve the solution. Figure 1 shows the pseudo-code of a generic GRASP algorithm.

During the construction phase, we iteratively build a solution by randomly selecting objects from a restricted candidate list (RCL). At each step, we form the RCL choosing those objects with the highest measure of attractiveness, we select a random object from the RCL, and adapt the greedy function to reflect the addition to the solution. Figure 4 shows the

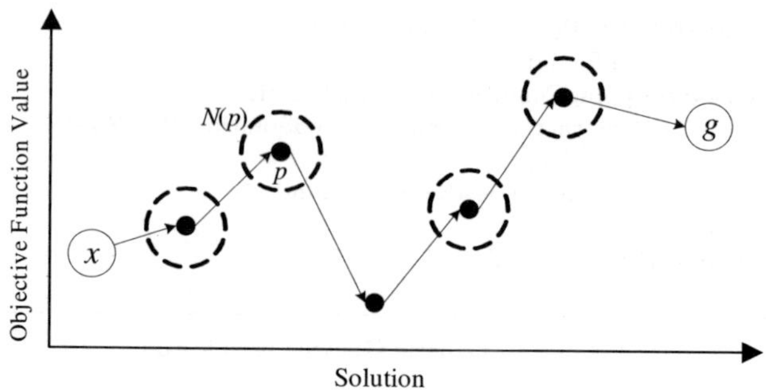

Fig. 2. A visualization of a path-relinking procedure. Given the initial solution s and the guiding solution g, we iteratively explore the trajectory linking these solutions, checking for any improvements on the way.

construction phase for our implementation.

The size of the list is typically restricted in one of two ways: by quality, when we choose the elements based on a threshold on the greedy function, or by cardinality. In the literature, they are often referred to as α and β, respectively. The size of the RCL controls the degree of greediness and randomness of the construction phase. A null RCL—i.e., of size 1—results in a purely greedy solution whereas a RCL size equal to the size of the problem yields a purely random solution.

After a solution is constructed, we attempt to improve it by performing a local search in an appropriately defined neighborhood. Given a solution x, we explore the neighborhood $N(x)$ aspiring to find a local (global) optimal solution. Although larger neighborhoods increase the probability of finding the global optimum, local search algorithms are often computationally expensive and thus careful consideration must be given to the election of the neighborhood. It is in this part of GRASP where the particular properties of the problem in question can be exploited to develop schemes that can provide an intensification of the search, while not compromising its running time. Finally, the best solution is updated if necessary with the newly found solution. The procedure is repeated until a stopping condition is met—for example, number of iterations, running time, etc.

Path-relinking was introduced by Glover and Laguna[11] as a method to integrate intensification and diversification to tabu search.

It generates new solutions by exploring routes that connect high-quality solutions by starting from one of these solutions, so-called *initiating solution,* and generating a path in the neighborhood space that leads toward the other solutions, a so-called *guiding solution.* This is completed by selecting moves that introduce attributes contained in the guiding solutions.

Path-relinking is a directly-focused instance of a strategy that seeks to fit in features of high quality solutions. On the other hand, instead of using an incentive that supports the inclusion of such attributes, the path relinking approach subordinates all other considerations to the goal of choosing moves that initiate the attributes of the guiding solutions, in order to generate a *good attribute composition* in the current solution. The composition at each step is determined by choosing the best move, using customary choice criteria, from the restricted set of moves that incorporate a maximum number of the attributes of guiding solutions.

Laguna and Martí[14] were the first to combine GRASP with a path-relinking procedure, essentially adding memory to a procedure that would otherwise be a multi-start algorithm.

3.2. *Proposed Algorithm*

In this section we propose a new heuristic that integrates GRASP with path-relinking for the linear ordering problem. Fig. 3 shows the pseudo-code for the implementation.

The measure of attractiveness for each object i consists of the difference between its row and column sums, given by

$$d_i = \sum_{j=1}^{n}(e_{ij} - e_{ji}), \quad i = 1, 2, \ldots, n. \tag{12}$$

In the context of the input-output tables, Eq. (12) represents the net flow of a sector. In earlier experimentations with a GRASP algorithm, we compared the use of Eq. (12) with the ratios as defined in Eq. (11). Although we have not observed significant differences in the quality of the solutions, adapting the greedy function when ratios are used is computationally more expensive. The use of Eq. (11) demands $\mathcal{O}(n^2)$ time to update the greedy function whereas Eq. (12) requires $\mathcal{O}(n)$ time.

Linear ordering problems usually have many alternative solutions—optimal and suboptimal—with the same objective function value. Therefore, it may occur at some point in the algorithm that the elite list becomes mostly populated by alternative solutions. Furthermore, it is increasingly difficult to enter a path-relinking as the best solution found approaches the optimal. We attempt to avoid such situations by expanding the size of the elite list and forcing a path-relinking procedure after a certain number of non-improving iterations. If an improvement is still not obtained and the elite list size reaches its limit, the elite list is deleted and a new one is constructed.

We now proceed to discuss in detail the different components of the algorithm.

3.2.1. *Construction Phase*

We initiate the procedure by creating the restricted candidate list (RCL) (see Fig. 4). The parameter β of GRASP determines the cardinality limit on the RCL—i.e., the number of elements in RCL. Larger values of β achieve greater diversity but at the cost of constructing many lower-quality solutions. The best value of β is usually determined by extensive testing.

After selecting an element at random from the RCL, we proceed to insert it in the partial solution. A conventional GRASP implementation would simply append the recently selected object s to the end of the partial solution. Instead, we added a procedure named 'Insert' (line 5 on Fig. 4) that seeks to insert the object in an optimal position. More precisely, let $T_k = (t_1, t_2, \ldots, t_k)$, $k = 1, 2, \ldots, n$, denote the current (partial) solution obtained after k steps. The Insert operation intercalates the most recently

```
Procedure GRASP(β, ρ, MaxIteration)
    BestSolutionFound = ∅;
    EliteList = ∅;
    nNonImprovingIt = 0;
    for k = 1,2,...,MaxIteration
        x = ConstructGreedyRandomizedSolution(β);
        LocalSearch(x);
        if x is better than worse solution in EliteList
            DoPathRelinking(EliteList,x);
            nNonImprovingIt = 0;
        else if nNonImprovingIt > γ
            ExpandEliteList(EliteList,ρ);
            DoPathRelinking(EliteList,x);
            nNonImprovingIt = 0;
        else nNonImprovingIt = nNonImprovingIt + 1;
        UpdateSolution(BestSolutionFound,x);
    end;
    return BestSolutionFound;
end;
```

Fig. 3. The GRASP pseudo-code

```
procedure ConstructGreedyRandomizedSolution(β)
    Solution = ∅, RCL = ∅;
    while |Solution| < N
        MakeRCL(RCL,β);
        s = SelectElementAtRandom(RCL);
        Insert(Solution,s);
        AdaptGreedyFunction(s);
    end;
end;
```

Fig. 4. The GRASP construction phase

selected object s in T_k in the position r that maximizes

$$\Delta(i, T_k) = \sum_{j=1}^{r-1} e_{t_j i} + \sum_{j=r}^{k} e_{i t_j} \tag{13}$$

breaking ties arbitrarily. First introduced by Chanas and Kobylański[4] as part of their heuristic, it can be considered as a very efficient local search procedure in a relatively small neighborhood. In fact, it can be implemented in $\mathcal{O}(k)$.

A step of the contruction phase finalizes with the task of adapting the greedy function. The row and column corresponding to the object s are removed from the matrix, and the attractiveness (12) of the objects is updated. We set $d_s = -M$, where M is a large positive value, and re-sort the top $n - k$ objects that have yet to be selected. The procedure continues until a solution is constructed. The overall complexity of the construction phase is $\mathcal{O}(n^2)$.

3.2.2. *Local Search*

We used a 2-exchange neighborhood for our local search. Given a solution π, its 2-exchange neighborhood $N(\pi)$ consists of all the solutions obtained by permuting the position of two objects in the ordering—i.e., if $\pi = (3, 1, 2)$, then $N(\pi) = \{(1, 3, 2), (2, 1, 3), (3, 2, 1)\}$. Clearly, for a problem of size n, $|N(\pi)| = \binom{n}{2}$. Consider a solution π and two objects $\pi(i)$ and $\pi(j)$ located in positions i and j respectively. For simplicity assume that $i < j$. The change in the objective function for an exchange of objects $\pi(i)$ and $\pi(j)$ is

$$\Delta Z(\pi, i, j) = -e'_{\pi(i)\pi(j)} - \sum_{k=i+1}^{j-1} (e'_{\pi(i)\pi(k)} + e'_{\pi(k)\pi(j)}) \tag{14}$$

where $e'_{ij} = e_{ij} - e_{ji}$. At completion, the local search would have exchanged the pair of objects that maximizes Eq. (14). The procedure of exploring the neighborhood and performing the exchange can be implemented in $\mathcal{O}(n^2)$.

3.2.3. *Path Relinking*

The solution provided by the local search procedure is used as the initial solution for the path-relinking. We randomly select a solution from the elite list as the guiding solution, determining the trajectory to be followed by the procedure. Figure 5 shows the pseudo-code for the path relinking procedure. The parameter ρ determines the size of the elite list as a fraction of the problem size.

```
procedure DoPathRelinking(EliteList,x)
    TempSolution = ∅;
    g = SelectSolutionAtRandom(EliteList);
    while x ≠ g do
        TempSolution = MakeNextMove(x, g);
        LocalSearch(TempSolution);
        if TempSolution is better than x or g then
            EliteList = EliteList ∪ TempSolution;
    end;
    AdjustEliteList(EliteList,ρ);
end;
```

Fig. 5. The GRASP Path-Relinking procedure

With the trajectory defined by the two end solutions, we proceed to perform a series of moves that will transform the initial solution into the guiding solution. In each iteration the algorithm performs a single move, thus creating a sequence of intermediate solutions (see Fig. 2). To add intensification to the process, we search the 2-exchange neighborhood of the intermediate solutions. The solutions obtained in this manner are added to the elite list if they are better than either the initial or the guiding solutions.

It should be noted that the search on the 2-exchange neighborhood may yield a previously examined solution in the path. However, this is not of concern in our implementation since we do not use the local minima during the procedure.

The moving process terminates when the algorithm reaches the guiding solution. At this point, the elite list size could have grown considerably due to the added solutions. The procedure 'AdjustEliteList' will discard the worst solutions, keeping the best ρn. The list is kept sorted at all times and therefore no sorting is needed. The complexity of the path-relinking is $\mathcal{O}(n^3)$.

4. Computational Results

In this section we discuss the computational results we obtained when applying our algorithm to two sets of problems:

(1) *LOLIB.* These are real-world instances of linear ordering problems that are publicly available on the internet.[22] They consist of 49 input-output tables for some European countries, with sizes up to 60 objects.

(2) *Mitchell.* This is a set of 30 random-generated instances by Mitchell,[18] with sizes ranging from 100 to 250 objects. Three different percentages of zero entries were used: 0, 10, and 20%, denoted by the last digit on the problem name. The generator as well as the instances are available at the author's web site.[19]

The optimal values for all instances are known. The generated instances are similar to those from LOLIB except for the numerical range of the entries—considerably larger for the latter. Despite attempts to replicate the characteristics of real-world instances such as those found in LOLIB, Mitchell's test set is significantly harder to solve.

All previous work on the linear ordering problem that included computational results predates the Mitchell instances, hence featuring only the LOLIB problems.

The algorithm was written in C++ and executed on a Pentium 4, 2.7 GHz, with 512 MB of memory. Empirically, we determined $\beta = 0.25$ and $\rho = 0.35$ as the best values for the parameters—e.g., for a problem of size n, the size of the RCL and the elite list are at most $0.25n$ and $0.35n$ respectively. The algorithm was executed five times for each problem instance, with a limit of 5000 GRASP iterations in its running time. We report the running time, number of iterations, and the gap between the best solution and the optimal solution. All times are reported in *seconds* and the gaps as *percentages.*

Figures 6 and 7 show the evolution of the gap as a function of the running time and the number of iterations for the LOLIB and Mitchell instances, respectively. Note that the units on the ordinates are percentage points. Tables 1 and 2 show the elapsed running time and gap values after 200 and 5000 iterations. The results reported are the averages of 5 runs for

Table 1. Results for the LOLIB Instances after 200 and 5000 iterations.

Instance	*Size*	*200 Iterations*		*5000 Iterations*	
		Gap(%)	*Time (s)*	*Gap(%)*	*Time (s)*
be75eec	50	0.0199	0.0282	0.0042	0.4062
be75np	50	0.0057	0.0188	0.0018	1.0406
be75oi	50	0.0459	0.0218	0.0125	0.7062
be75tot	50	0.0056	0.0218	0.0002	0.5156
stabu1	60	0.0124	0.0406	0.0024	1.3030
stabu2	60	0.0059	0.0468	0.0051	2.1624
stabu3	60	0.0329	0.0562	0.0090	1.9280
t59b11xx	44	0.0012	0.0156	0.0012	0.8250
t59d11xx	44	0.0082	0.0156	0.0000	0.5720
t59f11xx	44	0.0202	0.0124	0.0000	0.6562
t59i11xx	44	0.0000	0.0156	0.0000	0.7686
t59n11xx	44	0.0515	0.0186	0.0079	0.6592
t65b11xx	44	0.0223	0.0124	0.0133	0.7624
t65d11xx	44	0.0003	0.0156	0.0000	0.5876
t65f11xx	44	0.0090	0.0156	0.0006	0.5842
t65i11xx	44	0.0137	0.0156	0.0007	0.5938
t65l11xx	44	0.0000	0.0156	0.0000	0.8032
t65n11xx	44	0.0124	0.0156	0.0000	0.4624
t65w11xx	44	0.0014	0.0156	0.0000	0.5594
t69r11xx	44	0.0122	0.0156	0.0002	0.6844
t70b11xx	44	0.0080	0.0124	0.0080	0.7280
t70d11xn	44	0.0288	0.0750	0.0073	2.7406
t70d11xx	44	0.0232	0.0156	0.0120	0.4938
t70f11xx	44	0.0030	0.0126	0.0012	0.4282
t70i11xx	44	0.0022	0.0126	0.0009	0.4970
t70k11xx	44	0.0000	0.0156	0.0000	0.7218
t70l11xx	44	0.0007	0.0250	0.0000	0.6124
t70n11xx	44	0.0260	0.0126	0.0019	0.3970
t70u11xx	44	0.0023	0.0126	0.0000	0.7938
t70w11xx	44	0.0012	0.0156	0.0000	0.3874
t70x11xx	44	0.0000	0.0126	0.0000	0.6906
t74d11xx	44	0.0125	0.0188	0.0042	0.5032
t75d11xx	44	0.0028	0.0218	0.0020	0.7936
t75e11xx	44	0.0048	0.0126	0.0000	0.4688
t75i11xx	44	0.0017	0.0126	0.0000	0.6220
t75k11xx	44	0.0000	0.0156	0.0000	0.7156
t75n11xx	44	0.0710	0.0156	0.0000	0.5062
t75u11xx	44	0.0007	0.0156	0.0003	0.6530
tiw56n54	56	0.0170	0.0374	0.0044	0.8436
tiw56n58	56	0.0127	0.0312	0.0000	1.0124
tiw56n62	56	0.0187	0.0282	0.0007	1.0970
tiw56n66	56	0.0179	0.0342	0.0043	0.5218
tiw56n67	56	0.0063	0.0344	0.0006	2.0312
tiw56n72	56	0.0078	0.0376	0.0013	1.0470
tiw56r54	56	0.0157	0.0312	0.0020	0.7062
tiw56r58	56	0.0046	0.0438	0.0020	1.2720
tiw56r66	56	0.0353	0.0406	0.0061	0.8688
tiw56r67	56	0.0086	0.0312	0.0007	1.9250
tiw56r72	56	0.0006	0.0312	0.0006	1.8970

Table 2. Results for the Mitchell Instances after 200 and 5000 iterations.

Instance	*200 Iterations*		*5000 Iterations*	
	Gap(%)	*Time (s)*	*Gap(%)*	*Time (s)*
r100a2	0.0268	0.2374	0.0047	2.5374
r100b2	0.0462	0.2594	0.0207	2.6032
r100c2	0.0416	0.3094	0.0302	2.6376
r100d2	0.0389	0.2220	0.0343	2.4906
r100e2	0.0270	0.2030	0.0165	6.9312
r150a0	0.0137	0.6594	0.0070	8.0562
r150a1	0.0263	0.7126	0.0226	8.3282
r150b0	0.0230	0.9720	0.0102	8.1970
r150b1	0.0224	1.3220	0.0194	9.1000
r150c0	0.0207	1.2844	0.0167	27.9032
r150c1	0.0369	0.9282	0.0191	9.0032
r150d0	0.0146	1.1970	0.0122	8.5656
r150d1	0.0212	1.0874	0.0191	34.6218
r150e0	0.0148	0.8126	0.0056	33.1438
r150e1	0.0388	1.4842	0.0289	9.5812
r200a0	0.0181	2.6062	0.0080	21.2470
r200a1	0.0409	2.8906	0.0305	22.4062
r200b0	0.0196	2.0156	0.0151	21.4780
r200b1	0.0333	3.6532	0.0271	23.6062
r200c0	0.0096	3.1938	0.0084	20.4376
r200c1	0.0244	2.7312	0.0219	21.5844
r200d0	0.0159	2.7720	0.0089	32.5562
r200d1	0.0345	1.9844	0.0262	22.5906
r200e0	0.0212	2.9656	0.0158	22.8470
r200e1	0.0297	2.3780	0.0244	35.9750
r250a0	0.0233	5.4750	0.0156	47.1220
r250b0	0.0189	3.8188	0.0102	43.5406
r250c0	0.0159	10.1312	0.0132	52.2126
r250d0	0.0184	8.3750	0.0118	49.1438
r250e0	0.0241	5.4844	0.0146	48.9250

Table 3. Summary of the results obtained for the LOLIB and Mitchell test sets after 200 and 5000 iterations.

Problem Set	*Measure*	*200 Iterations*		*5000 Iterations*	
		Gap(%)	*Time (s)*	*Gap(%)*	*Time (s)*
LOLIB	*Average*	0.0125	0.0235	0.0024	0.8685
	Std. Dev.	0.0146	0.0131	0.0035	0.5188
	Maximum	0.0710	0.0750	0.0133	2.7406
Mitchell	*Average*	0.0254	2.4055	0.0173	21.9791
	Std. Dev.	0.0096	2.3638	0.0081	15.6266
	Maximum	0.0462	10.1312	0.0343	52.2126

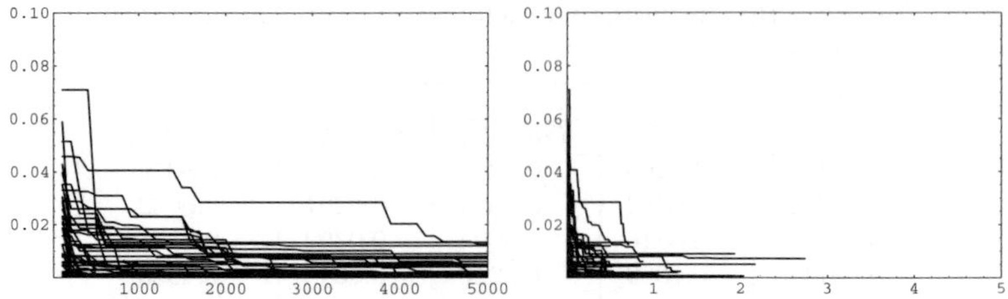

Fig. 6. LOLIB Instances: the gap from the optimal solution as a percentage is shown as a function of the number of iterations (left) and time (right). Time is in seconds.

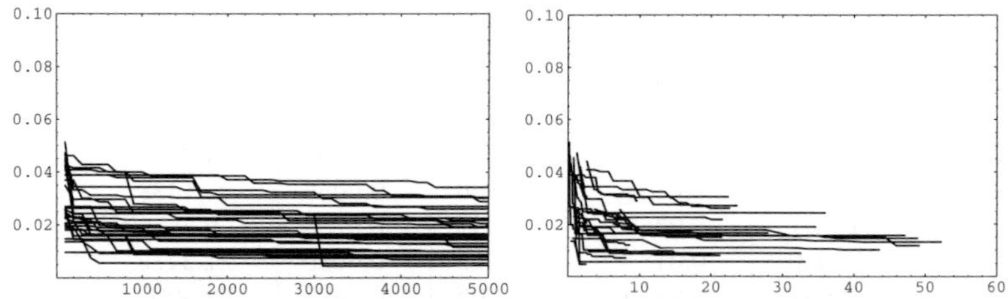

Fig. 7. Mitchell Instances: the gap from the optimal solution as a percentage is shown as a function of the number of iterations (left) and time (right). Time is in seconds.

each problem instance. The averages of the values presented in these tables are shown in Table 3.

The algorithm found optimal solutions for 47 out of 49 LOLIB instances, 17 of which were consistently solved to optimality. The average gap for the remaining LOLIB instances was 0.0127%. The average running time for these real-world instances was 0.87 seconds. Although none of the Mitchell instances were solved to optimality, the average gap after 5000 iterations was 0.0173% with an average running time of 21.98 seconds.

5. Concluding Remarks

In this chapter we implemented a new heuristic algorithm for the triangulation of input-output tables in economics. The algorithm is based on a greedy randomized adaptive search procedure (GRASP) with the addition of a path-relinking phase to further intensify the search.

The algorithm was tested on two sets of problems, exhibiting a remarkably robust performance as shown in Table 3. For all instances we obtained optimality gaps of less than 0.05% within 200 iterations and times ranging from 0.02 to 2.40 seconds on the average. We found optimal solutions for most of the LOLIB instances. No optimal solutions were obtained for the Mitchell instances, however, the average gap at termination for this set was 0.0173%. The results confirm the benefit of embedding GRASP with a path-relinking procedure.

Researchers in economics often use simulations in which many triangulation problems need to be solved in limited time. The efficiency and high-quality performance of our algorithm makes it a superior candidate for such application. Furthermore, since the algorithm is based upon the modelling of the triangulation problem as a linear ordering problem (LOP), it can be used for any application that accepts an LOP formulation such as those mentioned in Section 2.1.

References

1. G. Bolotashvili, M. Kovalev, and E. Girlich. New facets of the linear ordering polytope. *SIAM Journal on Discrete Mathematics*, 12(3):326–336, 1999.
2. R.E. Burkard, E. Çela, P.M. Pardalos, and L.S. Pitsoulis. The quadratic assignment problem. In P.M. Pardalos and D.-Z. Du, editors, *Handbook of Combinatorial Optimization*, pages 241–338. Kluwer Academic Publishers, 1998.
3. Vicente Campos, Fred Glover, Manuel Laguna, and Rafael Martí. An experimental evaluation of a scatter search for the linear ordering problem. *Journal of Global Optimization*, 21(4):397–414, December 2001.
4. Stefan Chanas and Przemysław Kobylański. A new heuristic algorithm solving the linear ordering problem. *Computational Optimization and Applications*, 6:191–205, 1996.
5. Hollis B. Chenery and Tsunehiko Watanabe. International comparisons of the structure of production. *Econometrica*, 26(4):487–521, October 1958.
6. Thomas Christof and Gerhard Reinelt. Low-dimensional linear ordering polytopes. 1997.
7. George B. Dantzig. Linear programming. *Operations Research*, 50(1):42–47, January 2002.
8. Thomas A. Feo and Mauricio G.C. Resende. Greedy randomized adaptive search procedures. *Journal of Global Optimization*, 2:1–27, 1995.
9. Paola Festa, Mauricio G.C. Resende, and Geraldo Veiga. Annotated bibliography of GRASP. http://www.research.att.com/~mgcr/grasp/annotated.
10. Michael R. Garey and David S. Johnson. *Computers and Intractability: A Guide to the Theory of NP-Completeness*. W.H. Freeman and Co., New York, USA, 1979.
11. Fred Glover and Manuel Laguna. *Tabu Search*. Kluwer Academic Publishers, Boston, 1997.
12. Martin Grötschel, Michael Jünger, and Gerhard Reinelt. A cutting plane algorithm for the linear ordering problem. *Operations Research*, 2(6):1195–1220, 1984.

13. Michael Jünger. *Polyhedral Combinatorics and the Acyclic Subdigraph Problem.* Number 7 in Research and Exposition in Mathematics. Heldermann Verlag, Berlin, 1985.
14. Manuel Laguna and Rafael Martí. GRASP and path relinking for 2-layer straight line crossing minimization. *INFORMS Journal on Computing*, 11(1):44–52, 1998.
15. Manuel Laguna, Rafael Martí, and Vicente Campos. Intensification and diversification with elite tabu search solutions for the linear ordering problem. *Computers & Operations Research*, 26:1217–1230, 1999.
16. Wassily Leontief. *Input-Output Economics.* Oxford University Press, New York, USA, 1986.
17. Rafael Martí. Multi-start methods. In Fred Glover and Gary A. Kochenberger, editors, *Handbook of Metaheuristics*, International Series in Operations Research & Management Sciences, chapter 12, pages 355–368. Kluwer Academic Publishers, 2003.
18. John E. Mitchell. Computational experience with an interior point cutting plane algorithm. Technical report, Mathematical Sciences, Rensellaer Polytechnic Intitute, Troy, NY 12180-3590, USA., 1997.
19. John E. Mitchell. Generating linear ordering problems, Dec. 2002. http://www.rpi.edu/~mitchj/generators/linord.
20. John E. Mitchell and Brian Borchers. Solving linear ordering problems with a combined interior point/simplex cutting plane algorithm. In H. Frenk et al., editor, *High Performance Optimization*, chapter 14, pages 345–366. Kluwer Academic Publishers, Dordrecht, The Netherlands, 2000.
21. Gerhard Reinelt. *The linear ordering problem: algorithms and applications.* Number 8 in Research and Exposition in Mathematics. Heldermann Verlag, Berlin, 1985.
22. Gerhard Reinelt. Linear ordering library (LOLIB), Dec. 2002. http://www.iwr.uni-heildelberg.de/iwr/comopt/soft/LOLIB/LOLIB.html.

CHAPTER 16

MINING ENCRYPTED DATA

B. Boutsinas

Department of Business Administration,
University of Patras Artificial Intelligence Research Center (UPAIRC),
University of Patras, GR-26500 Rio, Patras, Greece,
Tel: +30-610-997845, Fax: +30-610-996327
E-mail: vutsinas@bma.upatras.gr

G. C. Meletiou

T.E.I. of Epirus, P.O. Box 110, GR-47100 Arta and UPAIRC, Greece,
Tel: +30-6810-26825, Fax: +30-6810-75839
E-mail: gmelet@teiep.gr

M. N. Vrahatis

Department of Mathematics, UPAIRC,
University of Patras, GR-26500 Patras, Greece,
Tel: +30-610-997374, Fax: +30-610-992965
E-mail: vrahatis@math.upatras.gr

Business and scientific organizations, nowadays, own databases containing confidential information that needs to be analyzed, through data mining techniques, in order to support their planning activities. The need for privacy is imposed due to, either legal restrictions (for medical and socio-economic databases), or the unwillingness of business organizations to share their data which are considered as a valuable asset. Despite the diffusion of data mining techniques, the key problem of confidentiality has not been considered until very recently. In this chapter we address the issue of mining encrypted data, in order to both protect confidential information and to allow knowledge discovery. More specifically, we consider a scenario where a company having private databases negotiates a deal with a consultant. The company wishes the consultant to analyze its databases through data mining techniques. Yet the company is not willing to disclose any confidential information.

Keywords: Data mining, cryptography, security, privacy.

1. Introduction

Nowadays business or scientific organizations collect and analyze data, orders of magnitude greater than ever before, in order to support their planning activities. Consequently, considerable attention has been paid in the development of methods that contribute to knowledge discovery in business or scientific databases, using data mining techniques.[11] The new generation of data mining techniques are now applied to a variety of real life applications ranging from recognizing scenes to stock market analysis. Specifically, mining financial data presents special challenges.

Usually, data mining rules can be used either to classify data into predefined classes that are described by a set of concepts-attributes (classification), or to partition a set of patterns into disjoint and homogeneous clusters (clustering), or to represent frequent patterns in data in the form of dependencies among concepts-attributes (associations). Data mining algorithms typically are based on systematic search in large hypotheses spaces.

Business or scientific databases contain confidential information. The need for privacy is either due to legal restrictions (for medical and socio-economic databases) or due to the unwillingness of business organizations to expose their data, which are considered a valuable asset.

Despite the diffusion of data mining techniques, the key problem of confidentiality has not been addressed until very recently. In Ref. 7 ways through which data mining techniques can be used in a business setting to provide business competitors with an advantage, are presented. In Ref. 8 a technique to prevent the disclosure of confidential information by releasing only samples of the original data, independently of any specific data mining algorithm, is provided. In Refs. 2, 10 the authors propose to prevent the disclosure of confidential information, when association rules are to be extracted, by artificially decreasing the significance of these rules. In Ref. 13, the authors consider the scenario in which two parties owning private databases wish to run a classification data mining algorithm on the union of their databases, without revealing any confidential information. Similarly, in Ref. 9 the author addresses the issue of privacy preserving in distributed data mining, where organizations may be willing to share data mining association rules, but not the source data.

In this chapter we address the issue of mining encrypted data, in order to both protect confidential information and to allow knowledge discovery. More specifically, we consider a scenario in which a company having private databases negotiates a deal with a consultant. The company wishes the consultant to analyze its databases using data mining techniques, yet it is unwilling to disclose any confidential information. We address the problem by encrypting the private data and allowing the consultant to apply the data mining techniques on the encrypted data. Then, the consultant provides the company with the extracted data mining rules. Finally, the company decrypts those rules before use. Note that the decrypted rules should be the same as the rules that would be extracted from the original data. We investigate the applicability of certain cryptography techniques to the above scenario when either classification, clustering or association rules are to be extracted.

It has to be mentioned that, in our approach, the sender coincides with the receiver. In other words the encoder and the decoder are the same. (The protocol: Alice composes the plaintext; Alice encrypts it; Alice sends the ciphertext for data mining processing; Alice receives the encrypted answer; Alice decrypts and recovers the answer). Thus, we intend to propose "an Alice to Alice" cryptography for privacy preserving data mining.

To this end, we encrypt the data by applying a proper cryptosystem. We focus on the main question, which is the choice of the appropriate cryptosystem by taking under consideration that each attribute value, no matter where it is located in the original table of data (plaintext) has to be encrypted with the same sequence of symbols in the ciphertext.

A direct solution is for the set of all possible attribute values to play the role of the alphabet and the cryptosystem to be "mono-alphabetic". Notice that, in real life applications, the plaintext has no content as well as the cardinality of the "alphabet" is very large which is a serious problem for the cryptoanalyst (enemy). Of course, in the case of a text in the English language, a mono-alphabetic cryptosystem is based just on a permutation of the 26 symbols and thus it cannot resist to a frequency analysis attack. However, the case of encrypting business data is complicated. Attribute values can represent customer characteristics, product codes, etc. The result of a frequency analysis attack is unpredictable. A mono-alphabetic cryptosystem may resist, but this is not guaranteed. Accepting this risk seems to be a bad premise to build on. On the other hand, the idea to develop a cryptosystem which is based on a permutation of k symbols ($1000 \leqslant k \leqslant 20000$) is primitive.

In this chapter, we propose an alternative methodology based on distributed data mining techniques. In particular, to each attribute from the original table of data (the plaintext) a set of possible encryptions is assigned $E_i = (c_{i1}, c_{i2}, \ldots, c_{ik})$. Of course $i \neq j$ implies $E_i \cap E_j = \emptyset$. Each time a_i changes to one of the $c_{i1}, c_{i2}, \ldots, c_{ik}$.

In the rest of the chapter we first present the proposed methodology and then we briefly discuss some preliminary issues concerning distributed data mining techniques. The chapter closes with some concluding remarks.

2. The Proposed Methodology

The main acting agents of the protocol are, "Alice" that represents a business or scientific organization and "Bob" that represents a data mining consultant who handles the data mining process. Alice owns a database with fields and field values that correspond to attributes and attribute values referred by the data mining rules. Attribute and attribute values may describe, for instance, a profit related behavior of the customers of a company that need to be classified/clustered or products sold together in a transaction that need to be examined for existing dependencies. Attribute values, irrespective of what they represent, have to be encrypted. We consider a great number of such attribute values denoted by $g_1, \cdots, g_M$ and we also set $G = \{g_1, \cdots, g_M\}$. Trivially, each g_i can be represented as an integer $i : 1 \leqslant i \leqslant M$ denoting its index. Alternatively, since each g_i has a label like "good customer" or "driver" or "tomato", this label can be transformed to an integer. For instance, a label can be transformed to a string of bits with the help of ASCII code, in turn, each string corresponds to a number (integer). As a result, in both of the above cases each g_i can be represented as a small integer.

The proposed methodology is as follows:

During the first step, Alice selects and preprocesses the appropriate data and organizes it into relational tables. A relational table is supposed to be two dimensional, however it can be represented as one dimensional considering it in a *row major order* or in a *column major order*.

During the second, third and fourth step, encryption takes place. We propose two different encryption techniques which are described in detail in subsections 2.1 and 2.2. Note that, both encryption techniques are based on symmetric keys $r_i, 1 \leqslant i \leqslant s$. The s different keys r_i are repeated periodically for every record of any Q_j. Thus, the m-th record of any Q_j is encrypted using the key r_i, where $i = m(\bmod s) + 1$. It is this characteristic

---

encrypted data mining algorithm {
1. Alice collects data organized into relational tables $Q_1, Q_2, \cdots$
2. Alice obtains a "small" number of symmetric keys (or random numbers) $r_i, 1 \leqslant i \leqslant s$, e.g. $s = 4$
3. Alice obtains either a public key E and a private key D or a secret key X
4. Alice encrypts relational tables $Q_1, Q_2, \cdots$ as follows:
$\{\ Q_j \rightarrow ENCRYPTION_{IorII}x(r_i(Q_j)) = C_j\ \}$
5. Alice sends $C_1, C_2, \cdots$ to Bob. Data mining performed. Bob returns the obtained rules to Alice
6. Alice decrypts the rules.
}

---

that, later, supports the decryption of the extracted rules using distributed data mining algorithms.

At the fifth step, Alice sends the encrypted tables to Bob. Bob applies the proper data mining algorithm to the encrypted tables and a number of data mining rules are extracted. Of course, attribute and attribute values appeared in the rules are encrypted. Then Bob returns these rules to Alice.

During the final step, Alice decrypts the rules. Of course, after decryption, there will be rules concerning the same attributes and attribute values which, however, extracted from different subsets of the initial table. Thus, Alice synthesizes the final set of rules combining the corresponding rules by using distributed data mining algorithms, as it will be described in subsection 2.3.

2.1. *Encryption Technique I – The RSA Cryptosystem*

The first encryption technique is based on the RSA cryptosystem.[16] Two large primes p and q are selected and their product $N = p \cdot q$ is computed. Then e and d, the public and private key, respectively, are selected. These keys have to satisfy the following relation:

$$e \cdot d \equiv 1 \bmod [(p-1) \cdot (q-1)].$$

By r_i, $1 \leqslant i \leqslant s$ we denote the s random numbers which are chosen for the encryption.

Assume that k is the least integer such that $g \leqslant 2^k$ for all $g \in G$. Then the r_i's have to satisfy:

(1) $0 < r_i \leqslant N - 1 - 2^k$,

(2) For all $i_1, i_2 : 1 \leqslant i_1, i_2 \leqslant s$ holds that $r_{i_1} \oplus r_{i_1} > 2^{k+1}$, where $\oplus$ denotes the exclusive or operator.

Encryption:

$$Q_j \longmapsto (r_i \oplus Q_j)^e \bmod N = C_j.$$

Decryption:

$$C_j \longmapsto (C_j^d \bmod N) \oplus r_i.$$

Remark 1: Condition (1) is required for the encryption and description processes to be invertible.

Remark 2: If $g_1, g_2 \in G$, $g_1 \neq g_2$ then the encryptions are different. On the contrary assume that:

$$(g_1 \oplus r_1)^e \equiv (g_2 \oplus r_2)^e \bmod N,$$

then

$$g_1 \oplus r_1 = g_2 \oplus r_2 \;\Rightarrow\; g_1 \oplus g_2 = r_1 \oplus r_2,$$

which is a contradiction since:

$$r_1 \oplus r_2 > 2^{k+1} \geqslant g_1 \oplus g_2.$$

2.2. *Encryption Technique II – Using a Symmetric Cryptosystem*

The second encryption technique is based on the Discrete Logarithm Problem.[14] Let p be a large prime, $g < p$ for $g \in G$. By x we denote Alice's secret key, $0 < x \leqslant p - 2$. By $r_i, 1 \leqslant i \leqslant s$ we denote s random symmetric keys, $0 < r_i \leqslant p - 2$.

In the case of Q_j being an entry of the table consider:

Encryption:

$$Q_j \longmapsto Q_j^{r_i \cdot x} \bmod p = C_j.$$

Decryption:

$$C_j \longmapsto C_j^{(r_i \cdot x)^{-1}} \bmod p = Q_j.$$

Remark 3: For α a primitive element $\bmod p$ consider the pair $(\alpha^r, Q^{r \cdot x})$. Although it contains some "partial" information related to the random key r, r cannot be recovered from α^r.

Remark 4: Assume that $Q_1 \neq Q_2$. Then

$$(\alpha^{r_1}, Q_1^{r_1 \cdot x}) \neq (\alpha^{r_2}, Q_2^{r_2 \cdot x}).$$

2.3. *Distributed Data Mining*

Currently, data mining systems focus on real life problems that usually involve huge volumes of data. Therefore, one of the main challenges of these systems is to devise means to handle data that are substantially larger than available main memory on a single processor. Several approaches to this problem, are reported in the literature.

An obvious approach is to parallelize the data mining algorithms. This approach requires the transformation of the data mining algorithm to an optimized parallel algorithm suited to a specific architecture (e.g. Refs. 1, 12, 19). Another approach is based on windowing techniques, where data miners are supplied with a small subset of data, having a fixed size, the *window* (e.g. Refs. 5, 15, 18). Iteratively, the window is updated with new data of the remaining set, until a predefined accuracy is met.

An alternative approach is based on partitioning the initial data set into subsets, applying a data mining algorithm in parallel to these subsets and synthesizing the final data mining rules from the partial results.

For instance, in Ref. 6 a classification technique is proposed, called meta-learning, that exploits different classifiers supplied with different subsets of the data, in parallel. Then, partial results are combined, in the sense that the predictions of these classifiers are used to construct the final prediction. In Ref. 4, a classification methodology is proposed, for combining partial classification rules. It is, actually, a two-phase process. First, a number of classifiers are trained, each with a different subset of the data. Then, the trained classifiers are used in the construction of a new training data set, substantially smaller than the initial one. The latter data set is used to train the final classifier through an iterative process, that is guided by thresholds concerning the size of this data set and the achieved accuracy. In Ref. 3 an iterative clustering process is proposed that is based on partitioning a sample of data into subsets. In a first phase, each subset is given as an input to a clustering algorithm. The partial results form a dataset that it is partitioned into clusters, the meta-clusters, during a second phase. Under certain circumstances, meta-clusters are considered as the final clusters. Finally, in Ref. 17 an algorithm for extracting association rules is presented that is based on a logical division of the database into non-overlapping partitions. The partitions are considered one at a time and all associations for that partition are generated. Then, these associations are merged to generate a set of all potential associations. Finally, the actual associations are identified.

The latter approach, based on partitioning the initial data set, can be applied during the last step of the proposed methodology that concerns the decryption of the encrypted data mining rules. As mentioned earlier, the m-th record of any Q_j is encrypted using the key r_i, where $i = m(\mathrm{mod}\, s) + 1$. Thus, every Q_j can be partitioned into s subsets, where in every subset any g_k is encrypted using the same r_i. Notice, also, that the encryption of any g_k of a subset is different its encryptions in different subsets. After decrypting the extracted data mining rules, the obtained rules, partitioned by the subset they originated from, are identical to the partial rules that would be obtained by partitioning the initial data set into s subsets and applying a data mining algorithm in parallel to these subsets. Therefore, the key idea is that the rules obtained from each subset, after decryption, can be combined in order to construct the final set of rules, by using the distributed data mining algorithms mentioned above. Thus, Alice will obtain the final set of rules without revealing any confidential information to Bob.

3. Conclusions and Future Research

We have proposed a novel methodology for mining encrypted data. Such a methodology is very useful when the owner of the data is wishes to prevent the disclosure of confidential information.

Various cryptosystems have been tested. The obvious solutions based on a mono-alphabetic cryptosystem may not resist a frequency analysis attack. We have proposed two alternatives that perfectly fit with the problem requirements. In both cases, the decryption phase is based on distributed data mining algorithms.

Notice that distributed data mining algorithms may not maintain the accuracy that would be achieved by a simple data mining algorithm supplied with all the data. In other words, Alice may obtain rules which are not so accurate as they would be if she was not using distributed data mining algorithms, for instance in the case of mono-alphabetic cryptosystems. However, the loss in accuracy (if any) is usually small enough. Thus, the proposed methodology is considered to be acceptable.

References

1. R. Agrawal and J.C. Shafer. Parallel Mining of Association Rules: Design, implementation and experience. *IEEE Trans. on Knowledge and Data Engineering*, **8**(6):962–969 (1996).
2. M. Atallah, E. Bertino, A. Elmagarmid, M. Ibrahim, and V. Verykios. Dis-

closure Limitation of Sensitive Rules. *Proceedings of the 1999 Workshop on Knowledge and Data Engineering Exchange*, Chicago, 45–52 (1999).

3. B. Boutsinas and T. Gnardellis. On Distributing the clustering process. *Pattern Recognition Letters*, Elsevier Science Publishers B.V., **23**(8), 999-1008 (2002).
4. B. Boutsinas, G. Prassas, and G. Antzoulatos. On scaling up classification algorithms, submitted (2002).
5. P.S. Bradley, U.M. Fayyad , and C. Reina. Scaling Clustering Algorithms to Large Databases. *Proceedings of the 4th Int. Conf. on Knowledge Discovery and Data Mining*, 9–15 (1998).
6. P. Chan and S. Stolfo. Meta-learning for multistrategy and parallel learning. *Proceedings of the 2nd Int. Work. Multistrategy learning*, 150–165 (1993).
7. C. Clifton and D. Marks. Security and Privacy Implication of Data Mining. *Proceedings of the 1996 ACM Workshop on Data Mining and Knowledge Discovery* (1996).
8. C. Clifton. Protecting against Data Mining through Samples. *Proceedings of the 13th IFIP Conference on Database Security*, Seattle, Washington (1999).
9. C. Clifton. *Privacy Preserving Distributed Data Mining.* (2001)
10. E. Dasseni, V. Verykios, A. Elmagarmid, and E. Bertino. Hiding Association Rules by Using Confidence and Support. *LNCS* 2137, 369-383 (2001).
11. U.M. Fayyad, G. Piatetsky-Shapiro and P. Smyth. *Advances in Knowledge Discovery and Data Mining.* AAAI Press/MIT Press (1996).
12. X. Li and Z. Fang. Parallel clustering algorithms. *Parallel Computing*, **11**, 275–290 (1989).
13. Y. Lindell and B. Pinkas. Privacy Preserving Data Mining. *Advances in Cryptology- CRYPTO '00, LNCS* 1880, 36-53 (2000).
14. S.C. Pohlig and M. Hellman. An Improved Algorithm for Computing Logarithms over $GF(p)$ and its Cryptographic Significance. *IEEE Transactions on Information Theory*, **24**, 106–110 (1978).
15. F. Provost and V. Kolluri. Scaling Up Inductive Algorithms: An Overview. *Proceedings of the 3rd Proceedings of the Knowledge Discovery and Data Mining*, 239–242 (1997).
16. R. Rivest, A. Shamir, and L. Adlemann. A Method for Obtaining Digital Signatures and Public-Key Cryptosystems. *Commun. ACM*, **21**, 120–126 (1978).
17. A. Savasere, E. Omiecinski, and S. Navathe. An Efficient Algorithm for Mining Association Rules in Large Databases. *Proceedings of the 21th IEEE International Conference on Very Large Databases* (1995).
18. H. Toivonen. Sampling large databases for finding association rules. *Proceedings of the 22th IEEE International Conference on Very Large Databases*, India, 134–145 (1996).
19. X. Zhang, M. Mckenna, J. Mesirov, and D. Waltz. An efficient implementation of the backpropagation algorithm on the connection machine CM-2. Tech. Rep. RL89-1, Thinking Machines Corp. (1989).

CHAPTER 17

EXCHANGE RATE FORECASTING THROUGH DISTRIBUTED TIME-LAGGED FEEDFORWARD NEURAL NETWORKS

N.G. Pavlidis

Department of Mathematics,
University of Patras Artificial Intelligence Research Center (UPAIRC),
University of Patras,
GR–26110 Patras, Greece.
E-mail: npav@math.upatras.gr

D.K. Tasoulis

Department of Mathematics, UPAIRC,
University of Patras,
GR–26110 Patras, Greece.
E-mail: dtas@math.upatras.gr

G.S. Androulakis

Computer Technology Institute (CTI),
UPAIRC,
University of Patras,
GR–26110 Patras, Greece.
E-mail: gsa@math.upatras.gr

M.N. Vrahatis

Department of Mathematics, UPAIRC,
University of Patras,
GR–26110 Patras, Greece.
E-mail: vrahatis@math.upatras.gr

Throughout the last decade, the application of Artificial Neural Networks in the areas of financial and economic time series forecasting has been rapidly expanding. The present chapter investigates the ability of Distributed Time Lagged Feedforward Networks (DTLFN), trained through a popular Differential Evolution (DE) algorithm, to forecast the short–term behavior of the daily exchange rate of the Euro against the US Dollar. Performance is contrasted with that of focused time lagged feedforward networks, as well as with DTLFNs trained through alternative algorithms.

Keywords: Artificial neural networks, differential evolution algorithms, time series prediction.

1. Introduction

A central problem of science is forecasting; how can knowledge of the past behavior of a system be exploited in order to determine its future evolution. As of 1997 the foreign exchange market constitutes the world's largest market with daily transactions surpassing 1 trillion US dollars on busy days. More than 95 percent of this volume is characterized as speculative trading, i.e. transactions performed in order to profit from the short term movement of the exchange rate.

Two schools of thought compete in the field of financial forecasting, fundamentalists and technical analysts. Fundamentalists hold the view that forecasting needs to be based on the identification of a model that approximates the true underlying exchange rate determination dynamics. Thus, models that take into account inflation and interest rate differentials, balance of payments accounts and numerous other economic indicators, are constructed and evaluated. A key limitation of this approach is that most of the involved quantities are known only *a posteriori*. The scope of this approach, therefore, lies more within the realm of justification rather than prediction. In contrast, technical analysts exploit information contained in the history of prices, in market news, as well as in different technical indicators in order to form their expectations about future prices. A central limitation of this approach stems from the well–known Efficient Market Hypothesis (EMH) that states that all information concerning future values of financial assets traded in competitive markets is already incorporated in current prices. The best forecast for the future value is therefore the current value. Yet, as it has been shown in Ref. 4 for market participants to be willing to participate in the market, prices cannot reflect all available information; paradoxically, if that was the case there would be no reason for a market to exist, thus, at least the strong form of the EMH has to be violated. This however does not alter the fact that inferring the future evolution of market prices is a particularly hard problem. Indeed had prices been predictable to a large extent, once more there would be no incentive for market participants to enter the market.

The literature on artificial neural network (ANN) applications in the fields of financial time series has been rapidly expanding throughout the past decade.[11] Several researchers have considered the application of ANNs on the problem of foreign exchange forecasting with promising results.[1,3,8,9,17] The ability of ANNs to outperform different linear models[19] as well as the random walk model, using solely publicly available information, may be taken to imply that a certain structure does in fact exist, and most importantly, that ANNs, if properly designed and trained, can identify and exploit this structure in order to produce accurate forecasts.

The present chapter contributes further in this line of research, by investigating the ability of a particular type of feedforward networks, Distributed Time Lagged Feedforward Networks (DTLFN)[7,18] to forecast the time series of the daily exchange rate of the Euro against the US Dollar (USD). Particular emphasis is attributed to the prediction of the direction of change of the spot exchange rate, since this information is sufficient to render speculative trading profitable. The task at hand is particularly difficult due to the limited number of available observations, the highly competitive nature of the particular market, the noise and finally, the nonstationarity present in the data. A key merit of ANNs is noise tolerance; i.e. the ability to identify patterns in data contaminated with noise. DTLFNs are further characterized by the highly desirable property that they can cope effectively with nonstationarity, unlike static multilayer feedforward networks. The best results obtained were achieved using a novel global optimization technique, called Differential Evolution (DE) algorithm, introduced in Ref. 16 and implemented in the context of ANN training in Ref. 10, in combination with a modified error performance function.

The chapter is organized as follows; Section 2 discusses focused and distributed time–lagged feedforward networks, and also provides a brief introduction to the workings of the DE algorithm. Section 3 presents the empirical results and the Section 4 is devoted to concluding remarks and future work.

2. Artificial Neural Networks

2.1. *Focused Time-Lagged Feedforward Neural Networks*

Artificial Neural Networks (ANNs) are parallel computational models comprised of densely interconnected, simple, adaptive processing units, and characterized by a natural propensity for storing experiential knowledge and rendering it available for use. ANNs resemble the human brain in two fundamental respects; firstly, knowledge is acquired by the network from its environment through a learning process, and secondly, interneuron connection strengths, known as synaptic weights are employed to store the acquired knowledge.[6,7]

The building block of an ANN is the *artificial neuron*, which consists of a number n of synapses, a weight vector w, an activation function f, and finally output y. Each element of the weight vector w_i corresponds to an input x_i, $i = 1, ..., n$ and the primary operation of an artificial neuron, consists of summing up the products of the inputs and their corresponding weight. This sum, also known as *excitation level*, constitutes the argument of the activation function, which in turn, determines the output of the neuron. It should be noted that a bias term is included by incorporating an additional weight b whose corresponding input is always set to $x_0 = 1$. In general, the operation of a single neuron receiving as input the vector x, is summarized by the following equations:

$$\xi = \sum_{i=0}^{n} w_i x_i, \qquad y = f(\xi).$$

The most frequently encountered types of activation functions are:

(a) the *hard limit* function:

$$f(\xi) = \begin{cases} 1, & \text{if} \quad \xi \geqslant h, \\ 0, & \text{if} \quad \xi < h, \end{cases}$$

(b) the *piecewise linear* function:

$$f(\xi) = \begin{cases} 1, & \text{if} \quad \xi \geqslant 1, \\ 0, & \text{if} \quad \xi \leqslant 0, \\ \xi, & \text{if} \quad 0 < \xi < 1, \end{cases}$$

(c) the *standard (logistic) logsig* function:

$$f(\xi) = \frac{1}{1 + e^{-\xi}},$$

(d) the *hyperbolic tangent* function:

$$f(\xi) = \tanh \frac{\xi}{2} = \frac{1 - e^{-\xi}}{1 + e^{-\xi}}.$$

Combining a set of neurons, extended ANNs with different topologies, can be created. *Feedforward Neural Networks* (FNNs), are ANNs in which neurons are organized in layers and no feedback connections are established.

In FNNs, inputs are assigned to sensory nodes that comprise the input layer, while the output of the network is produced by the neurons that form the output layer. All other neurons are assigned to intermediate, hidden, layers. The output of a neuron in layer j becomes an input of all the neurons that belong to the next layer, $j+1$, as shown in Figure 1.

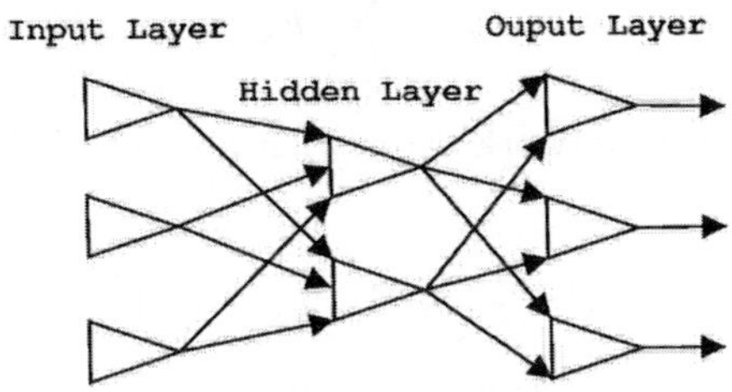

Fig. 1. Feedforward Network

The operation of such networks consists of a series of iterative steps. At the beginning the states of the input layer neurons are assigned to the elements of the input vector, also called pattern vector, and the remaining hidden and output layer neurons are passive. In the next step the neurons of the first hidden layer collect and sum their inputs and compute their output. This procedure is propagated forward through the layers of the FNN until the final network output is computed.

The computational power of FNNs is based on the fact that they can adapt to a specific training set. According to the *universal approximation theorem*,[2] a FNN composed of neurons with nonlinear activation functions and a single hidden layer is sufficient to approximate an arbitrary continuous function. Assuming, as in the case of time series prediction, that the input vector for the network consists of a number of delayed observations of the time series, and the target is the next value, then the *universal myopic mapping theorem*[14,15] states that any shift-invariant map can be approximated arbitrarily well by a structure consisting of a bank of linear filters feeding a static ANN. This type of ANN is known as Focused Time–Lagged Feedforward Neural Networks (FTLFN).

In this context, a training set T of P patterns is defined as:

$$T = \left\{ (x_k, d_k) \left| \begin{array}{l} x_k = (x_t, \ldots, x_{t-n}) \in R^n \\ d_k = x_{t+1} \in R \end{array} \right. , \quad k = 1, \ldots, P \right\},$$

where x_k represents the kth training pattern, and d_k, the desired response vector for this pattern. The purpose of training is to assign to the free parameters of the network W, values that will minimize the discrepancy between network output and desired response. The training process starts by presenting all the patterns to the network and computing a total error function E, defined as:

$$E = \sum_{k=1}^{P} E_k,$$

where E_k is the partial network error with respect of the kth training pattern, defined as:

$$E_k = \frac{1}{2}\Big(y_i(W, x_k) - d_{ki}\Big)^2.$$

Each full pass of all the patterns is called a training epoch. If the error performance value drops below the desired accuracy, then it is obvious that the aim of the training algorithm has been fulfilled and the algorithm is terminated. Thus supervised training is a non-trivial minimization problem.

$$\min_W E(W)$$

Training a FNN, the information content of the training set is stored in the synaptic weights, forming the long term memory of the network. Once a network is trained the acquired knowledge can be readily exploited. The ability of an ANN to respond correctly to patterns not encountered during training is called generalization. A satisfactory performance on the training set coupled with a poor generalization performance is known as overfitting, or, overtraining. Overfitting implies that the network has adjusted too much on the training set, thereby capturing potential inaccuracies and errors that might exist. Clearly, for the task of time series prediction generalization is of critical importance, since satisfactory performance on the known part of the time series is of little use by itself.

2.2. *Distributed Time–Lagged Feedforward Neural Networks*

The universal myopic theorem previously stated, is limited to maps that are shift invariant. An immediate implication of this limitation is that FTLFN are suitable for modeling stationary processes[7]. As discussed in the following section, using FTLFNs to forecast the future behavior of the exchange rate of the Euro against the USD, frequently results in overfitting and rarely produces a satisfactory performance. A possible alternative in order to overcome this limitation is to distribute the impact of time throughout the network and not only at the input end. To this end, *Distributed Time Lagged Feedforward Networks* (DTLFN)[18] are considered. The construction of such a network is based on the model of an artificial neuron known as *multiple input neuronal filter*. In this setting the output of each neuron in layer j is inserted in a tapped delay line memory of order m. In effect this line reproduces the m previous values of the output of the neuron and transmits them, as well as the current value, to all the processing units in layer $j+1$. An alternative way to envisage this operation is to imagine that the output of the tapped delay line and the current output are fed to a linear *Finite Impulse Response* filter whose output is in turn transmitted unaltered to neurons in the succeeding layer. The operation of a multiple input neuronal filter is summarized by the following equations:

$$y = f(\sum_{i=1}^{z}\sum_{l=0}^{m} w_{il}x_{i-l} + b).$$

where z stands for the number of inputs of the neuron, and m denotes the order of the tapped delay line.

2.3. *Differential Evolution Training Algorithm*

In a recent work, Storn and Price [16] have presented a novel minimization method, called Differential Evolution (DE), designed to handle non–differentiable, nonlinear and multimodal objective functions. DE exploits a *population* of potential solutions to probe the search space. At each iteration of the algorithm, mutation and crossover are applied in order to obtain more accurate approximations to a solution.

To apply DE to neural network training the approach introduced in Ref. 10 is adopted. Primarily, a number (NP) of N–dimensional weight vectors is specified. Weight vectors are initialized using a random number generator. At each iteration of the algorithm, called *generation*, new weight vectors are generated through the combination of randomly selected members of the existing population. This is the *mutation* operation. The resulting weight vectors are then mixed with a predetermined weight vector, called the *target* weight vector. This stage of the algorithm is called *crossover*. The outcome of the mutation and crossover operation yields the *trial* weight vector. The trial weight vector is accepted if and only if it reduces the value of the error function E. The final operation is known as *selection*. Subsequently, a brief outline of the workings of the two DE main operations is presented.

The first DE operator is the *mutation operator*. Specifically, for each weight vector, a new vector v^i_{g+1}, called mutant vector, is generated according to the relation:

$$v^i_{g+1} = w^i_g + \mu(w^{best}_g - w^i_g) + \mu(w^{r1}_g - w^{r2}_g).$$

where w^i_g, $i = 1, ..., NP$ represents the i–th weight vector of the population, g stands for the current generation index, and w^{best}_g denotes the best member of the previous generation. The constant parameter, $\mu > 0$, is a real number, called mutation constant, which controls the amplification of the difference between the two weight vectors. Finally, w^{r1}_g and w^{r2}_g are two randomly selected weight vectors of the current generation, different from w^i_g. To stimulate further the diversity among members of the new population, the *crossover operator* is applied. For each component of $j = 1, ..., n$ of the mutant weight vector a randomly selected real number $r \in [0, 1]$. If $r \leq \rho$, where $\rho > 0$ is the crossover constant, then the j–th component of the trial vector is replaced by the j–th component of the mutant vector. Otherwise the j–th component of the target vector is selected.

3. Empirical Results

The data set used in the present study is provided by the official website of the European Central Bank (ECB), and consists of the daily exchange rate of the Euro against the USD, starting from the introduction of the Euro on January 1st 1999 and extending to October 10th 2001. Observations after October 11th 2001 were excluded, due to the international turmoil which had a major impact on the international foreign exchange rate markets. Clearly, no method can provide reliable forecasts once exogenous factors never previously encountered, exert a major impact on the formulation of market prices. The total number of observations included in present the study was therefore limited to 619.

The time series of daily exchange rates, illustrated in Figure 2, is clearly nonstationary. An approach frequently encountered in the literature, to overcome the problem of nonstationarity is to consider the first differences of the series, or the first differences of the natural logarithms. Both of these approaches transform the original, nonstationary, time series, to a stationary one. If, however, the original series is contaminated with noise, there is a danger that both of these transformations will accentuate the presence of noise, in other words increase the noise–to–signal ratio, while at the same time eliminate valuable information. Our experience indicates that for the task of one–step–ahead prediction both of these transformations inhibit the training process and effectively impose the use of larger network architectures, while at the same time, there is no indication of superior forecasting ability. Thus, the original time series of daily exchange rates was considered, and for the purpose of training ANNs with nonlinear transfer functions the data was normalized within the range $[-1, 1]$. The data set was divided into a test set, containing the last 30 observations, and a training set containing all previous data points. Input patterns consisted of a number of time lagged values, $x_k = (x_t, ..., x_{t-n})$, whereas the desired response for the network was set to the next day's exchange rate, $d_k = x_{t+1}$.

Numerical experiments were performed using the Neural Network Toolbox version 4.0 for Matlab 6.0, as well as a Neural Network C++ Interface built under the Linux Operating system using the g++ compiler. Using the Neural Network Toolbox version 4.0 provided the opportunity to apply several well–known deterministic training algorithms. More specifically, the following algorithms were considered:

* Standard Back Propagation (BP),[13]
* Back Propagation with Adaptive Stepsize (BPAS),
* Resilient Back Propagation (RPROP),[12]
* Conjugate Gradient Algorithms (CG), and
* Levenberg–Marquardt (LM).[5]

The first three training algorithms, BP, BPAS, and RPROP, exploit gradi-

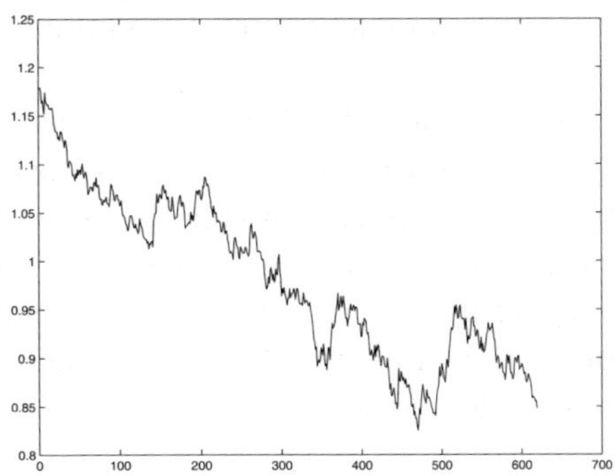

Fig. 2. Time Series of the Daily Euro/USD Exchange Rate

ent information to determine the update of synaptic weights, whereas, CG algorithms and the LM algorithm also incorporate an approximation of the Hessian matrix, thereby exploiting second order information. The DTLFNs trained through the DE algorithm were implemented using a Neural Network C++ Interface built under the Linux Operating system. This interface was selected since it greatly reduced the time required to train ANNs.

All training methods considered were extensively tested with a wide range of parameters. The BP and BPAS algorithms frequently encountered grave difficulties in training. Relative to speed measures RPROP proved to be the fastest. A crucial difference was relative to the reliability of performance were the LM and DE algorithms proved to be the most reliable, in the sense that ANNs with the same topology and activation functions, trained through these algorithms tended to exhibit small variability in performance on the test set. This finding however was contingent on the size of the network; as the number of hidden neurons and layers increased this highly desirable property from the viewpoint of financial forecasting, tended to vanish.

Network topology has been recognized as a critical determinant of network performance. Numerical experiments performed in the context of the present study conform to this finding. Unfortunately, the problem of identifying the "optimal" network topology for a particular task is very hard and currently remains an open research problem. To find a suitable network we proceed according to the following heuristic method. Starting with a network with a single hidden layer and a minimum number of hidden neurons, usually two, we proceed to add neurons and layers as long as performance on both the test and training set is improving. For the purposes of the present study ANNs with a single hidden layer proved to be sufficient. It is worth noting that adding more layers tends to inhibit the generalization ability of the network.

To evaluate the performance of different predictors, several measures have been proposed in the literature.[1,3,8,9,19] The primary focus of the present study was to create a system capable of capturing the direction of change of daily exchange rates, i.e. whether tomorrow's rate will be lower or higher relative to today's rate. To this end a measure called *sign prediction* was applied. Sign prediction measures the percentage of times for which the following inequality holds on the test set:

$$(\widehat{x_{t+1}} - x_t) * (x_{t+1} - x_t) > 0$$

where, $\widehat{x_{t+1}}$ represents the prediction generated by the ANN, x_{t+1} refers to the true value of the exchange rate at period $t+1$ and, finally, x_t stands for the value of the exchange rate at the present period, t. If the above inequality holds, then the ANN has correctly predicted the direction of change of the exchange rate.

As previously mentioned the EMH states that the best possible forecast of tomorrow's rate is today's rate, $\widehat{x_{t+1}} = x_t$. We refer to this forecast as the *naive predictor*, since it requires knowledge of only the last value of the time series. Despite the fact that EMH has been theoretically questioned and different researchers have been capable of outperforming the naive predictor, comparing the accuracy of forecasts with that of the naive predictor remains a benchmark for comparison. To evaluate the performance of different ANNs with respect to the naive predictor a measure called *acrnn* is devised. The acrnn measure captures the percentage of times for which the absolute deviation between the true value and the value predicted by the ANN is smaller than the absolute deviation between the true value and the value predicted by the naive predictor. In other words, acrnn measures the percentage of times for which the following inequality holds on the test set.

$$|x_{t+1} - \widehat{x_{t+1}}| < |x_{t+1} - x_{t+1}^{naive}|$$

where $x_{t+1}^{naive} = x_t$.

Initially, different ANNs were trained on the training set and consequently, their ability to produce accurate predictions for the entire test set was evaluated. Overall, the obtained results were unsatisfactory. Irrespective of the ANN type, topology and, training function applied, the accuracy of the predictions generated by the trained ANNs for the observations that belong to the test set, was not significantly better than the naive predictor. In other words, an acrnn measure consistently above 50% was not achieved by any network. Moreover, with respect to sign prediction, no ANN was capable of consistently predicting the direction of change with accuracy exceeding 55%. Plotting predictions generated by ANNs against the true evolution of the time series it became evident that the predictions produced were very similar to a time–lagged version of the true time series. Alternatively stated, the ANN approximated closely the behavior of the naive predictor. Due to the fact that daily variations were indeed very small compared to the value of the exchange rate this behavior resulted to very small mean squared error on the training set. Performance did not improve as the number of time lags provided as inputs to the network, n, increased. Indeed, the best results were obtained for values of n higher than 2, and lower than 10. Despite the fact that the task of training different ANNs became easier and performance on the training set was improving, as n was increasing, performance on the test set showed no signs of improvement. This behavior indicated that instead of capturing the underlying dynamics of the system, increasing the free parameters of the network *per se* rendered ANNs prone to overfitting. Similar results were obtained as the number of hidden neurons was increased above 5. To avoid overfitting early stopping was applied. Early stopping implies the division of the data set into a training, a validation and a test set. At each training epoch, the error on the validation set is computed but no information from this measurement is included in the update of synaptic weights. Training is terminated once the error on the validation set increases as training proceeds beyond a critical point. Incorporating early stopping did not produce a significant improvement of performance. This outcome can be justified by the presence of nonstationarity which implies that the selection of an appropriate validation set is not trivial. In effect the patterns selected to comprise the validation set need to bear a structural similarity with those that comprise the test set, a prerequisite that cannot be guaranteed to hold if a set of patterns just before the test set is selected as a validation set for the particular task.

A possible explanation for the evident inability of different ANNs to produce accurate one–step–ahead forecasts of the daily exchange rate of the Euro against the USD is that the market environment in the particular foreign exchange market is rapidly changing. This is a reasonable assumption taking into consideration the intense competition among market participants. If this claim is valid, then the knowledge stored in the synaptic weights of trained ANNs becomes obsolete for the task of prediction, as patterns further in the test set are considered. To evaluate the validity of this claim an alternative approach is considered. More specifically, ANNs are trained on the training set, predictions for a test set consisting of the five patterns that immediately follow the end of the training set are generated and, network performance is evaluated. Subsequently, the first pattern that belongs to the test set is assigned to the training set. A test set consisting of five patterns immediately following the end of the new training set is selected, and the process is repeated until forecasts for the entire test set consisting of 30 observations are generated. To promote the learning process and avoid the phenomenon of generated predictions being a time–lagged version of the true time series a modified error performance function was implemented for DTLFNs trained through the DE algorithm.

This function assigns an error value of zero for the kth pattern as long as the DTLFN accurately predicts the direction of change for the particular pattern, otherwise error performance is computed as in the standard mean squared error performance function.

$$E_k = \begin{cases} 0 & \text{if } (\widehat{x_{t+1}} - x_t) * (x_{t+1} - x_t) > 0 \\ \\ \frac{1}{2}\sum(x_{t+1} - \widehat{x_{t+1}})^2 & \text{otherwise} \end{cases}$$

To generate predictions for the entire test set, the training and evaluation processes were repeated 25 times with the maximum number of training epochs set to 150. The performance a FTLFN and a DTLFN trained though the LM algorithm using a mean squared error performance function, as well as that of a DTLFN trained through the DE algorithm using the modified error performance function, is reported in Table 1. The generalization ability of the first two networks is clearly unsatisfactory. The accuracy of predictions is inferior to that of the naive predictor and average sign prediction is considerably lower than 50%. This is not however the case for the last DTLFN. The predictions generated by the DTLFN that was trained using the DE algorithm and the modified error performance function, clearly outperform the naive predictor, with an average acrnn value of 59.2%. Most importantly average sign prediction assumes a value of 68%, which is substantially above 50% and substantially higher than the sign prediction achieved in Ref. 3. A significant advantage due to the incorporation of the modified error performance function and the DE training algorithm, which is not evident from Table 1, is the fact that network performance on the training set became a much more reliable measure of generalization. In other words, a reduction of the error on the training was most frequently associated with superior performance on the test set. At the same time, the phenomenon of deteriorating performance on the test as training proceeded was very rarely witnessed.

4. Concluding Remarks

The ability of Distributed Time–Lagged Neural Networks (DTLFNs), trained using a Differential Evolution (DE) algorithm and a modified error performance function, to accurately forecast the direction of change of the daily exchange rate of the Euro against the US Dollar has been investigated. To this end only the history of previous values has been exploited. Attention was focused on sign prediction, since it is sufficient to produce a speculative profit for market participants, but results were also contrasted with the naive predictor. Comparing the out–of–sample performance of the

Table 1. Test Set Performance

Topology: 5 * 5 * 1						
Iteration	FTLFN		DTLFN LM		DTLFN DE	
	sign	acrnn	sign	acrnn	sign	acrnn
1	100%	40%	40%	0%	80%	60%
2	40%	0%	40%	0%	80%	80%
3	60%	0%	40%	0%	60%	60%
4	20%	20%	60%	0%	80%	80%
5	80%	20%	60%	0%	80%	80%
6	20%	0%	40%	0%	40%	40%
7	60%	20%	40%	0%	80%	80%
8	40%	20%	40%	0%	60%	20%
9	20%	0%	40%	0%	80%	60%
10	60%	20%	80%	0%	60%	60%
11	40%	40%	20%	0%	80%	80%
12	20%	0%	20%	0%	80%	80%
13	80%	20%	60%	20%	80%	80%
14	20%	0%	60%	20%	80%	80%
15	20%	20%	60%	0%	60%	60%
16	80%	20%	40%	0%	80%	80%
17	0%	0%	60%	0%	80%	80%
18	60%	20%	40%	0%	80%	80%
19	80%	40%	100%	20%	100%	100%
20	40%	20%	40%	20%	60%	40%
21	40%	20%	20%	0%	60%	40%
22	40%	0%	60%	0%	40%	20%
23	40%	0%	20%	0%	40%	40%
24	60%	20%	40%	0%	40%	0%
25	20%	0%	20%	0%	40%	0%

DTLFNs trained through the proposed approach with that of DTLFNs and FTLFNs trained through different deterministic algorithms and using the mean squared error performance function, the proposed approach proved to be superior for the particular task. The results from the numerical experiments performed were promising, with correct sign prediction reaching an average of 68% and the average percentage of times for which the predictions were more accurate than the naive predictor being 59.2%. A further advantage of this approach was the fact that network performance on the training set became a more reliable indicator of generalization ability.

Further work will include the application of the present approach to different financial time series as well as the consideration of alternative training methods based on evolutionary and swarm intelligence methods.

References

1. A.S. Andreou, E.F. Georgopoulos and S.D. Likothanassis. Exchange–Rates Forecasting: A Hybrid Algorithm Based on Genetically Optimized Adaptive Neural Networks. *Computational Economics*, in press.
2. G. Cybenko. Approximation by superpositions of a sigmoidal function. *Mathematical Control Signals Systems*, **2**:303–314 (1989).
3. C.L. Giles, S. Lawrence and A.C. Tsoi. Noisy Time Series Prediction using a Recurrent Neural Network and Grammatical Inference. *Machine Learning*, **44**(1/2):161–183 (2001).
4. S.J. Grossman and J. Stiglitz. On the Impossibility of Informationally Efficient Markets. *American Economic Review*, **70**:393–408 (1980).
5. M.T. Hagan and M. Menhaj. Training Feedforward Networks with the Marquardt Algorithm. *IEEE Transactions on Neural Networks*, 5(6):989–993 (1994).
6. M.H. Hassoun. *Fundamentals of Artificial Neural Networks.* MIT Press (1995).
7. S. Haykin. *Neural Networks: A Comprehensive Foundation.* Macmillan College Publishing Company, New York (1999).
8. C.M. Kuan and T. Liu. Forecasting Exchange Rates Using Feedforward and Recurrent Neural Networks. *Journal of Applied Econometrics*, 10:347–364 (1995).
9. M.T. Leung, A.S. Chen and H. Daouk. Forecasting Exchange Rates using General Regression Neural Networks. *Computers & Operations Research*, 27:1093-1110 (2000).
10. V.P. Plagianakos and M.N. Vrahatis. Training Neural Networks with Threshold Activation Functions and Constrained Integer Weights. *Proceedings of the IEEE International Joint Conference on Neural Networks (IJCNN)* (2000).
11. A. Refenes. *Neural Networks in the Capital Markets.* John Wiley and Sons (1995).

12. M. Riedmiller and H. Braun. A Direct Adaptive Method for Faster Backpropagation Learning: The RPROP Algorithm. *Proceedings of the IEEE International Conference on Neural Networks*, San Francisco, CA, pp. 586–591 (1993).
13. D.E. Rumelhart, G.E. Hinton and R.J. Williams. Learning Internal Representations by Error Propagation. In D.E. Rumelhart and J.L. McClelland (Eds.) *Parallel distributed processing: Explorations in the microstructure of cognition.* Cambridge, MA, MIT Press, 1:318–362 (1986).
14. I.W. Sandberg and L. Xu. Uniform Approximation of Multidimensional Myopic Maps. *IEEE Transactions on Circuits and Systems*, 44:477–485 (1997).
15. I.W. Sandberg and L. Xu. Uniform Approximation and Gamma Networks. *Neural Networks*, 10:781–784 (1997).
16. R. Storn and K. Price. Differential Evolution – A Simple and Efficient Heuristic for Global Optimization over Contituous Spaces. *Journal of Global Optimization*, 11:341–359 (1997).
17. W. Verkooijen. A Neural Network Approach to Long–Run Exchange Rate Prediction. *Computational Economics*, 9:51–65 (1996).
18. E. Wan. Time Series Prediction using a Connectionist Network with Internal Delay Lines. in *Time Series Prediction: Forecasting the Future and Understanding the Past*, A.S. Weigend and N.A. Gershenfeld. Reading, MA: Addison–Wesley, pp. 195–217 (1993).
19. B. Wu. Model–free Forecasting for Nonlinear Time Series (with Application to Exchange Rates). *Computational Statistics & Data Analysis*, 19:433–459 (1995).

CHAPTER 18

NETWORK FLOW PROBLEMS WITH STEP COST FUNCTIONS

R. Yang
Department of Industrial and Systems Engineering,
University of Florida,
303 Weil Hall, Gainesville, FL 32611, USA
E-mail: maggiey@ufl.edu

P.M. Pardalos
Department of Industrial and Systems Engineering,
University of Florida,
303 Weil Hall, Gainesville, FL 32611, USA
E-mail: pardalos@ufl.edu

Network flow problems are widely studied, especially for those having convex cost functions, fixed-charge cost functions, and concave functions. However, network flow problems with general nonlinear cost functions receive little attention. The problems with step cost functions are important due to the many practical applications. In this paper, these problems are discussed and formulated as equivalent mathematical mixed 0-1 linear programming problems. Computational results on randomly generated test beds for these exact approached solution procedure are reported in the paper.

Keywords:
Nonconvex network problem; lot sizing; minimum-cost network flows.

1. Introduction

Given a finite time horizon T and positive demands for a single item, production, inventory, and transportation schedules should be determined to minimize the total cost, including production cost, inventory cost, and transportation cost, on the condition that the demand only be satisfied from production at multiple facilities in the current period or by inventory from the previous periods (Backlogging is not allowed). For convex network flow problems if the local optimal solution is found, then it is also the global optimum. Even large scale problems are still tractable in the convex case. However, it is well known that concave network flow problems are NP-hard problems [11]. In the worst case, it need to enumerate all the local optimal solution before the global optimum is found. Many heuristic algorithms are used to solve concave network flow problems. Kim and Pardalos[1,2,3] solved the fixed charge network flow problem by a series of linear underestimate functions to approximate the cost function dynamically and recursively. Fontes *et al.* [4] exploited the optimal property of the spanning tree structure, and then used the swap algorithm to find an upper bound. Ortega [5] introduced the concepts of demand and supply super-nodes, found a feasible solution based on super-nodes, then applied a branch-and-cut algorithm, which is extended from Kim's slope scaling algorithm. For problems with non-convex functions or non-concave functions, which are also NP-hard problems, by exploiting some specific properties, some heuristic methods based on local search have been proposed. Chan et al. [6] considered the function having the properties (i) nondecreasing and (ii) the variable cost is nonincreasing. They exploited the ZIO (Zero Inventory Ordering) property and designed an algorithm with an upper bound no more than $\frac{4}{3}$ times the optimal cost and $\frac{5.6}{4.6}$ when the ordering cost function is constant. D. Shaw [8] presented a pseudo-polynomial algorithm for network problems with piecewise linear production costs and general holding costs by dynamic programming. Lamar [7] considered general nonlinear arc cost functions and converted it to an "equivalent" concave function on an extended network, and then applied any concave function minimization method to solve it.

However, in the previous studies, some practical applications are overlooked. We know that in the United States, trucking is the dominant mode of freight transportation and accounts for over 75 percent of the nation's freight bill. There are two ways to charge in trucking industry: full truckload (TL) and less than full truckload (LTL). Only LTL has been studied in Ref. 6. The way of TL operation is to charge for the full truck independent of the quantity shipped. The costs increase according to the incremental transportation capacity. In some capital-intensive industries, for example, the semiconductor industry, setup costs are so huge compared to the expenses of daily maintenance operations that the production functions are very close to staircase functions.

In this paper, we focus on the network flow problems with staircase cost functions and give its mathematical programming formulation. Based on variable redefinition, a tighter reformulation of these problems is also provided, which greatly reduced the problem size, including the number of binary variables and the number of constraints.

2. Problem Description

Over a fixed, finite planning horizon of T periods, a class of optimization models is proposed to coordinate production, transportation, and inventory decisions in a special supply chain network. It is not allowed the products transported between facilities. The products are only kept in factories. Assume that there is no inventory in retailers. Backlogging is also forbidden. The production and transportation costs are nondecreasing step functions, and the inventory holding cost is linear to the number of items. Without loss of generality, the model assumes no starting inventory. The objective is to minimize the total cost by assigning the appropriate production, inventory and transportation quantities to fulfill demands while the production, inventory, and transportation costs are step functions that can vary from period to period and from facility to facility. The multi-facility lot-sizing problem can be formulated using the following notation:

Parameters

T	number of periods in the planning horizon
P	number of factories
R	number of retailers
t	index of periods, $t \in \{1, \ldots, T\}$
p	index of factories, $p \in \{1, \ldots, P\}$
r	index of retailers, $r \in \{1, \ldots, R\}$
h_{tp}	unit inventory holding cost at factory p in period t
$Q_{tp}(q_{tp})$	production cost function of q_{tp} items produced at factory p in period t
$C_{tpr}(x_{tpr})$	transportation cost function of x_{tpr} items delivered from factory p to retailer r in period t
d_{tr}	demand at retailer r in period t
Uq_{tp}	production capacity at factory p in period t
Uc_{tpr}	transportation capacity from factory p to retailer r in period t

DECISION VARIABLES

q_{tp} number of items produced at factory p in period t

x_{tpr} number of items transported from factory p to retailer r at the end of period t

I_{tp} number of items in inventory at factory p at the end of period t

This problem can be formulated as a capacitated network flow problem:

$$\textit{minimize} \sum_{t=1}^{T}\sum_{p=1}^{P}\{Q_{tp}(q_{tp}) + h_{tp}I_{tp}\} + \sum_{t=1}^{T}\sum_{p=1}^{P}\sum_{r=1}^{R} C_{tpr}(x_{tpr})$$

subject to

$$I_{t-1,p} + q_{tp} = \sum_{r=1}^{R} x_{tpr} + I_{tp} \qquad \forall p, t \tag{1}$$

$$\sum_{p=1}^{P} x_{tpr} = d_{tr} \qquad \forall t, r \tag{2}$$

$$I_{p0} = 0 \qquad \forall p \tag{3}$$

$$0 \leq q_{tp} \leq Uq_{tp} \qquad \forall p, t \tag{4}$$

$$0 \leq x_{tpr} \leq Uc_{tpr} \qquad \forall p, t, r \tag{5}$$

$$I_{tp} \geq 0 \qquad \forall p, t \tag{6}$$

Where

$$Q_{tp}(q_{tp}) = \begin{cases} 0 & q_{tp} = 0 \\ Q_{tpk} & b_{tpk} < q_{tp} \leq b_{tpk+1} \end{cases}$$

$$C_{tpr}(x_{tpr}) = \begin{cases} 0 & x_{tpr} = 0 \\ C_{tprl} & e_{tprl} < x_{tprl} \leq e_{tprl+1} \end{cases}$$

are step functions.

The first term of the objective function describes the total production cost at every factory in all periods given the number of its products. The second term shows the total inventory cost incurred at every factory at the end of each period. The third term is the total transportation cost when products are delivered from factories to retailers. Here, as we mentioned above, it is assumed that the over-demand products are stored in factories. Constraints (1) and (2) are the flow conservation constraints at the production and demand points respectively. Without loss of generality, constraints

(3) assume that initial inventory in every facility is zero. Constraints (4) and (5) are the production capacity and transportation capacity constraints respectively. Constraints (6) are the general nonnegative constraints.

Obviously, this model is a nonlinear program. Staircase functions are neither convex nor concave. When the value switches from one level to the next, it results in a higher unit cost rather than a nonincreasing discount rate.

First, we reformulate it as a mixed 0-1 linear programming. We need more decision variables.

Additional Parameters of the Problem

K_{tp}	number of production price levels at factory p in period t
L_{tpr}	number of transportation price levels from factory p to retailer r in period t
k	index of K_{tp}, $k \in \{1, \ldots, K_{tp}\}$
l	index of L_{tpr}, $l \in \{1, \ldots, L_{tpr}\}$
Q_{tpk}	k_{th} production price level at factory p in period t
b_{tpk}	the capacity of the k_{th} production price level at factory p in period t
C_{tprl}	l_{th} transportation price level from facility p to retailer r in period t
e_{tprl}	the capacity of the l_{th} transportation price level from factory p to retailer r in period t

DECISION VARIABLES

q_{tp}	number of items produced at factory p in period t
x_{tpr}	number of items transported from factory p to retailer r at the end of period t
I_{tp}	number of items in inventory at factory p at the end of period t
y_{tpk}	if $q_{tp} \in k_{th}$ level, then $y_{tpk} = 1$; otherwise, 0
z_{tprl}	if $x_{tpr} \in l_{th}$ level, then $z_{tpkl} = 1$; otherwise, 0
ω_{tpki}	if $\omega_{tpk1} \leq b_{tpk+1}$, then $\omega_{tpk1} = 1$; otherwise, 0 if $\omega_{tpk2} > b_{tpk}$, then $\omega_{tpk2} = 1$; otherwise, 0
μ_{tprli}	if $\mu_{tprl1} \leq e_{tprl+1}$, then $\mu_{tprl1} = 1$; otherwise, 0 if $\mu_{tprk2} > e_{tprl}$, then $\mu_{tprl2} = 1$; otherwise, 0

Now, the problem is reformulated as a mixed-integer programming.

Problem (DMIP):

$$Min \sum_{t=1}^{T}\sum_{p=1}^{P}\sum_{k=1}^{K_{tp}} Q_{tpk}y_{tpk} + \sum_{t=1}^{T}\sum_{p=1}^{P}\sum_{r=1}^{R}\sum_{l=1}^{L_{tpr}} C_{tprl}z_{tprl} + \sum_{t=1}^{T}\sum_{p=1}^{P} h_{tp}I_{tp}$$

subject to

$$I_{t-1,p} + q_{tp} = \sum_{r=1}^{R} x_{tpr} + I_{tp} \qquad \forall p, t \tag{7}$$

$$\sum_{p=1}^{P} x_{tpr} = d_{tr} \qquad \forall r, t \tag{8}$$

$$I_{p0} = 0 \qquad \forall p \tag{9}$$

$$q_{tp} + M\omega_{tpk1} \geq b_{tpk+1} + 1 \qquad \forall p, t; \forall k \cap k < K_{tp} \tag{10}$$

$$-q_{tp} + M\omega_{tpk2} \geq -b_{tpk} \qquad \forall p, t; \forall k \cap k < K_{tp} \tag{11}$$

$$\omega_{tpk1} + \omega_{tpk2} - My_{tpk} \leq 1 \qquad \forall p, t; \forall k \cap k < K_{tp} \tag{12}$$

$$x_{tpr} + M\mu_{tprl1} \geq e_{tprl+1} + 1 \qquad \forall p, t, r; \forall l \cap l < L_{tpr} \tag{13}$$

$$-x_{tpr} + M\mu_{tprl2} \geq -e_{tprl} \qquad \forall p, t, r; \forall l \cap l < L_{tpr} \tag{14}$$

$$\mu_{tprl1} + \mu_{tprl2} - Mz_{tprl} \leq 1 \qquad \forall p, t, r; \forall l \cap l < L_{tpr} \tag{15}$$

$$0 \leq q_{tp} \leq b_{tpK_{tp}} \qquad \forall p, t \tag{16}$$

$$0 \leq x_{tpr} \leq e_{tprL_{tpr}} \qquad \forall p, t, r \tag{17}$$

$$I_{tp} \geq 0 \qquad \forall p, t \tag{18}$$

$$y_{tpr} \in \{0, 1\} \qquad \forall p, t, r, \tag{19}$$

$$z_{tprl} \in \{0, 1\} \qquad \forall p, t, r, l \tag{20}$$

$$\omega_{tpki} \in \{0, 1\} \qquad \forall p, t, k; i = 1, 2 \tag{21}$$

$$\mu_{tprli} \in \{0, 1\} \qquad \forall p, t, r, l; i = 1, 2 \tag{22}$$

Here, M is a very large positive number.

The first term in the objective function still demonstrates the total production cost. When the number of products is decided, the corresponding index y_{tpk} is set to one. Thus, the value of production cost is singled out from the step function. Similarly, the second term shows the total transportation cost from factories to retailers in the planning horizon. The third term describes the inventory cost incurred in the factories.

Constraints (7) and (8) are the classical flow conservation constraints at the production and demand points respectively. Constraints (9) assume that initial inventory in every factory is zero. Constraints (10) demonstrate that if the quantity of production q_{tp} is less than some certain capacity level b_{tpk+1}, then its corresponding index ω_{tpk1} will be set to one. On the contrary, constraints (11) show that if the number of items q_{tp} is more than or equal to some certain capacity level b_{tpk}, then the corresponding index ω_{tpk2} will be assigned to one. Constraints (12) describe that if the quantity of production q_{tp} belongs to some segment $(b_{tpk}, b_{tpk+1}]$, the index of whether or not to produce (y_{tpk}) is set to one. Thus the corresponding cost value will be singled out. In a similar way, constraints (13), (14), and (15) describe the same situation in transportation that if the quantity of products delivered is within a certain range in segments, then the corresponding index will be chosen. Constraints (16) and (17) represent production capacity and transportation capacity constraints respectively. Constraints (18) are the general nonnegative constraints. Constraints (19), (20), (21), and (22) are the general binary constraints.

In this model, for every segment we add three extra binary variables. In all the total number of binary variables is three times as many as the total number of segments. Let us define the number of total segments including production segments and transportation segments as L^{Sum}. Then the number of total variables is $2TP + TPR + 3L^{Sum}$. The number of constraints is $T(P + R) + 3L^{Sum}$.

For this formulation, we can see that many binary variables needed. Besides, the constraints (10), (11), (14), and (15) are not very tight. If we consider the quantity produced in one period and satisfied part of demand in just one period, which is the current period or the following periods, a tighter model will be formulated.

3. A Tighter Formulation

The quantity of items (q_{tp}) produced at facility p in period t is split into several parts $(q_{pt\tau})$ according to which period it supplies, where $\tau \in \{t, \dots, T\}$. A series of decision variables $u_{pt\tau}$ are defined, which is the percentage of $(q_{pt\tau})$ in the total demand in period τ. If we aggregate the demand in every period and represent it by D_t, it is clear that $q_{tp} = \sum_{\tau=t}^{T} u_{pt\tau} D_t$. In this problem, we need to consider the capacity levels. Since $q_{pt\tau}$ implies that the items should be stored at factory p from period t until period τ, we need the unit inventory holding cost during this interval. We also consider step functions as an accumulation of a series of minor setup costs if we switch the capacity to the next level, then we need pay more which equals to difference between both levels which is called minor setup cost. Next, the derived parameters are given.

Derived Parameters

D_t	the total demand in period t. $D_t = \sum_{r=1}^{R} d_{tr}$
$H_{pt\tau}$	unit inventory holding cost at factory p from period t to period τ. $H_{pt\tau} = \sum_{t'=t}^{\tau-1} h_{t'p}$
F_{tp}^k	the additional production cost from price level k to $k+1$ at factory p in period t. Assume $F_{tp}^1 = Q_{tp1}$
G_{tpr}^l	the additional transportation cost from price level l to $l+1$ when products are delivered from factory p to retailer r in period t. Assume $G_{tpr}^1 = C_{tpr1}$
b_{tp}^k	the difference between the k_{th} and $k+1_{st}$ production capacity level at factory p in period t
e_{tpr}^l	the difference between the l_{th} and $l+1_{st}$ transportation capacity level from factory p to retailer r in period t
k	index of K_{tp}, $k \in \{1, \ldots, K_{tp}\}$
l	index of L_{tpr}, $l \in \{1, \ldots, L_{tpr}\}$

DECISION VARIABLES

$u_{pt\tau}^k$	the percentage of items produced at factory p in period t to supply period τ in the demand of period τ at the k_{th} level, where $\tau \in \{t, \ldots, T\}$. Clearly $q_{tp}^k = \sum_{\tau=t}^{T} u_{pt\tau}^k D_\tau$
y_{tp}^k	if factory p produces in period t at price level k, then $y_{tp}^k = 1$; otherwise, 0
v_{tpr}^l	the percentage of items delivered from factory p to retailer r in period t at level l
z_{tpr}^l	if items are delivered from factory p to retailer r in period t at level l, then $z_{tpr}^l = 1$; otherwise, 0

Now, the problem can be reformulated as following.

Problem (ETFP):

$$Min \sum_{t=1}^{T}\sum_{p=1}^{P}\sum_{k=1}^{K_{tp}} F_{tp}^k y_{tp}^k + \sum_{t=1}^{T}\sum_{p=1}^{P}\sum_{r=1}^{R}\sum_{l=1}^{L_{tpr}} G_{tpr}^l z_{tpr}^l + \sum_{p=1}^{P}\sum_{t=1}^{T}\sum_{\tau=t}^{T}\sum_{k=1}^{K_{tp}} H_{pt\tau} u_{pt\tau}^k D_\tau$$

subject to

$$\sum_{p=1}^{P}\sum_{\tau=1}^{t}\sum_{k=1}^{K_{\tau p}} u_{p\tau t}^{k} = 1 \qquad \forall t \tag{23}$$

$$\sum_{\tau=t}^{T} u_{pt\tau}^{k} D_{\tau} \le b_{tp}^{k} \qquad \forall p, t, k \tag{24}$$

$$u_{pt\tau}^{k} \le y_{tp}^{k} \qquad \forall p, t, k, \tau \tag{25}$$

$$y_{tp}^{k} \ge y_{tp}^{k+1} \qquad \forall p, t; \forall k \cap k < K_{tp} - 1 \tag{26}$$

$$\sum_{p=1}^{P}\sum_{l=1}^{L_{tpr}} v_{tpr}^{l} = 1 \qquad \forall r, t \tag{27}$$

$$v_{tpr}^{l} d_{tr} \le e_{tpr}^{l} \qquad \forall p, t, l, r \tag{28}$$

$$v_{tpr}^{l} \le z_{tpr}^{l} \qquad \forall p, t, l, r \tag{29}$$

$$z_{tpr}^{l} \ge z_{tpr}^{l+1} \qquad \forall p, t, r; \forall l \cap l < L_{tpr} - 1 \tag{30}$$

$$\sum_{\tau=1}^{t}\sum_{k=1}^{K_{\tau p}} u_{p\tau t}^{k} D_{t} = \sum_{r=1}^{R}\sum_{l=1}^{L_{tpr}} v_{tpr}^{l} d_{tr} \qquad \forall p, t \tag{31}$$

$$u_{pt\tau}^{k} \ge 0 \qquad \forall k, p, t, \tau = t, ...T \tag{32}$$

$$y_{tp}^{k} \in \{0, 1\} \qquad \forall p, t, k \tag{33}$$

$$v_{tpr}^{l} \ge 0 \qquad \forall p, t, r, l \tag{34}$$

$$z_{tpr}^{l} \in \{0, 1\} \qquad \forall p, t, r, l \tag{35}$$

In the objective function, the first term demonstrates the total production cost. The difference from previous model is that if y_{tp}^{k} equals one, then $y_{tp}^{1} \cdots y_{tp}^{k-1}$ are all assigned to one since the F_{tp}^{k} is the minor setup cost. Since the cost of lower level segments is less than that of higher level segments, the lower level capacities are filled first. In the previous model, if y_{tpk} is one, then the other index of factory p in period t must be zero. In a similar way, the second term shows the total transportation cost from factories to retailers at the planning horizon. The third term describes the inventory cost incurred in the factories, where $u_{pt\tau}^{k} D_{\tau}$ is the items produced at factory p in period t to fulfill demands in period τ.

Constraints (23) demonstrate the demand in period t is supplied by items produced from period 1 to period t at all levels. Constraints (24) describe the items produced qualifying for a certain price level is bounded above. Constraints (25) indicate that if there are some items to satisfy the demand, the corresponding indicator will be set to one. Constraints (26) demonstrate that if items are produced at the k_{th} price level then they should be produced at all of proceeding levels. Similarly, constraints (27) demonstrate demand in retailer r is supplied by all the plants at all price levels. Constraints (28) describe the items transported qualifying for a certain price level is bounded above. Constraints (29) indicate that if there are some items to satisfy the demand, the corresponding indicator will be set to one. Constraints (30) demonstrate that if items are transported at the l_{th} price level then they should be produced at all of proceeding levels. Constraints (31) are the flow conservation constraints, items produced should be equal to those delivered. Constraints (32), (33), (34), and (35) are nonnegative and binary constraints.

We have already known the property that y_{tp}^k are consecutive one followed by a series of zeros. According to this, Constraints (25) can be converted to

$$\sum_{\tau=t}^{T} u_{pt\tau}^k \leq (T-t)y_{tp}^k \qquad \forall t, p$$

This formulation reduces the binary variables from $3L^{Sum}$ to L^{Sum}. And the number of variables is $O(L^{Sum})$. The total number of constraints is decreased to $T(R+1) - TPR + 3L^{Sum}$ now.

4. Experimental Design and Computational Results

In this section, we report the performance of the two formulations on a Pentium 4 2.8GHz personal computer with 512MB RAM. First, an interactive data-generated program is developed. The number of periods, factories, and retailers need input. And the number of price levels in every staircase cost function also need to be given. However, the default is that every production cost function has two segments and every transportation cost function has three segments since in reality the flexibility of production capacity is less than that of transportation. Product demands are assumed to be uniformly distributed between [20, 80] in every period. Unit inventory holding costs are randomly generated from a uniform distribution between [3.00, 7.00] with a mean of \$5.00 and deviation of \$1.33. Manufacture's capacity increases as a uniform distribution with a mean of 0.9 times that of total demand and a standard deviation of 0.36 times that of total demand. While the production cost increases as a uniform distribution between 0 and 700 times the mean of unit inventory cost. The transportation capacity is assigned up by a uniform distribution between 0 and 8 times the mean of demand.

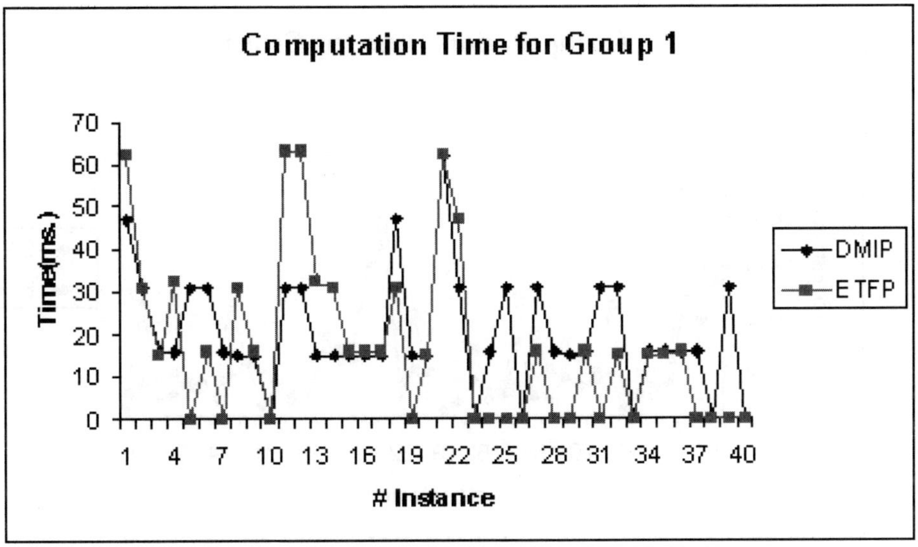

Fig. 1. CPU time for Group 1

Table 1. Problem Characteristics

Group	Period	Plant	Retailer	Total Steps	Nodes	Arcs
1	2	3	2	48	13	53
2	10	3	2	240	61	269
3	20	3	2	480	121	539
4	10	4	5	680	101	718

Table 2. Size of both formulations

Group	# Constraints		# Total Var.		# Binary Var.	
	DMIP	ETFP	DMIP	ETFP	DMIP	ETFP
1	154	138	168	102	144	48
2	770	690	840	750	720	240
3	1540	1380	1680	2100	1440	480
4	2130	1900	2320	1720	2040	680

We can change the problem characteristics to get more problem classes, such as changing the length of the time horizon, the number of plants and retailers, and the segment number of cost functions. Four groups are generated and in each group 40 instances are included. We coded the two formulations in Microsoft Visual C++ 7.0 and called CPLEX callable libraries

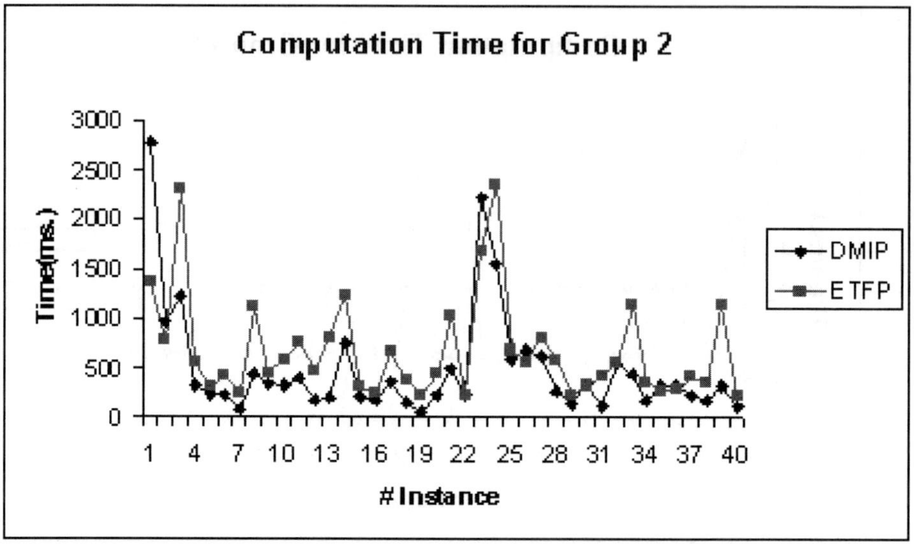

Fig. 2. CPU time for Group 2

8.0 to solve them. In the Problem (ETFP), since the nonbinary variables are all between 0 and 1, to avoid the round error by digital limitation they are scaled up by one hundred.

Table 3. Results of problems

Group	# Opt.		Rel. Er.(%)		Avg. Times (ms.)		Max. Time (ms.)	
	DMIP	ETFP	DMIP	ETFP	DMIP	ETFP	DMIP	ETFP
1	40	40	0	0	807	688	62	63
2	38	33	1.26	0.62	464.736	676.447	2781	2359
3	34	20	0.15	0.81	1662.618	6215.450	4688	28063
4	33	24	–	0.36	3442.485	43663.417	11000	202610

The problem characteristics are shown in Table 1. First, a small size problem is provided, then we enlarge time periods. The number of segments is increased in proportion. Next, we test the performance by a little larger set of facilities. Table 2 gives the problem sizes of the two formulations, including the number of constraints, variables, and binary variables. It is easy to see that even for a small set of facilities and a couple of periods, when we consider the effect of step cost functions, the problem size

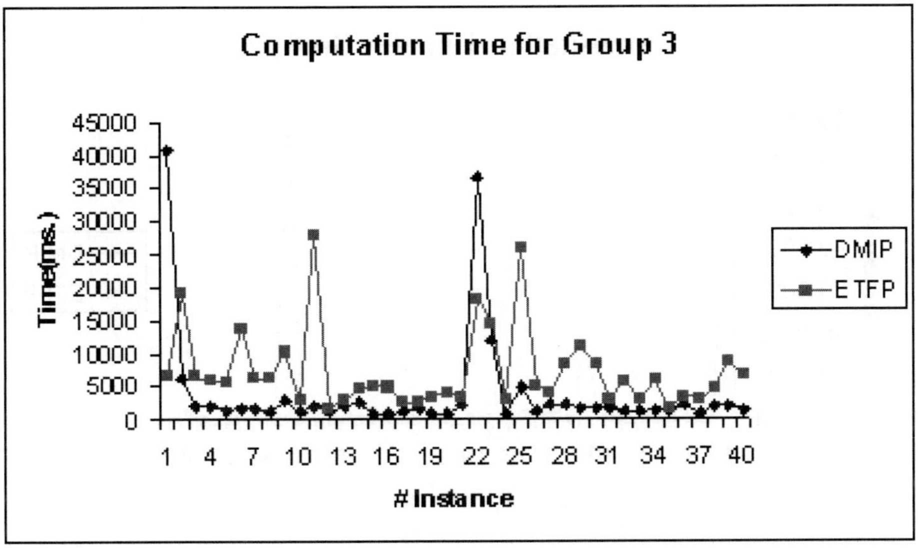

Fig. 3. CPU time for Group 3

exponentially increases. Some computational results are offered in Table 3. The second column gives the number of optimum obtained by the two formulations respectively. The third column shows the average relative error of the near optimal solutions. Here, average relative error is defined as the absolute value of difference between the near optimal solution and the optimum over the optimum. The meaning of average time and maximum time is straightforward. However, here we only consider those instances having optimal solutions. Figures 1 – 4 visualize the computation time for four groups of the two formulations, respectively. For Group 4, since those achieved a near optimal solution in DMIP also obtained a near optimal solution in ETFP, the relative error is unavailable. So we put "–" in that cell.

We observed from Table 3 that ETFP has a tendency to need more time to solve problems. However, compared to DMIP, it needs less time to obtain a near optimal solution. Figure 4 illustrates this property. This is expected since ETFP has tighter upper bounds on production arcs, which reduces the feasible region to search for the optimal solution. This property can be exploited in heuristic algorithms for large-scale problems expected

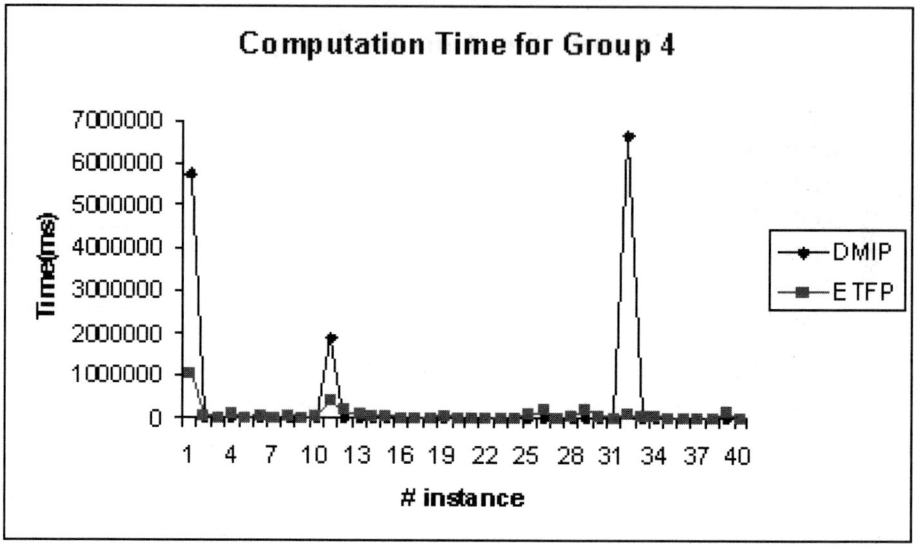

Fig. 4. CPU time for Group 4

to be significantly beneficial.

5. Conclusion

In this paper, we have considered network flow problems with step cost functions. First, an equivalent mathematical formulation is given as a mixed 0-1 linear programming problem. The computational experience suggests that a tighter formulation is possible. It is promising to exploit this property to find practical heuristic algorithms for large size problems. In the test bed of the problem with 20 periods, four plants, and five retailers with total 1360 segments, some instances failed to be optimized by CPLEX. More than half of the instances only can get the near optimal solutions. Thus, it is important to extend the proposed formulations to develop practical algorithms to solve large problems in further research.

References

1. D.Kim and P.M.Pardalos, A solution approach to the fixed charge network flow problem using a dynamic slope scaling procedure, *Operations Research Letters* **24**, pp. 195-203 (1999).

2. D.Kim and P.M.Pardalos, Dynamic Slope Scaling and Trust Interval Techniques for solving Concave Piecewise Linear Network Flow Problems, *Networks* **35**, pp. 216-222 (2000).
3. D.Kim and P.M.Pardalos, A Dynamic Domain Contraction Algorithm for Nonconvex Piecewise Linear Network Flow Problems, *Journal of Global Optimization* **17**, pp. 225-234 (2000).
4. D. M. Fontes, E. Hadjiconstantinou, and N. Christofides, Upper Bounds for Single-source Uncapacitated Concave Minimum-Cost Network Flow Problems, *Networks* **41**, pp. 221-228 (2003).
5. F. Ortega, L. A. Wolsey, A Branch-and-Cut Alg. for the Single-commodity, Uncapacitated, Fixed-charge Network Flow Problem, *Networks* **41**, pp. 143-158 (2003).
6. L. Chan, A. Muriel, Z. Shen, D. Simchi-levi, On the effectiveness of zero-inventory-ordering policies for the economic lot-sizing model with a class of piecewise linear cost structures, *Operations Research* **50**, pp. 1058-1067 (2002).
7. B. W. Lamar, A method for solving Network Flow Flow Problems with General Nonlinear Arc Costs, in *Network optimization Problems*, D.-Z. Du and P. M. Pardalos(eds.), pp. 147-167, World Scientific (1993).
8. D. X. Shaw, A. P. M. Wagelmans, An algorithm for Single-Item Capacitated Economic Lot Sizing with Piecewise Linear Production Costs and General Holding Costs, *Management Science* **44**, pp. 831-838 (1998)
9. G. M. Guisewhite and P.M.Pardalos, Single-Source Uncapacitated Minimum Concave Cost Network Flow Problems, in H.E. Bradley(ed.), *Operational Research '90*, Pergamon Press, Ocford, England, pp. 703-713 (1990).
10. G. M. Guisewhite and P.M.Pardalos. Minimum Concave-Cost Network Flow Problems: Applications, Complexity, and Algorithms, *Annals of Operations Research* **25**, pp. 75-100 (1990).
11. G. M. Guisewhite and P.M.Pardalos. Algorithms for the Single-Source Uncapacitated Minimum Concave-Cost Network Flow Problems, *Journal of Global Optimization* **1**, pp. 309-330 (1991).
12. D.B. Khang and O. Fujiwara. Approximate Solutions of Capacitated Fixed-Charge Minimum Cost Network Flow Problems, *Networks* 21, pp. 689-704 (1991).

CHAPTER 19

MODELS FOR INTEGRATED CUSTOMER ORDER SELECTION AND REQUIREMENTS PLANNING UNDER LIMITED PRODUCTION CAPACITY

K. Taaffe

Industrial and Systems Engineering Department,
University of Florida, PO Box 116595, Gainesville, FL 32611
E-mail: taafe@ufl.edu

J. Geunes

Industrial and Systems Engineering Department,
University of Florida, PO Box 116595, Gainesville, FL 32611
E-mail: geunes@ise.ufl.edu

Manufacturers regularly face the challenge of determining the best allocation of production resources to customer orders in make-to-order systems. Past research on dynamic requirements planning problems has led to models and solution methods that help production planners to effectively address this challenge. These models typically assume that the orders the production facility must meet are exogenously determined and serve as input parameters to the model. In contrast, we approach the problem by allowing the production planning model to implicitly decide which among all outstanding orders a production facility should satisfy in order to maximize the contribution to profit from production. The *order selection* models we provide generalize classical capacitated lot-sizing problems by integrating order-selection and production-planning decisions under limited production capacities. Building on prior analysis of an uncapacitated version of the problem, this chapter studies strong problem formulations and develops heuristic solution algorithms for several capacitated versions. Using a broad set of more than 3,000 randomly generated test problems, these heuristic solution methods provided solutions that were, on average, within 0.67% of the optimal solution value.

1. Introduction

Firms that produce made-to-order goods often make critical order acceptance decisions prior to planning production for the orders they ultimately accept. These decisions require the firm's representatives (typically sales/marketing personnel in consultation with manufacturing management) to determine which among all customer orders the firm will satisfy. In certain contexts, such as those involving highly customized goods, the customer works closely with sales representatives to define an order's requirements and, based on these requirements, the status of the production system, and the priority of the order, the firm quotes a lead time for order fulfillment, which is then accepted or rejected by the customer (see Yano [28]). In other competitive settings, the customer's needs are more rigid and the customer's order must be fulfilled at a precise future time. The manufacturer can either commit to fulfilling the order at the time requested by the customer, or decline the order based on several factors, including the manufacturer's capacity to meet the order and the economic attractiveness of the order. These "order acceptance and denial" decisions are typically made prior to establishing future production plans and are most often made based on the collective judgment of sales, marketing, and manufacturing personnel, without the aid of the types of mathematical decision models typically used in the production planning decision process.

When the manufacturing organization is highly capacity constrained and customers have firm delivery date requirements, it is often necessary to satisfy a subset of customer orders and to deny an additional set of potentially profitable orders. In some contexts, the manufacturer can choose to employ a rationing scheme in an attempt to satisfy some fraction of each customer's demand (see Lee, Padmanabhan, and Whang [18]). In other settings, such a rationing strategy cannot be implemented, i.e., it may not be desirable or possible to substitute items ordered by one customer in order to satisfy another customer's demand. In either case, it may be necessary for the firm to deny certain customer orders (or parts of orders) so that the manufacturer can meet the customer-requested due dates for the orders it accepts. Assessing the profitability of an order in isolation, prior to production planning, leads to myopic decision rules that fail to consider the best set of actions from an overall profitability standpoint. The profitability of an order, when gauged solely by the revenues generated by the order and perceived customer priorities, neglects the impacts of important operations cost factors, such as the opportunity cost of manufacturing capacity consumed by the order, as well as economies of scale in production. Decisions on the collective set of orders the organization should accept can be a critical determinant of the firm's profitability. Past operations modeling literature has not fully addressed integrated customer order selection and production planning decisions in make-to-order systems. This chapter partially fills this gap by developing modeling and solution approaches for integrating these decisions in single-stage systems with dynamic demand and production capacity limits.

Wagner and Whitin [27] first addressed the basic uncapacitated Economic Lot-Sizing Problem (ELSP), and numerous extensions and generalizations of this problem have subsequently been addressed in the literature (e.g., Zangwill,[29] Love,[20] Thomas,[25] Afentakis, Gavish, and Karmarkar,[2] and Afentakis and Gavish [1]). A substantial amount of research on the capacitated version of the lot-sizing problem (CLSP) also exists, beginning with the work of Florian and Klein [11] (see also Baker, Dixon, Magazine, and Silver,[4] and Florian, Lenstra, and Rinnooy Kan [12]). The development and application of strong valid inequalities for the mixed integer programming formulation of the CLSP (beginning in the 1980s) has allowed researchers to solve large problem instances in acceptable computing time (e.g., Barany, Van Roy, and Wolsey,[6] Pochet,[22] and Leung, Magnanti, and Vachani [19]). See Lee and Nahmias,[17] Shapiro [23], and Baker [3] for detailed discussions on dynamic requirements planning problems.

Geunes, Romeijn, and Taaffe [14] addressed the uncapacitated requirements planning problem with *order selection flexibility*. Given a set of outstanding customer orders over a finite horizon, fixed plus variable production costs, and end-of-period (variable) holding costs in each period, they develop a model that determines the order selection, production quantity, and inventory holding decisions in each period that maximize net contribution to profit. In this chapter we generalize this model to account for time-varying, finite production capacities in each period. While the uncapacitated version is solvable in polynomial time, as we later discuss, the capacitated version is NP-Hard and therefore requires customized heuristic solution approaches. The main contributions of this chapter include the generalization of the class of order selection problems (see Geunes, Romeijn, and Taaffe [14]) to address settings with limited production capacities, and the development of optimization-based modeling and solution methods for this class of problems. We extend a tight formulation of the uncapacitated version of the problem to a capacitated setting, which often allows solving general capacitated instances via branch-and-bound in reasonable computing time. For those problems that cannot be solved via branch-and-bound in reasonable time, we provide a set of three effective heuristic solution methods. Computational test results indicate that the proposed solution methods for the general capacitated version of the problem are very effective, producing solutions within 0.67% of optimality, on average, for a broad set of 3,240 randomly generated problem instances.

Lee, Çetinkaya, and Wagelmans [16] recently considered contexts in which demands can be met either earlier (through early production and delivery) or later (through backlogging) than specified without penalty, provided that demand is satisfied within certain *demand time windows*, for the uncapacitated, single-stage lot-sizing problem. Their model still assumes ultimately, however, that all demand must be filled during the planning horizon, while our approach does not consider the notion of time windows. Charnsirisakskul, Griffin, and Keskinocak [8] also consider a context allowing flexibility in order deliveries. Their model emphasizes the benefits of producer flexibility in setting lead times for individual orders. Integrating lead-time quotation, order selection and production planning decisions, they determine under what conditions lead-time flexibility is most useful for increasing the producer's profits. In many industries, the producer may not enjoy the flexibility to choose, within certain limits, the lead time for product delivery. Our model considers this more restrictive case, emphasizing algorithmic approaches for efficiently solving the new problem class we define.

The remainder of this chapter is organized as follows. Section 2 presents a formal definition and mixed integer programming formulation of the general capacitated production planning problem with order selection flexibility. In Section 3 we consider various mixed integer programming formulations of this problem, along with the advantages and disadvantages of each formulation strategy. We also provide several heuristic solution approaches for capacitated problem instances. Section 4 provides a summary of a set of computational tests used to gauge the effectiveness of the formulation strategies and heuristic solution methods described in Section 3. Section 5 concludes with a summary and directions for future research.

2. Order Selection Problem Definition and Formulation

Consider a producer who manufactures a good to meet a set of outstanding orders over a finite number of time periods, T. Producing the good in any time period t requires a production setup at a cost S_t and each unit costs an additional p_t to manufacture. We let $M(t)$ denote the set of all orders that request delivery in period t (we assume zero delivery lead time for ease of exposition; the model easily extends to a constant delivery lead time without loss of generality), and let m denote an index for orders. The manufacturer has a capacity to produce C_t units in period t, $t = 1, \ldots, T$. We assume that that no shortages are permitted, i.e., no planned backlogging[a], and that items can be held in inventory at a cost of h_t per unit remaining at the end of period t. Let d_{mt} denote the quantity of the good requested by order m for period t delivery, for which the customer will pay r_{mt} per unit, and suppose the producer is free to choose any quantity between zero and d_{mt} in satisfying order m in period t (i.e., rationing is possible, and the customer will take as much of the good as the supplier can provide, up to d_{mt}). The producer thus has the flexibility to decide which orders it will choose to satisfy in each period and the quantity of demand it will satisfy for each order. If the producer finds it unprofitable to satisfy a certain order in a period, they can choose to reject the order at the beginning of the planning horizon. The manufacturer incurs a fixed shipping cost for delivering order m in period t equal to F_{mt} (any variable shipping cost can be subtracted from the revenue term, r_{mt}, without loss of generality). The producer, therefore, wishes to maximize net profit over a T-period horizon, defined as the total revenue from orders satisfied minus total production (setup + variable), holding, and delivery costs incurred over the horizon. To formulate this problem we define the following decision variables:

$$x_t = \text{Number of units produced in period } t,$$

$$y_t = \begin{cases} 1, \text{ if we setup for production in period } t, \\ 0, \text{ otherwise,} \end{cases}$$

$$I_t = \text{Producer's inventory remaining at the end of period } t,$$

$$v_{mt} = \text{Proportion of order } m \text{ satisfied in period } t,$$

$$z_{mt} = \begin{cases} 1, \text{ if we satisfy any positive fraction of order } m \text{ in period } t, \\ 0, \text{ otherwise.} \end{cases}$$

We formulate the Capacitated Order Selection Problem (OSP) as follows.
[OSP]

$$\text{maximize} \sum_{t=1}^{T} \left(\sum_{m \in M(t)} (r_{mt} d_{mt} v_{mt} - F_{mt} z_{mt}) - S_t y_t - p_t x_t - h_t I_t \right) \quad (1)$$

subject to:

[a]Extending our models and solution approaches to allow backlogging at a per unit per period backlogging cost is fairly straightforward. We have chosen to omit the details for the sake of brevity.

Inventory Balance:

$$I_{t-1} + x_t = \sum_{m \in M(t)} d_{mt} v_{mt} + I_t, \; t = 1, \ldots, T, \tag{2}$$

Capacity/Setup Forcing:

$$0 \leq x_t \leq C_t y_t, \; t = 1, \ldots, T, \tag{3}$$

Demand Bounds:

$$0 \leq v_{mt} \leq z_{mt}, \; t = 1, ..., T, m \in M(t), \tag{4}$$

Nonnegativity:

$$I_0 = 0, I_t \geq 0, \; t = 1, \ldots, T, \tag{5}$$

Integrality:

$$y_t, z_{mt} \in \{0, 1\}, \; t = 1, ..., T, m \in M(t). \tag{6}$$

The objective function (1) maximizes net profit, defined as total revenue less fixed shipping and total production and inventory holding costs. Constraint set (2) represents inventory balance constraints, while constraint set (3) ensures that no production occurs in period t if we do not perform a production setup in the period. If a setup occurs in period t, the production quantity is constrained by the production capacity, C_t. Constraint set (4) encodes our assumption regarding the producer's ability to satisfy any proportion of order m up to the amount d_{mt}, while (5) and (6) provide nonnegativity and integrality restrictions on variables. Observe that we can force any order selection (z_{mt}) variable to one if qualitative and/or strategic concerns (e.g., market share goals) require satisfying an order regardless of its profitability.

In this chapter we investigate not only the OSP model as formulated above, but also certain special cases and restrictions of this model that are of both practical and theoretical interest. In particular, we consider the special case in which no fixed delivery charges exist, i.e., the case in which all fixed delivery charge (F_{mt}) parameters equal zero. We denote this version of the model as the OSP-NDC. We also explore contexts in which customers do not permit partial demand satisfaction, i.e., a restricted version of the OSP in which the continuous v_{mt} variables must equal the binary delivery-charge forcing (z_{mt}) variable values, and can therefore be substituted out of the formulation; let OSP-AND denote this version of the model (where AND implies all-or-nothing demand satisfaction). Observe that for the OSP-AND model we can introduce a new revenue parameter

$R_{mt} \equiv r_{mt}d_{mt}$, where the total revenue from order m in period t must now equal $R_{mt}z_{mt}$. Table 1 defines our notation with respect to the different variants of the OSP problem.

Table 1. Classification of model special cases and restrictions.

Model	Fixed Delivery Charges Exist	Partial Order Satisfaction Allowed
OSP	Y	Y
OSP-NDC	N	Y
OSP-AND	U	N

Y = Yes; N = No.
U: Model and solution approaches unaffected by this assumption.

We distinguish between these model variants not only because they broaden the model's applicability to different contexts, but also because they can substantially affect the model's formulation size and complexity, as we next briefly discuss. Let $M_\Sigma = \sum_{t=1}^{T} |M(t)|$ denote the total number of customer orders over the T-period horizon, where $|M(t)|$ is the cardinality of the set $M(t)$. Note that the [OSP] formulation contains $M_\Sigma + T$ binary variables and $M_\Sigma + 2T$ constraints, not including the binary and non-negativity constraints. The OSP-NDC model, on the other hand, in which $F_{mt} = 0$ for all order-period (m, t) combinations, allows us to replace each z_{mt} variable on the right-hand-side of constraint set (4) with a 1, and eliminate these variables from the formulation. The OSP-NDC model contains only T binary variables and therefore requires M_Σ fewer binary variables than the [OSP] formulation, a significant reduction in problem size and complexity. In the OSP-AND model, customers do not allow partial demand satisfaction, and so we require $v_{mt} = z_{mt}$ for all order-period (m, t) combinations; we can therefore eliminate the continuous v_{mt} variables from the formulation. While the OSP-AND, like the OSP, contains $M_\Sigma + T$ binary variables, it requires M_Σ fewer total variables than the [OSP] formulation as a result of eliminating the v_{mt} variables. Table 2 summarizes the size of each of these variants of the OSP with respect to the number of constraints, binary variables, and total variables.

Based on the information in this table, we would expect the OSP and OSP-AND to be substantially more difficult to solve than the OSP-NDC. As we will show in Section 4, the OSP-AND actually requires the greatest amount of computation time on average, while the OSP-NDC requires the least.

Table 2. Problem Size Comparison for Capacitated Versions of the OSP.

	OSP	**OSP-NDC**	**OSP-AND**
Number of Constraints *	$M_\Sigma + 2T$	$M_\Sigma + 2T$	$2T$
Number of Binary Variables	$M_\Sigma + T$	T	$M_\Sigma + T$
Number of Total Variables	$2M_\Sigma + 3T$	$M_\Sigma + 3T$	$M_\Sigma + 3T$

*Binary restriction and nonnegativity constraints are not included.

Note that the OSP-AND is indifferent to whether fixed delivery charges exist, since we can simply reduce the net revenue parameter, $R_{mt} \equiv r_{mt}d_{mt}$, by the fixed delivery-charge value F_{mt}, without loss of generality. In the OSP-AND then, the net revenue received from an order equals $R_{mt}z_{mt}$, and we thus interpret the z_{mt} variables as binary "order selection" variables. In contrast, in the OSP, the purpose of the binary z_{mt} variables is to force us to incur the fixed delivery charge if we satisfy any fraction of order m in period t. In this model we therefore interpret the z_{mt} variables as fixed delivery-charge forcing variables, since their objective function coefficients are fixed delivery cost terms rather than net revenue terms, as in the OSP-AND. Note also that since both the OSP-NDC and the OSP-AND require only one set of order selection variables (the continuous v_{mt} variables for the OSP-NDC and the binary z_{mt} variables for the OSP-AND), their linear programming relaxation formulations will be identical (since relaxing the binary z_{mt} variables is equivalent to setting $z_{mt} = v_{mt}$). The OSP linear programming relaxation formulation, on the other hand, explicitly requires both the v_{mt} and z_{mt} variables, resulting in a larger LP relaxation formulation than that for the OSP-NDC and the OSP-AND. These distinctions will play an important role in interpreting the difference in our ability to obtain strong upper bounds on the optimal solution value for the OSP and the OSP-AND in Section 4.3. We next discuss solution methods for the OSP and the problem variants we have presented.

3. OSP Solution Methods

To solve the OSP, we must decide which orders to select and, among the selected orders, how much of the order we will satisfy while obeying capacity limits. We can show that this problem is NP-Hard through a reduction from the capacitated lot-sizing problem as follows. If we consider the special case of the OSP in which $\sum_{t=1}^{j} C_t \geq \sum_{t=1}^{j} \sum_{m \in M(t)} d_{mt}$ for $j = 1, \ldots, T$ (which implies that satisfying all orders is feasible) and $\min_{t=1,\ldots,T, m \in M(t)} \{r_{mt}\} \geq \max_{t=1,\ldots,T} \{S_t\} + \max_{t=1,\ldots,T} \{p_t\} + \sum_{t=1}^{T-1} h_t$ (which implies that it is profitable to satisfy all orders in every period), then total revenue is fixed and the problem is equivalent to a capacitated lot-sizing problem, which is an NP-Hard optimization problem (see Florian and Klein [11]).

Given that the OSP is NP-Hard, we would like to find an efficient method for obtaining good solutions for this problem. As our computational test results in Section 4 later show, we were able to find optimal solutions using branch-and-bound for many of our randomly generated test instances. While this indicates that the majority of problem instances we considered were not terribly difficult to solve, there were still many instances in which an optimal solution could not be found in reasonable computing time. Based on our computational test experience in effectively solving problem instances via branch-and-bound using the CPLEX 6.6 solver, we focus on strong LP relaxations for the OSP that provide quality upper bounds on optimal net profit quickly, and often enable solution via branch-and-bound in acceptable computing time. For those problems that cannot be solved via branch-and-bound, we employ several customized heuristic methods, which we discuss in Section 3.2. Before we discuss the heuristics used to obtain lower bounds for the OSP, we first present our reformulation strategy, which helps to substantially improve the upper bound provided by the linear programming relaxation of the OSP.

3.1. *Strengthening the OSP Formulation*

This section presents an approach for providing good upper bounds on the optimal net profit for the OSP. In particular, we describe two LP relaxations for the OSP, both of which differ from the LP relaxation obtained by simply relaxing the binary restrictions of the [OSP] formulation (constraint set (6)) in Section 2. We will refer to this simple LP relaxation of the [OSP] formulation as OSP-LP, to distinguish this relaxation from the two LP relaxation approaches we provide in this section.

The two LP relaxation formulations we next consider are based on a reformulation strategy developed for the uncapacitated version of the OSP, which we will refer to as the UOSP. Geunes, Romeijn, and Taaffe [15] provide a "tight" formulation of the UOSP for which they show that the optimal linear programming relaxation solution value equals the optimal (mixed integer) UOSP solution value (note that a similar approach for the basic ELSP was developed by Wagelmans, van Hoesel, and Kolen [26]). We next discuss this reformulation strategy in greater detail by first providing a tight linear programming relaxation for the UOSP. We first note that for the UOSP, an optimal solution exists such that we never satisfy part of an order, i.e., v_{mt} equals either zero or d_{mt}; thus we can substitute the v_{mt} variables out of the [OSP] formulation by setting $v_{mt} = z_{mt}$ for all t and $m \in M(t)$. Next observe that since $I_t = \sum_{j=1}^{t} x_j - \sum_{j=1}^{t} \sum_{m \in M(j)} d_{mj} z_{mj}$, we can eliminate the inventory variables from the [OSP] formulation via substitution. After introducing a new variable production and holding cost parameter, c_t, where $c_t \equiv p_t + \sum_{j=t}^{T} h_j$, the objective function of the UOSP can be rewritten as:

$$\text{maximize} \sum_{j=1}^{T} \sum_{m \in M(j)} R_{mj} z_{mj} +$$

$$\sum_{t=1}^{T} h_t \left(\sum_{j=1}^{T} \sum_{m \in M(j)} d_{mj} z_{mj} \right) - \sum_{t=1}^{T} (S_t y_t + c_t x_t) \tag{7}$$

We next define ρ_{mt} as an adjusted revenue parameter for order m in period t, where $\rho_{mt} = d_{mt} \sum_{j=t}^{T} h_j + R_{mt}$. Our reformulation procedure requires capturing the exact amount of production in each period allocated to every order. We thus define x_{mtj} as the number of units produced in period t used to satisfy order m in period j, for $j \geq t$, and replace each x_t with $\sum_{j=t}^{T} \sum_{m \in M(j)} x_{mtj}$. The following formulation provides the tight linear programming relaxation of the UOSP.

[UOSP]

$$\text{maximize} \sum_{j=1}^{T} \sum_{m \in M(j)} \rho_{mj} z_{mj} - \sum_{t=1}^{T} \left(S_t y_t + c_t \sum_{j=t}^{T} \sum_{m \in M(j)} x_{mtj} \right) \tag{8}$$

subject to:

$$\sum_{t=1}^{j} x_{mtj} - d_{mj} z_{mj} = 0, \; j = 1, ..., T, \; m \in M(j), \tag{9}$$

$$\left(\sum_{m \in M(j)} d_{mj} \right) y_t - \sum_{m \in M(j)} x_{mtj} \geq 0, \; t = 1, ..., T, \; j = t, ..., T, \tag{10}$$

$$-z_{mj} \geq -1, \; j = 1, ..., T, \; m \in M(j), \tag{11}$$

$$y_t, x_{mtj}, z_{mj} \geq 0,\ t = 1, ..., T,\ j = t, \ldots, T,\ m \in M(j). \qquad (12)$$

Note that since a positive cost exists for setups, we can show that the constraint $y_t \leq 1$ is unnecessary in the above relaxation, and so we omit this constraint from the relaxation formulation. It is straightforward to show that the above [UOSP] formulation with the additional requirements that all z_{mj} and y_t are binary variables is equivalent to our original OSP when production capacities are infinite.

To obtain the LP relaxation for the OSP (when production capacities are finite), we add finite capacity constraints to [UOSP] by forcing the sum of x_{mtj} over all $j \geq t$ and all $m \in M(j)$ to be less than the production capacity C_t in period t. That is, we can add the following constraint set to [UOSP] to obtain an equivalent LP relaxation for the OSP:

$$\sum_{j=t}^{T} \sum_{m \in M(j)} x_{mtj} \leq C_t,\ t = 1, \ldots, T. \qquad (13)$$

Note that this LP relaxation approach is valid for all three variants of the OSP, the general OSP, the OSP-NDC, and the OSP-AND. Observe that the above constraint can be strengthened by multiplying the right-hand-side by the setup forcing variable y_t. To see how this strengthens the formulation, note that constraint set (10) in the [UOSP] formulation implies that

$$\sum_{j=t}^{T} \sum_{m \in M(j)} x_{mtj} \leq \sum_{j=t}^{T} \left(\sum_{m \in M(j)} d_{mj} \right) y_t,\ t = 1, ..., T.$$

To streamline our notation, let $X_{t,T} = \sum_{j=t}^{T} \sum_{m \in M(j)} x_{mtj}$ and $D_{t,T} = \sum_{j=t}^{T} \left(\sum_{m \in M(j)} d_{mj} \right)$ for $t = 1, \ldots, T$ denote aggregated production variables and order amounts, respectively. Constraint set (13) can be rewritten as

$$X_{t,T} \leq C_t,\ t = 1, \ldots, T,$$

and the aggregated demand forcing constraints (10) can now be written as $X_{t,T} \leq D_{t,T} \cdot y_t$. If we do not multiply the right-hand-side of capacity constraint set (13) by the forcing variable y_t, the formulation allows solutions, for example, such that $X_{t,T} = C_t$ for some t, while $X_{t,T}$ equals only a fraction of $D_{t,T}$. In such a case, the forcing variable y_t takes the fractional value $\frac{X_{t,T}}{D_{t,T}}$, and we only absorb a fraction of the setup cost in period t. Multiplying the right-hand-side of (13) by y_t, on the other hand, would force $y_t = \frac{X_{t,T}}{C_t} = 1$ in such a case, leading to an improved upper bound on the optimal solution value. We can therefore strengthen the LP relaxation solution that results from adding constraint set (13) by instead using the following capacity forcing constraints.

$$X_{t,T} \leq \min\{C_t, D_{t,T}\} \cdot y_t, \; t = 1, \ldots, T. \tag{14}$$

Note that in the capacitated case we now explicitly require stating the $y_t \leq 1$ constraints in the LP relaxation, since it may otherwise be profitable to violate production capacity in order to satisfy additional orders. We refer to the resulting LP relaxation with these aggregated setup forcing constraints as the [ASF] formulation, which we formulate as follows.

[ASF]

$$\text{maximize} \sum_{j=1}^{T} \sum_{m \in M(j)} \rho_{mj} z_{mj} - \sum_{t=1}^{T} \left(S_t y_t + c_t \sum_{j=t}^{T} \sum_{m \in M(j)} x_{mtj} \right)$$

subject to:

Constraints (9–12, 14)

$$y_t \leq 1, \; t = 1, \ldots, T. \tag{15}$$

We can further strengthen the LP relaxation formulation by disaggregating the demand forcing constraints (10) (see Erlenkotter [9], who uses this strategy for the uncapacitated facility location problem). This will force y_t to be at least as great as the maximum value of $\frac{x_{mtj}}{d_{mj}}$ for all $j = t, \ldots, T$ and $m \in M(j)$. The resulting Disaggregated Setup Forcing (DASF) LP relaxation is formulated as follows.

[DASF]

$$\text{maximize} \sum_{j=1}^{T} \sum_{m \in M(j)} \rho_{mj} z_{mj} - \sum_{t=1}^{T} \left(S_t y_t + c_t \sum_{j=t}^{T} \sum_{m \in M(j)} x_{mtj} \right)$$

subject to:

Constraints (9, 11, 12, 14, 15)

$$x_{mtj} \leq d_{mj} y_t, \; t = 1, ..., T, \; j = t, ..., T, \; m \in M(j). \tag{16}$$

Each of the LP relaxations we have described provides some value in solving the capacitated versions of the OSP. Both the OSP-LP and ASF relaxations can be solved very quickly, and they frequently yield high quality solutions. The DASF relaxation further improves the upper bound on the optimal solution value. But as the problem size grows (i.e., the number of orders per period or the number of time periods increases), [DASF] becomes intractable, even via standard linear programming solvers. We present results for each of these relaxation approaches in Section 4. Before doing this, however, we next discuss methods for determining good feasible solutions, and therefore lower bounds, for the OSP via several customized heuristic solution procedures.

3.2. *Heuristic Solution Approaches for OSP*

While the methods discussed in the previous subsection often provide strong upper bounds on the optimal solution value for the OSP (and its variants), we cannot guarantee the ability to solve this problem in reasonable computing time using branch-and-bound due to the complexity of the problem. We next discuss three heuristic solution approaches that allow us to quickly generate feasible solutions for OSP. As our results in Section 4 report, using a composite solution procedure that selects the best solution among those generated by the three heuristic solution approaches provided feasible solutions with objective function values, on average, within 0.67% of the optimal solution value. We describe our three heuristic solution approaches in the following three subsections.

3.2.1. *Lagrangian relaxation based heuristic*

Lagrangian relaxation (Geoffrion [13]) is often used for mixed integer programming problems to obtain stronger upper bounds (for maximization problems) than provided by the LP relaxation. As we discussed in Section 3.1, our strengthened linear programming formulations typically provide very good upper bounds on the optimal solution value of the OSP. Moreover, as we later discuss, our choice of relaxation results in a Lagrangian subproblem that satisfies the so-called integrality property (see Geoffrion [13]). This implies that the upper bound provided by our Lagrangian relaxation scheme will not provide better bounds than our LP relaxation. Our purpose for implementing a Lagrangian relaxation heuristic, therefore, is strictly to obtain good feasible solutions using a Lagrangian-based heuristic. Because of this we omit certain details of the Lagrangian relaxation algorithm and implementation, and describe only the essential elements of the general relaxation scheme and how we obtain a heuristic solution at each iteration of the Lagrangian algorithm.

Under our Lagrangian relaxation scheme, we add (redundant) constraints of the form $x_t \leq My_t$, $t = 1, \ldots, T$ to the [OSP] formulation (where M is some large number), eliminate the forcing variable y_t from the right-hand side of the capacity/setup forcing constraints (3), and then relax the resulting modified capacity constraint (3) (without the y_t multiplier on the right-hand side) in each period. The resulting Lagrangian relaxation subproblem is then simply an uncapacitated OSP (or UOSP) problem. Although the Lagrangian multipliers introduce the possibility of negative unit production costs in the Lagrangian subproblem, we retain the convexity of the objective function, and all properties necessary for solving the UOSP problem via a Wagner-Whitin [27] based shortest path approach still hold (for details on this shortest path solution approach, please see Geunes, Romeijn, and Taaffe [14]). We can therefore solve the Lagrangian subproblems in polynomial time. Because we have a tight formulation of the UOSP, this implies that the Lagrangian relaxation satisfies the integrality property, and the Lagrangian solution will not provide better upper bounds than the LP relaxation. We do, however, use the solution of the Lagrangian subproblem at each iteration of a subgradient optimization algorithm (see Fisher [10]) as a starting point for heuristically generating a feasible solution, which serves as a candidate lower bound on the optimal solution value for OSP.

Observe that the subproblem solution from this relaxation will satisfy all constraints of the OSP except for the relaxed capacity constraints (3). We therefore call a feasible solution generator (FSG) at each step of the subgradient algorithm, which can take any starting capacity-infeasible solution and generate a capacity-feasible solution. (We also use this FSG in our other heuristic solution schemes, as we later describe.) The FSG works in three main phases. Phase I first considers performing additional production setups (beyond those prescribed by the starting solution) to try to accommodate the desired production levels and order selection decisions provided in the starting solution, while obeying production capacity limits. That is, we consider shifting production from periods in which capacities are violated to periods in which no setup was originally planned in the starting solution. It is possible, however, that we still violate capacity limits after

Phase I, since we do not eliminate any order selection decisions in Phase I.

In Phase II, after determining which periods will have setups in Phase I, we consider those setup periods in which production still exceeds capacity and, for each such setup period, index the orders satisfied from production in the setup period in nondecreasing order of contribution to profit. For each period with violated capacity, in increasing profitability index order, we shift orders to an earlier production setup period, if the order remains profitable and such an earlier production setup period exists with enough capacity to accommodate the order. Otherwise we eliminate the order from consideration. If removing the order from the setup period will leave excess capacity in the setup period under consideration, we consider shifting only part of the order to a prior production period; we also consider eliminating only part of the order when customers do not require all-or-nothing order satisfaction. This process is continued for each setup period in which production capacity is violated until total production in the period satisfies the production capacity limit. Following this second phase of the algorithm, we will have generated a capacity-feasible solution. In the third and final phase, we scan all production periods for available capacity and assign additional profitable orders that have not yet been selected to any excess capacity if possible. The Appendix contains a detailed description of the FSG algorithm.

3.2.2. *Greatest unit profit heuristic*

Our next heuristic solution procedure is motivated by an approach taken in several well-known heuristic solution approaches for the ELSP. In particular, we use a similar "myopic" approach to those used in the Silver-Meal [24] and Least Unit Cost (see Nahmias [21]) heuristics. These heuristics proceed by considering an initial setup period, and then determining the number of consecutive period demands (beginning with the initial setup period) that produce the lowest cost per period (Silver-Meal) or per unit (Least Unit Cost) when allocated to production in the setup period. The next period considered for a setup is the one immediately following the last demand period assigned to the prior setup; the heuristics proceed until all demand has been allocated to some setup period. Our approach differs from these approaches in the following respects. Since we are concerned with the profit from orders, we take a greatest profit rather than a lowest cost approach. We also allow for accepting or rejecting various orders, which implies that we need only consider those orders that are profitable when assigning orders to a production period. Moreover, we can choose not to perform a setup if no selection of orders produces a positive profit when allocated to the setup period. Finally, we apply our "greatest unit profit" heuristic in a capacitated setting, whereas a modification of the Silver-Meal and Least Unit Cost heuristics is required for application to the capacitated lot-sizing problem.

Our basic approach begins by considering a setup in period t (where t initially equals 1) and computing the maximum profit per unit of demand satisfied in period t using only the setup in period t. Note that, given a setup in period t, we can sort orders in periods $t, \ldots, T$ in non-increasing order of contribution to profit based solely on the variable costs incurred when assigning the order to the setup in period t (for the OSP when fixed delivery charges exist we must also subtract this cost from each order's contribution to profit). Orders are then allocated to the setup in non-increasing order of contribution to profit until either the setup capacity is exhausted or no additional attractive orders exist. After computing the maximum profit per unit of demand satisfied in period t using only the setup in period t, we then compute the maximum profit per unit satisfied in periods $t, \ldots, t+j$ using only the setup in period t, for $j = 1, \ldots, j'$, where period j' is the first period in the sequence such that the maximum profit per unit in periods $t, \ldots, t+j'$ is greater than or equal to the maximum profit per unit in periods $t, \ldots, t+j'+1$. The capacity-feasible set of orders that leads to the greatest profit per unit in periods $t, \ldots, j'$ using the setup in period t is then assigned to production in period t, assuming the maximum profit per unit is positive. If the maximum profit per unit for any given setup period does not exceed zero, however, we do not assign any orders to the setup and thus eliminate the setup.

Since we consider a capacity-constrained problem, we can either consider period $j'+1$ (as is done in the Silver-Meal and Least Unit Cost heuristics) or period $t+1$ as the next possible setup period following period t. We use both approaches and retain the solution that produces higher net profit. Note that if we consider period $t+1$ as the next potential setup period following period t, we must keep track of those orders in periods $t+1$ and higher that are already assigned to period t (and prior) production, since these will not be available for assignment to period $t+1$ production. Finally, after applying this greatest unit profit heuristic, we apply Phase III of the FSG algorithm (see the Appendix) to the resulting solution, in an effort to further improve the heuristic solution value by looking for opportunities to effectively use any unused setup capacity.

3.2.3. *Linear programming rounding heuristic*

Our third heuristic solution approach uses the LP relaxation solution as a starting point for a linear programming rounding heuristic. We focus on rounding the setup (y_t) and order selection (z_{mt}) variables that are fractional in the LP relaxation solution (rounding the order selection variables is not, however, relevant for the OSP-NDC problem, since the z_{mt} variables do not exist in this special case). We first consider the solution that results by setting all (non-zero) fractional y_t and z_{mt} variables from the LP relaxation solution to one. We then apply the second and third phases of our FSG algorithm (described previously in Section 3.2.1 and provided in the Appendix) to ensure a capacity feasible solution, and to search for unselected orders to allocate to excess production capacity in periods where the setup variable was rounded to one.

We also use an alternative version of this procedure, where we round up the setup variables with values greater than or equal to 0.5 in the LP relaxation solution, and round down those with values less than 0.5. Again we subsequently apply Phases II and III of the FSG algorithm to generate a good capacity-feasible solution (if the maximum setup variable value takes a value between 0 and 0.5, we round up only the setup variable with the maximum fractional variable value and apply Phases II and III of the FSG algorithm). Finally, based on our discussion in Section 3.1, note that we have a choice of three different formulations for generating LP relaxation starting solutions for the rounding procedure: formulation [OSP] (Section 2) and the [ASF] and [DASF] formulations (Section 3.1). As our computational results later discuss, starting with the LP relaxation solution from the [DASF] formulation provides solutions that are, on average, far superior to those provided using the other LP relaxation solutions. However, the size of this LP relaxation also far exceeds the size of our other LP relaxation formulations, making this formulation impractical as problem sizes become large. We use the resulting LP relaxation solution under each of these formulations and apply the LP rounding heuristic to all three of these initial solutions for each problem instance, retaining the solution that provides the highest net profit.

4. Computational Testing Scope and Results

This section discusses a broad set of computational tests intended to evaluate our upper bounding and heuristic solution approaches. Our results focus on gauging both the ability of the different LP relaxations presented in Section 3.1 to provide tight upper bounds on optimal profit, and the performance of the heuristic procedures discussed in Section 3.2 in providing good feasible solutions. Section 4.1 next discusses the scope of our computational tests, while Sections 4.2 and 4.3 report results for the OSP, OSP-NDC, and OSP-AND versions of the problem.

4.1. *Computational Test Setup*

This section presents the approach we used to create a total of 3,240 randomly generated problem instances for computational testing, which consist of 1,080 problems for each of the OSP, OSP-NDC, and OSP-AND versions of the problem. Within each problem version (OSP, OSP-NDC, and OSP-AND), we used three different settings for the number of orders per period, equal to 25, 50, and 200. In order to create a broad set of test instances, we considered a range of setup cost values, production capacity limits, and per unit order revenues.[b] Table 3 provides the set of distributions used for randomly generating these parameter values in our test cases. The total number of combinations of parameter distribution settings shown in Table 3 equals 36, and for each unique choice of parameter distribution settings we generated 10 random problem instances. This produced a total of 360 problem instances for each of the three values of the number of orders per period (25, 50, and 200), which equals 1,080 problem instances for each problem version.

As the distributions used to generate production capacities in Table 3 indicate, we maintain a constant ratio of average production capacity per period to average total demand per period. That is, we maintain the same average order size (average of d_{mt} values) across each of these test cases, but the average capacity per period for the 200-order problem sets is four times that of the 50-order problem sets and eight times that of the 25-order problems. Because the total number of available orders per period tends to strongly affect the relative quality of our solutions (as we later discuss), we report performance measures across all test cases and also individually within the 25, 50, and 200 order problem sets.

In order to limit the scope of our computational tests to a manageable size, we chose to limit the variation of certain parameters across all of the test instances. The per unit production cost followed a distribution of U[20, 30] for all test instances (where U$[a, b]$ denotes a Uniform distribution on

[b]These three parameters appeared to be the most critical ones to vary widely in order to determine how robust our solution methods were to problem parameter variation.

Table 3. Probability distributions used for generating problem instance parameter values.

Parameter	Number of Distribution Settings	Distributions used for Parameter Generation*
Setup cost (varies from period-to-period)	3	U[350, 650]
		U[1750, 3250]
		U[3500, 6500]
Per unit per period holding cost**	2	$0.15 \times p/50$
		$0.25 \times p/50$
Production capacity in a period (varies from period-to-period)***	3	U[$d/3 - .05d$, $d/3 + .05d$]
		U[$d/2 - .1d$, $d/2 + .1d$]
		U[$d - .15d$, $d + .15d$]
Per unit order revenue (varies from order-to-order)	2	U[28,32]
		U[38,42]

* U[a, b] denotes a uniform distribution on the interval [a, b].
** p denotes the variable production cost. We assume 50 working weeks in one year.
*** d denotes the expected per-period total demand, which equals the mean of the distribution of order sizes multiplied by the number of orders per period.

the interval $[a, b]$), and all problem instances used a 16-period planning horizon. We also used an order size distribution of U[10, 70] for all test problems (i.e., the d_{mt} values follow a uniform distribution on [10, 70]). For the OSP, the distribution used for generating fixed delivery charges was U[100, 600].[c] By including a wide range of levels of production capacity, setup cost, and order volumes, we tested a set of problems which would fairly represent a variety of actual production scenarios.

Observe that the two choices for distributions used to generate per unit order revenues use relatively narrow ranges. Given that the distribution used to generate variable production cost is U[20, 30], the first of these per unit revenue distributions, U[28, 32], produces problem instances in which the contribution to profit (after subtracting variable production cost) is quite small—leading to fewer attractive orders after considering setup and holding costs. The second distribution, U[38, 42], provides a more profitable set of orders. We chose to keep these ranges very narrow because our preliminary test results showed that a tighter range, which implies less per unit revenue differentiation among orders, produces more difficult problem

[c]We performed computational tests with smaller per-order delivery charges, but the results were nearly equivalent to those presented for the OSP-NDC in Table 4, since the profitability of the orders remained essentially unchanged. As we increased the average delivery charge per order, more orders became unprofitable, creating problem instances that were quite different from the OSP-NDC case.

instances. Those problem instances with a greater range of per unit revenue values among orders tended to be solved in CPLEX via branch-and-bound much more quickly than those with tight ranges, and we wished to ensure that our computational tests reflected more difficult problem instances.

A tighter range of unit revenues produces more difficult problem instances due to the ability to simply 'swap' orders with identical unit revenues in the branch-and-bound algorithm, leading to alternative optimal solutions at nodes in the branch-and-bound tree. For example, if an order m in period t is satisfied at the current node in the branch-and-bound tree, and some other order m' is not satisfied, but $r_{mt} = r_{m't}$ and $d_{mt} = d_{m't}$, then a solution which simply swaps orders m and m' has the same objective function as the first solution, and no improvement in the bound occurs as a result of this swap. So, we found that when the problem instance has less differentiation among orders, the branch-and-bound algorithm can take substantially longer, leading to more difficult problem instances. Barnhart *et al.* [7] and Balakrishnan and Geunes [5] observed similar swapping phenomena in branch-and-bound for machine scheduling and steel production planning problems, respectively.

All linear and mixed integer programming (MIP) formulations were solved using the CPLEX 6.6 solver on an RS/6000 machine with two PowerPC (300MHz) CPUs and 2GB of RAM. We will refer to the best solution provided by the CPLEX branch-and-bound algorithm as the *MIP solution.* The remaining subsections summarize our results. Section 4.2 reports the results of our computational experiments for the OSP-NDC and the OSP, and Section 4.3 presents the findings for the OSP-AND (all-or-nothing order satisfaction) problem. For the OSP-AND problem instances discussed in Section 4.3, we assume that the revenue parameters provided represent revenues in excess of fixed delivery charges (since we always satisfy all or none of the demand for the OSP-AND, this is without loss of generality).

4.2. *Computational Results for the OSP and the OSP-NDC*

Recall that the OSP assumes that we have the flexibility to satisfy any proportion of an order in any period, as long as we do not exceed the production capacity in the period. Because of this, when no fixed delivery charges exist, the only binary variables in the OSP-NDC correspond to the T binary setup variables, and solving these problem instances to optimality using CPLEX's MIP solver did not prove to be very difficult. The same is not necessarily true of the OSP-AND, as we later discuss in Section 4.3.

Surprisingly, the OSP (which includes a binary fixed delivery-charge forcing (z_{mt}) variable for each order-period combination) was not substantially computationally challenging either. All of the OSP-NDC and all but two of the OSP instances were solved optimally using branch-and-bound within the allotted branch-and-bound time limit of one hour. Even though we are able to solve the OSP and OSP-NDC problem instances using CPLEX with relative ease, we still report the upper bounds provided by the different LP relaxations for these problems in this section. This allows us to gain insight regarding the strength of these relaxations as problem parameters change, with knowledge of the optimal mixed integer programming (MIP) solution values as a benchmark.

Table 4 presents optimality gap measures based on the solution values resulting from the LP (OSP-LP) relaxation upper bound, the aggregated setup forcing (ASF) relaxation upper bound, and the disaggregated setup forcing (DASF) relaxation upper bound for the OSP-NDC and OSP problem instances. The last row of the table shows the percentage of problem instances for which CPLEX was able to find an optimal solution via branch-and-bound. As Table 4 shows, for the OSP-NDC, all three relaxations provide good upper bounds on the optimal solution value, consistently producing gaps of less than 0.25%, on average. As expected, the [ASF] formulation provides better bounds than the simple OSP-LP relaxation, and the [DASF] formulation provides the tightest bounds. We note that as the number of potential orders and the per-period production capacities increase, the relative performance of the relaxations improves, and the optimality gap decreases. Since an optimal solution exists such that at most one order per period will be partially satisfied under any relaxation, as the problem size grows, we fulfill a greater proportion of orders in their entirety. So the impact of our choice of which order to partially satisfy diminishes with larger problem sizes. Note also, however, that a small portion of this improvement is attributable to the increased optimal solution values in the 50- and 200-order cases.

For the OSP, we have non-zero fixed delivery costs and cannot therefore eliminate the binary z_{mt} variables from formulation [OSP]. In addition, since formulation [OSP] includes the continuous v_{mt} variables, it has the highest number of variables of any of the capacitated versions we consider. This does not necessarily, however, make it the most difficult problem class for solution via CPLEX, as a later comparison of the results for the OSP and OSP-AND indicates.

The upper bound optimality gap results reported in Table 4 for the OSP are significantly larger than those for the OSP-NDC.[d] This is because this formulation permits setting fractional values of the fixed delivery-charge forcing (z_{mt}) variables, and therefore does not necessarily charge the entire fixed delivery cost when meeting a fraction of some order's demand. For this problem set the [DASF] formulation provides substantial value in obtaining strong upper bounds on the optimal net profit although, as shown in Table 5, the size of this formulation makes solution via CPLEX substantially more time consuming as the number of orders per period grows to 200.

Table 4. OSP-NDC and OSP Problem Optimality Gap Measures.

	OSP-NDC				**OSP**			
	Orders per Period			Overall	Orders per Period			Overall
Gap Measure	25	50	200	Average	25	50	200	Average
OSP-LP v. MIP*	0.24%	0.14%	0.05%	0.14%	9.26%	6.09%	0.57%	5.31%
ASF vs. MIP**	0.18	0.12	0.04	0.11	9.21	6.07	0.56	5.28
DASF v. MIP***	0.11	0.07	0.03	0.07	1.58	0.35	0.10	0.68
% Optimal******	100	100	100	100	100	99.7	99.7	99.8

Note: Entries in each "orders per period" class represent average among 360 test instances.
* (OSP-LP − MIP)/MIP × 100%.
** (ASF − MIP)/MIP × 100%.
*** (DASF − MIP)/MIP × 100%.
**** % of problems for which CPLEX branch-and-bound found an optimal solution.

Table 5 summarizes the solution times for solving the OSP-NDC and the OSP. The MIP solution times reflect the average time required to find an optimal solution for those problems that were solved to optimality in CPLEX (the two problems that CPLEX could not solve to optimality are not included in the MIP solution time statistics). We used the OSP-LP formulation as the base formulation for solving all mixed integer programs. The table also reports the times required to solve the LP relaxations for each of our LP formulations (OSP-LP, ASF, and DASF). We note that the [ASF] and [DASF] LP relaxations often take longer to solve than the mixed integer problem itself. The [DASF] formulation, despite providing the best upper bounds on solution value, quickly becomes less attractive as the problem size grows because of the size of this LP formulation. Nonetheless, the relaxations provide extremely tight bounds on the optimal solution as shown in the table. As we later show, however, solving the problem to

[d]For the two problems that could not be solved to optimality via branch-and-bound using CPLEX due to memory limitations, the MIP solution value used to compute the upper bound optimality gap is the value of the best solution found by CPLEX.

optimality in CPLEX is not always a viable approach for the restricted OSP-AND discussed in the following section.

Table 5 reveals that the MIP solution times for the OSP were also much greater than for the OSP-NDC. This is due to the need to simultaneously track the binary (z_{mt}) and continuous (v_{mt}) variables for the OSP with non-zero fixed delivery costs. As expected, the average and maximum solution times for each relaxation increased with the number of orders per period. As we noted previously, the percentage optimality gaps, however, substantially decrease as we increase the number of orders per period.

Table 5. OSP-NDC and OSP Solution Time Comparison.

	OSP-NDC Orders per Period			**OSP** Orders per Period		
Time Measure (CPU seconds)	25	50	200	25	50	200
Average MIP Solution Time	0.1	0.1	0.2	3.3	19.1	129.4
Maximum MIP Solution Time	0.1	0.1	0.3	44.8	541.3	3417.2
Average OSP-LP Solution Time	0.1	0.1	0.3	0.1	0.1	0.3
Maximum OSP-LP Solution Time	0.1	0.2	0.5	0.1	0.1	0.5
Average ASF Solution Time	0.5	1.5	14.0	0.4	1.0	8.3
Maximum ASF Solution Time	0.7	2.2	25.2	0.6	1.6	15.4
Average DASF Solution Time	5.3	27.3	727.2	3.3	15.7	333.8
Maximum DASF Solution Time	18.4	64.3	1686.7	12.1	47.1	1251.9

Note: Entries represent average/maximum among 360 test instances.
LP relaxation solution times include time consumed applying the LP rounding heuristic to the resulting LP solution, which was effectively negligible.

We next present the results of applying our heuristic solution approaches to obtain good solutions for the OSP and OSP-NDC. We employ the three heuristic solution methods discussed in Section 3.2, denoting the Lagrangian-based heuristic as LAGR, the greatest unit profit heuristic as GUP, and the LP rounding heuristic as LPR. Table 6 provides the average percentage deviation from the best upper bound (as a percentage of the best upper bound) for each heuristic solution method. Note that since we found an optimal solution for all but two of the OSP and OSP-NDC problem instances, the upper bound used in computing the heuristic solution gaps is nearly always the optimal mixed integer solution value. The last row in Table 6 shows the resulting lower bound gap from our composite solution procedure, which selects the best solution among all of the heuristic methods applied. The average lower bound percentage gap is within 0.06% of optimality for the OSP-NDC, while that for the OSP is 1.69%, indicating that overall, our heuristic solution methods are quite effective.

As the table indicates, the heuristics perform much better in the absence of fixed delivery costs. For the Lagrangian-based and LP rounding heuristics, we can attribute this in part to the difficulty in obtaining good relaxation upper bounds for the OSP as compared to the OSP-NDC. Observe that as the upper bound decreases, i.e., as the number of orders per period increases, these heuristics tend to improve substantially. The GUP heuristic, on the other hand, appears to have difficulty identifying a good combination of setup periods in the presence of fixed delivery charges. Although it appears, based on average performance, that the LPR heuristic dominates the LAGR and GUP heuristics, the last row of the table reveals that this is not universally true. Each of our heuristic approaches provided the best solution value for some nontrivial subset of the problems tested.

Table 6. OSP and OSP-NDC Heuristic Solution Performance Measures.

	OSP-NDC				**OSP**			
	Orders per Period			Overall	Orders per Period			Overall
Gap Measure	25	50	200	Average	25	50	200	Average
LAGR v. UB*	1.34%	0.58%	0.32%	0.75%	6.35%	4.07%	2.16%	4.19%
GUP v. UB**	1.00	0.69	0.44	0.71	7.27	6.91	5.39	6.52
LPR v. UB***	0.25	0.15	0.05	0.15	8.32	5.31	0.96	4.86
Best LB****	0.10	0.07	0.02	0.06	3.08	1.55	0.44	1.69

Note: Entries in each "orders per period" class represent average among 360 test instances.
* (LAGR − UB)/UB × 100%.
** (GUP − UB)/UB × 100%.
*** (LPR − UB)/UB × 100%.
**** Uses the best heuristic solution value for each problem instance.

4.3. *Computational Results for the OSP-AND*

We next provide our results for the OSP-AND where, if we choose to accept an order, we must satisfy the entire order, i.e., no partial order satisfaction is allowed. Finding the optimal solution to the OSP-AND can be much more challenging than for the OSP, since we now face a more difficult combinatorial "packing" problem (i.e., determining the set of orders that will be produced in each period is similar to a multiple knapsack problem).

Table 7 provides upper bound optimality gap measures based on the solution values resulting from our different LP relaxation formulations, along with the percentage of problem instances that were solved optimally via the CPLEX branch-and-bound algorithm. Observe that the upper bound optimality gap measures are quite small and only slightly larger than those observed for the OSP-NDC. The reason for this is that the LP relaxation formulations are identical in both cases (as discussed in Section 2), and the optimal LP relaxation solution violates the all-or-nothing requirement for at most one order per period. Thus, even in the OSP-NDC case, almost all orders that are selected are fully satisfied in the LP relaxation solution. In contrast to the [OSP] formulation, the binary z_{mt} variables in the OSP-AND model now represent "order selection" variables rather than fixed delivery-charge forcing variables. That is, since we net any fixed delivery charge out of the net revenue parameters R_{mt}, and the total revenue for an order in a period now equals $R_{mt}z_{mt}$ in this formulation, we have strong preference for z_{mt} variable values that are either close to one or zero. In the [OSP] formulation, on the other hand, the z_{mt} variables are multiplied by the fixed delivery-charge terms (F_{mt}) in the objective function, leading to a strong preference for low values of the z_{mt} variables and, therefore, a weaker upper bound on optimal net profit. Note also that as the number of possible orders increases (from the 25-order case to the 200-order case), the influence of the single partially satisfied order in each period on the objective function value diminishes, leading to a reduced optimality gap as the number of orders per period increases. As the last row of Table 7 indicates, we were still quite successful in solving these problem instances to optimality in CPLEX. The time required to do so, however, was substantially greater than that for either the OSP or OSP-NDC, because of the complexities introduced by the all-or-nothing order satisfaction requirement.

Table 8 summarizes the resulting solution time performance for the OSP-AND. We note here that our relaxation solution times are quite reasonable, especially as compared to the MIP solution times, indicating that quality upper bounds can be found very quickly. Again, the MIP solution times reflect the average time required to find an optimal solution for those problems that were solved to optimality in CPLEX (those problems which CPLEX could not solve to optimality are not included in the MIP solution time statistics). The table does not report the time required to solve our different LP relaxation formulations, since the OSP-AND LP relaxation is identical to the OSP-NDC LP relaxation, and these times are therefore shown in Table 5.

Unlike our previous computational results for the OSP and the OSP-NDC, we found several problem instances of the OSP-AND in which an optimal solution was not found either due to reaching the time limit of one hour or because of memory limitations. For the problem instances we were able to solve optimally, the MIP solution times were far longer than those for the OSP problem. This is due to the increased complexity resulting from the embedded "packing problem" in the OSP-AND problem. Interestingly, however, in contrast to our previous results for the OSP, the average and maximum MIP solution times for the OSP-AND were *smaller* for the 200-order per period problem set than for the 25 and 50-order per period problem sets. The reason for this appears to be because of the nearly non-existent integrality gaps of these problem instances, whereas these gaps increase when the number of orders per period is smaller.

Table 7. OSP-AND Optimality Gap Measures.

	Orders per Period			
Gap Measurement	25	50	200	**Overall Average**
OSP-LP vs. MIP Solution*	0.34%	0.20%	0.06%	0.20%
ASF vs. MIP Solution**	0.28	0.18	0.05	0.17
DASF vs. MIP Solution***	0.21	0.10	0.03	0.11
% Optimal******	96.7	94.2	100	97

Note: Entries within each "orders per period" class represent average among 360 test instances.
* (OSP-LP – MIP)/MIP × 100%.
** (ASF – MIP)/MIP × 100%.
*** (DASF – MIP)/MIP × 100%.
**** % of problems for which CPLEX branch-and-bound found an optimal solution.

Table 8. OSP-AND Solution Time Comparison

	Orders per Period		
Time Measure (CPU seconds)	25	50	200
Average MIP Solution Time	42.0	67.9	21.9
Maximum MIP Solution Time	1970.1	1791.8	1078.8

Table 9 shows that once again our composite heuristic procedure performed extremely well on the problems we tested. The percentage deviation from optimality in our solutions is very close to that of the OSP-NDC, and much better than that of the OSP, with an overall average performance within 0.25% of optimality. We note, however, that the best heuristic solution performance for both the OSP-NDC and the OSP-AND occurred using the LP rounding heuristic applied to the DASF LP relaxation solution. As Table 5 showed, solving the DASF LP relaxation can be quite time consuming as the number of orders per period grows, due to the size of this formulation. We note, however, that for the OSP-NDC and OSP-AND, applying the LP rounding heuristic to the ASF LP relaxation solution produced results very close to those achieved using the DASF LP relaxation solution in much less computing time. Among all of the 3,240 OSP, OSP-NDC, and OSP-AND problems tests, the best heuristic solution value was within 0.67% of optimality on average, indicating that overall, the heuristic solution approaches we presented provide an extremely effective method for solving the OSP and its variants.

Table 9. OSP-AND Heuristic Solution Performance Measures

	OSP-AND			
	Orders per Period			Overall
Gap Measurement	25	50	200	Average
LAGR vs. UB*	3.95%	3.92%	0.33%	2.73%
GUP vs. UB**	1.85	0.83	0.46	1.04
LPR vs. UB***	0.80	0.31	0.12	0.41
Best LB******	0.49	0.19	0.06	0.25

Note: Entries within each "orders per period" class represent average among 360 test instances.
* (LAGR – UB)/UB × 100%.
** (GUP – UB)/UB × 100%.
*** (LPR – UB)/UB × 100%.
**** Uses the best heuristic solution value for each problem instance.

5. Summary and Directions for Future Research

When a producer has discretion to accept or deny production orders under limited capacity, determining the best set of orders to accept based on both revenue and production/delivery cost implications can be quite challenging. We have proposed several capacitated versions of a combined order selection and production planning model that addresses this challenge. We considered variants of the problem both with and without fixed delivery charges, as well as contexts that permit the producer to satisfy any chosen fraction of any order quantity, thus allowing the producer to ration its capacity. We provided three linear programming relaxations that produce strong upper bound values on the optimal net profit from integrated order selection and production planning decisions. We also provided a set of three effective heuristic solution methods for the OSP. Computational tests performed on a broad set of randomly generated problems demonstrated the effectiveness of our heuristic methods and upper bounding procedures. Problem instances in which the producer has the flexibility to determine any fraction of each order it will supply, and no fixed delivery charges exist, were easily solved using the MIP solver in CPLEX. When fixed delivery charges are present, however, the problem becomes more difficult, particularly as the number of available orders increases. Optimal solutions were still obtained, however, for nearly all test instances within one hour of computing time when partial order satisfaction was allowed. When the producer must take an all-or-nothing approach, satisfying the entire amount of each order it chooses to satisfy, the problem becomes substantially more challenging, and the heuristic solutions we presented become a more practical approach for solving such problems.

The models we presented can serve as a starting point for future research on more general models. Suppose that instead of picking and choosing individual orders by period, the producer must satisfy a given customer's orders in *every* period if the producer satisfies that customer's demand in any single period. In other words, a customer cannot be served only when it is desirable for the producer, since this would result in poor customer service. We might also pose a slightly more general version of this problem, which requires serving a customer in some contiguous set of time periods, if we satisfy any portion of the customer's demand. This would correspond to contexts in which the producer is free to begin serving a market at any time and can later stop serving the market at any time in the planning horizon; however, full service to the market must continue between the start and end period chosen. Future research might also consider varying degrees of producer flexibility, where certain minimum order fulfillment requirements must be met. Finally, we might also consider a situation in which the producer can acquire additional capacity at a cost in order to accommodate more orders than current capacity levels allow. This generalization of the order selection models can potentially further increase net profit from integrated order selection, capacity planning, and production planning decisions.

Appendix A. Description of Feasible Solution Generator (FSG) Algorithm for OSP

This appendix describes the Feasible Solution Generator (FSG) algorithm, which takes as input a solution that is feasible for all OSP problem constraints except the production capacity constraints, and produces a capacity-feasible solution. Note that we present the FSG algorithm as it applies to the OSP, and that certain straightforward modifications must be made for the OSP-AND version of the problem.

Phase I: Assess attractiveness of additional setups

0) Let j denote a period index, let $p(j)$ be the most recent production period prior to and including period j, and let $s(j)$ be the next setup after period j. If no production period exists prior to and including j, set $p(j) = 0$. Set $j = T$ and $s(j) = T + 1$ and let X_j denote the total planned production (in the current, possibly capacity-infeasible solution) for period j.

1) Determine the most recent setup $p(j)$ as described in Step 0. If $p(j) = 0$, go to Phase II. If $X_{p(j)} \leq C_{p(j)}$, set $s(p(j) - 1) = p(j)$ and $j = p(j) - 1$ and repeat Step 1 (note that we maintain $s(j) = j + 1$). Otherwise, continue.

2) Compare the desired production in period $p(j)$, $X_{p(j)}$, with actual capacities over the next $s(j) - p(j)$ periods. If $X_{p(j)} > \sum_{t=p(j)}^{s(j)-1} C_t$, and the sum of the revenues for all selected orders for period j exceed the setup cost in period j, then add a production setup in period j and transfer all selected orders in period j to the new production period j. Otherwise do not add the setup in period j. Set $s(p(j) - 1) = p(j), j = p(j) - 1$, and return to Step 1.

Phase II: Transfer/remove least profitable production orders

0) Let $d_{m,p(j),j}$ denote the amount of demand from order m in period j to be satisfied by production in period $p(j)$ in the current (possibly capacity-infeasible) plan. When reading in the problem data, all profitable order and production period combinations were determined. Based on the solution, we maintain a list of all orders that were satisfied, and this list is kept in *non-decreasing* order of per-unit profitability. Per-unit profitability is defined as follows: $\Pi_{m,p(j),j} = r_{mj} - p_{p(j)} - \sum_{t=p(j)}^{j-1} h_t - \frac{F_{mj}}{d_{mj}}$. We will use this list to determine the least desirable production orders to maintain.

1) If no periods have planned production that exceeds capacity, go to Phase III. While there are still periods in which production exceeds capacity, find the next least profitable order period combination, $(m^*, p(j^*), j^*)$, in the list.

2) If $X_{p(j^*)} > C_{p(j^*)}$, consider shifting or removing an amount equal to $d^* = \min\{d_{m^*,p(j^*),j^*}, X_{p(j^*)} - C_{p(j^*)}\}$ from production in period $p(j^*)$ (otherwise, return to Step 1). If an earlier production period $\tau < p(j^*)$ exists such that $X_\tau < C_\tau$, then move an amount equal to $\min(d^*, C_\tau - X_\tau)$ to the production in period τ, i.e., $d_{m^*,\tau,j^*} =$

min $(d^*, C_\tau - X_\tau)$. Otherwise, reduce the amount of production in period $p(j^*)$ by d^* and set $d_{m^*,p(j^*),j^*} = d_{m^*,p(j^*),j^*} - d^*$.

3) Update all planned production levels and order assignments and update the number of periods in which production exceeds capacity. Return to Step 1.

Phase III: Attempt to increase production in under-utilized periods

0) Create a new list for each period of all profitable orders <u>not</u> fulfilled. Each list is indexed in *non-increasing* order of per-unit profitability, as defined earlier. Let j denote the first production period.

1) If $j = T + 1$, STOP with a feasible solution. Otherwise, continue.

2) If $C_{p(j)} > X_{p(j)}$, excess capacity exists in period $p(j)$. Choose the next most profitable order from period j, and let m^* denote the order index for this order. Let $d_{m^*,p(j),j} = \min\{d_{m^*,j}, C_{p(j)} - X_{p(j)}\}$, and assign an additional $d_{m^*,p(j),j}$ to production in period $p(j)$.

3) If there is remaining capacity and additional profitable orders exist for period j, the repeat Step 2. Otherwise, set $j = j + 1$ and return to Step 1.

References

1. P. Afentakis and B. Gavish. Optimal Lot-Sizing Algorithms for Complex Product Structures. *Oper. Res.* **34**, 237-249 (1986).
2. P. Afentakis, B. Gavish, and U. Karmarkar. Computationally Efficient Optimal Solutions to the Lot-Sizing Problem in Multistage Assembly Systems. *Management Sci.* **30**, 222-239 (1984).
3. K.R. Baker. Requirements Planning. Ch. 11 in *Handbooks in Operations Research and Management Science v4, Logistics of Production and Inventory* (S.C. Graves, A.H.G. Rinnooy Kan, and P.H. Zipkin, eds.), North-Holland, Amsterdam (1993).
4. K.R. Baker, P. Dixon, M. Magazine, and E. Silver. An Algorithm for the Dynamic Lot Size Problem with Time-Varying Production Capacity Constraints. *Management Sci.* **24**, 1710-1720 (1978).
5. A. Balakrishnan, and J. Geunes. Production Planning with Flexible Product Specifications: An Application to Specialty Steel Manufacturing. *Oper. Res.* 51(1), 94-112 (2003).
6. I. Barany, T.J. Van Roy, and L.A. Wolsey. Strong Formulations for Multi-item Capacitated Lot Sizing. *Management Sci.* **30**(10), 1255-1261 (1984).
7. C. Barnhart, E.L. Johnson, G.L. Nemhauser, M.W.P. Savelsbergh, and P.H. Vance. Branch-And-Price: Column Generation for Solving Huge Integer Programs. *Oper. Res.* **46**(3), 316-329 (1998).

8. K. Charnsirisakskul, P. Griffin, and P. Keskinocak. Order Selection and Scheduling with Lead-Time Flexibility. 2002 IIE Research Conference Proceedings (2002).
9. D. Erlenkotter. A Dual-Based Procedure for Uncapacitated Facility Location. *Oper. Res.* **26**, 992-1009 (1978).
10. M.L. Fisher. Lagrangian Relaxation Method for Solving Integer Programming Problems. *Management Sci.* **27**(1), 1-18 (1981).
11. M. Florian and M. Klein. Deterministic Production Planning with Concave Costs and Capacity Constraints. *Management Sci.* 18, 12-20 (1971).
12. M. Florian, J. Lenstra, and A.H.G. Rinnooy Kan. Deterministic Production Planning: Algorithms and Complexity. *Management Sci.* 26, 669-679 (1980).
13. A.M. Geoffrion. Lagrangian Relaxation for Integer Programming. *Math. Prog. Study* **2**, 82-114 (1974).
14. J. Geunes, H.E. Romeijn, and K. Taaffe. Models for Integrated Production Planning and Order Selection. 2002 IIE Research Conference Proceedings (2002).
15. J. Geunes, H.E. Romeijn, and K. Taaffe. Requirements Planning with Dynamic Pricing and Order Selection Flexibility. Working Paper, ISE Department, University of Florida, Gainesville Florida (2003).
16. C.Y. Lee, S. Çetinkaya, and A.P.M. Wagelmans. A Dynamic Lot Sizing Model with Demand Time Windows. *Management Sci.* **47**(10), 1384-1395 (2001).
17. H.L. Lee and S. Nahmias. Single-Product, Single-Location Models. Ch. 1 in *Handbooks in Operations Research and Management Science v4, Logistics of Production and Inventory* (S.C. Graves, A.H.G. Rinnooy Kan, and P.H. Zipkin, eds.), North-Holland, Amsterdam (1993).
18. H.L. Lee, V. Padmanabhan, and S. Whang. Information Distortion in a Supply Chain: The Bullwhip Effect. *Management Sci.* 43(4), 546-558 (1997).
19. J. Leung, T.L. Magnanti, and R. Vachani. Facets and Algorithms for Capacitated Lot Sizing. *Math. Programming* 45, 331-359 (1989).
20. S. Love. A Facilities in Series Inventory Model with Nested Schedules. *Management Sci.* **18**, 327-338 (1972).
21. S. Nahmias. *Production and Operations Analysis*, 4th ed., McGraw-Hill Irwin, Boston (2001).
22. Y. Pochet. Valid inequalities and separation for capacitated economic lot sizing. *Oper. Res. Letters*, **7**, 109-116 (1988).
23. J.F. Shapiro. Mathematical Programming Models and Methods for Production Planning and Scheduling. Ch. 8 in *Handbooks in Operations Research and Management Science v4, Logistics of Production and Inventory* (S.C. Graves, A.H.G. Rinnooy Kan, and P.H. Zipkin, eds.), North-Holland, Amsterdam (1993).
24. E.A. Silver and H.C. Meal. A Heuristic for Selecting Lot Size Requirements for the Case of a Deterministic Time-Varying Demand Rate and Discrete Opportunities for Replenishment. *Prod. and Inv. Mgt.* 14(2), 64-74 (1973).
25. J. Thomas. Price-Production Decisions with Deterministic Demand. *Management Sci.* 16(11), 747-750 (1970).

26. A.P.M. Wagelmans, S. van Hoesel, and A. Kolen. Economic Lot Sizing: An $O(n \log n)$-algorithm that Runs in Linear Time in the Wagner-Whitin Case. *Oper. Res.* **40**, S145-S156 (1992).
27. H. Wagner and T. Whitin. Dynamic Version of the Economic Lot Size Model. *Management Sci.* **5**, 89-96 (1958).
28. C.A. Yano. Setting Planned Leadtimes in Serial Production Systems with Tardiness Costs. *Management Sci.* 33(1), 95-106 (1987).
29. W. Zangwill. A Backlogging Model and a Multi-Echelon Model Of a Dynamic Economic Lot Size Production System. *Management Sci.* **15**, 506-527 (1969).